MySQL 8（中文版） Cookbook

MySQL 8 Cookbook

[美] Karthik Appigatla 著

周彦伟　孟治华　王学芳 译

電子工業出版社
Publishing House of Electronics Industry
北京•BEIJING

内 容 简 介

MySQL 8.0 的发布是 MySQL 发展历史上的一个重要里程碑，也是开源数据库领域内的一个大事件。针对这个版本，MySQL 官方团队做了太多的工作，从查询优化到集群架构，从参数调整到特性支持，MySQL 都有了革命性的变化。本书基于 MySQL 8.0，以基础知识为入手点，以讲解技术特性为目标，以案例作为理论的补充，详细介绍了 MySQL 的方方面面，提供了超过 150 个高性能数据库查询与管理技巧，是 MySQL 入门者和管理者的必读之作。

版权贸易合同登记号　图字：01-2018-5009

图书在版编目（CIP）数据

MySQL 8 Cookbook（中文版）/（美）卡西克·阿皮加特拉（Karthik Appigatla）著；周彦伟，孟治华译. —北京：电子工业出版社，2018.11
书名原文：MySQL 8 Cookbook
ISBN 978-7-121-35010-8

Ⅰ. ①M… Ⅱ. ①卡… ②周… ③孟… Ⅲ. ①SQL 语言－程序设计 Ⅳ. ①TP311.132.3

中国版本图书馆 CIP 数据核字(2018)第 207742 号

策划编辑：张春雨
责任编辑：许　艳
印　　刷：北京盛通商印快线网络科技有限公司
装　　订：北京盛通商印快线网络科技有限公司
出版发行：电子工业出版社
　　　　　北京市海淀区万寿路 173 信箱　邮编 100036
开　　本：787×980　1/16　印张：28　字数：560 千字
版　　次：2018 年 11 月第 1 版
印　　次：2023 年 2 月第 10 次印刷
定　　价：115.00 元

凡所购买电子工业出版社图书有缺损问题，请向购买书店调换。若书店售缺，请与本社发行部联系，联系及邮购电话：（010）88254888，88258888。
质量投诉请发邮件至 zlts@phei.com.cn，盗版侵权举报请发邮件至 dbqq@phei.com.cn。
本书咨询联系方式：010-51260888-819，faq@phei.com.cn。

推荐语

自 2010 年加入 MySQL 原厂推广 MySQL 技术以来，我深刻地体会到一本优秀的 MySQL 专业书籍对 MySQL 技术人来说是多么重要，一本结构完善、内容深入浅出的 MySQL 书籍又是多么难得。而这本书正是这样一本少见的好书，因为原著作者是顶尖的 MySQL 专家，更为难得的是，两位主要译者都具备不错的英文能力和深厚的 MySQL 专业背景。周彦伟在大中华区的 MySQL 专业人士中名声响亮，由于他在 MySQL 技术领域取得的成就，以及多年来对 MySQL 社区发展的领导，Oracle 公司特别授予他 MySQL 技术人最高等级的荣誉——Oracle ACE Director（MySQL）。第二译者是孟治华先生，他在美国顶尖大学取得计算机专业硕士学位后，在美国的世界 500 强公司积累了十多年的数据库开发和管理经验。他们精心挑选并翻译了由 Karthik Appigatla 所著的 *MySQL 8 Cookbook* 一书，这本书对于 MySQL 入门者来说，可以作为由浅入深地全面了解 MySQL 的学习宝典；对于 MySQL 专家来说，可以作为解决疑难杂症的工具书。

当我畅快淋漓地拜读完本书后，深刻感觉到译者对一本好的翻译著作的贡献不亚于原著作者的贡献。由于译者们都是顶尖的 MySQL 专家，对原文所谈的专业技术和数据库管理场景都有深刻理解，所以能够贴切地把原著的意思以中文表达出来，并使之符合国内读者的阅读习惯。本书的翻译质量远高于一般技术书的翻译质量，很高兴看到这么一本 MySQL 技术书问世，也期望它能造福更多 MySQL 专业人士以及有兴趣成为 MySQL 专业人的小伙伴们。

杜修文

MySQL 全球事业部技术顾问群北区经理

听说彦伟兄又有数据库新书推出，真是欣喜若狂，他居然能从创业之初的繁重公务中抽出身来，亲历亲为，为大家再献上一本数据库领域的上乘之作。此书与市面上已有的

MySQL 书籍不同，它是关于 MySQL 的最新版本 MySQL 8.0 的维护与管理工作的，展示了 MySQL 全新的知识与内容。MySQL 8.0 于 2018 年 4 月发布，距今已半年有余，但相关的书籍却寥寥无几，这对开源数据库在中国的推广非常不利。本书的出版将弥补这一缺憾。

一个行业的进步离不开那些默默前行的负重者，更需要登高疾呼的感召者。感谢 Karthik Appigatla 为我们编写了如此前卫实用的内容，感谢周彦伟、孟治华和王学芳把如此精彩难得的书籍带给中国读者，为国内的 MySQL 使用者再点一盏明灯。

刘启荣

中国计算机行业协会开源数据库专业委员会副会长，京东金融运维总监

MySQL 官方最新发布的 MySQL 8 与以往的版本相比变化很大，市面上尚无相关的中文书籍，可以说此书的出版应时应景。

当年严复提出，翻译要做到信、达、雅，这实际上是非常难的事情，不是所有精通外语的人都能翻译好书的。作为一个读者，我一口气读完了本书，确实没有出现卡壳现象，对内容的理解效率有明显提升。

对老司机来说，此书能帮助快速了解新功能，让大家放心大胆地继续探索；对新手来说，此书能作为使用手册，帮助快速找到问题的解决方法。

田发明

中国计算机行业协会开源数据库专业委员会秘书长

我个人学习 MySQL 技术是从 MySQL 5.5 版本开始的，相比之前的 MySQL 版本，MySQL 5.5 的改进已经非常大。而多年后的今天，继 MySQL 5.6、MySQL 5.7 之后，MySQL 推出了令人期待和振奋的 MySQL 8 版本。MySQL 8 的更新幅度在 MySQL 历史上可谓空前，它增添了很多重磅的新特性，比如业内广受欢迎的“数据字典”，这个更新取消了已存在长达几十年的 frm 文件，将插件式数据库的重心偏向了 InnoDB，极大地提高了 meta data 的访问性能，并且支持了 DDL 的原子性，将来还可以进一步实现 DDL 的回滚等操作；同时支持了隐藏索引，这使得 DBA 能够更加灵活地对数据库调优；在统计分析方面，还支持了被大家期待已久的通用表表达式、窗口函数等，这涉及一些在数据库层面可以实现的复杂

计算，更新简单方便；当然还包括已经正式发布一年多的MGR，这个架构的推出可以说是MySQL划时代的变革。

新功能不一而足，值得了解。以前，想要学习和使用 MySQL 的同学可能只能看英文手册。现在就不同了，从 MySQL 8 正式发布到现在时间并不长，在这短短的几个月里，这本 MySQL 8 的中文书籍就要面世了，这是多么值得称赞的一件事情。

本人阅读了几章，可以感受到译者对 MySQL 的深刻理解和准确把握，该书可以很好地帮助对 MySQL 8 感兴趣的同学，强烈推荐。

王竹峰

去哪儿网数据库总监，Oracle MySQL ACE

《MySQL 运维内参》作者

关于译者

周彦伟，Oracle ACE Director（MySQL），中国计算机行业协会开源数据库专业委员会会长，中国 MySQL 用户组（ACMUG）主席，曾在去哪儿网、人人网等互联网公司工作多年，专注于 MySQL 数据库的技术和推广，著有《MySQL 运维内参》，该书被业内从业者视为 MySQL 面试宝典。2018 年创办极数云舟，提供 MySQL 技术产品和顾问服务。

孟治华，美国波士顿大学计算机专业硕士，拥有 14 年以上的数据库相关项目经验，曾多年在美国为 Walmart、CVS Health 等世界 500 强公司提供数据仓库和商业智能项目咨询服务。回国后积极参与 ACMUG 等知名开源社区活动，并致力于开源数据库知识的推广和传播。

王学芳，中国矿业大学（北京）计算机技术专业在读研究生，对数据库很感兴趣，一直关注 MySQL 社区的发展动态、技术革新等内容。

谨以此书献给我的母亲，A.S Gayathri，以及我的父亲 A.S.R.V.S.N Murty，向他们无私的奉献和矢志不渝的精神致敬。

——*Karthik Appigatla*

关于作者

Karthik Appigatla 是一位备受尊敬的数据库架构师，他在性能调优领域闻名于世。他为世界各地的许多公司提供设计咨询、性能调优、数据库架构设计和培训服务。在过去十年中，他曾供职于雅虎、Pythian 和 Percona 等公司。目前，他任职于 LinkedIn，在那里他发明了一种新的分析查询方法，并于 2017 年在都柏林的 SRECon 上发表了关于这个新发明的演讲。

我要感谢我的妻子 Lalitha 和我的兄弟 Kashyap 的鼓励。此外，没有女儿 Samhita 的配合，这本书恐怕也无法完成。

关于审校者

Marco Ippolito 是一位意大利的软件工程师，现为 Imagining IT 公司的软件开发总监。Marco 在牛津大学完成了软件工程专业的研究生课程，曾供职于英特尔、惠普、谷歌、戴尔和甲骨文公司（在被收购的 MySQL 团队中），以及`@platformsh`等初创公司和拥有大量 MySQL 用户的公司（如`@bookingcom`）。可以通过`marco.ippolito@imaginingit.com`与他联系。与他一起工作（远程的或现场的）过的团队来自世界各地，讲着各种不同的语言：意大利语、英语、西班牙语、巴西葡萄牙语、德语和法语。

Kedar Mohaniraj Vaijanapurkar 是一名拥有十多年经验的 MySQL 数据库顾问，专注于从编程到数据库管理的领域。他的目标是通过使用 MySQL 数据库系统传播快乐。除了使用 MySQL 和相关的开源技术，他还探索了云、自动化和 NoSQL 技术。他与家人一起生活在印度文化城市瓦多达拉（Vadodara）。可以通过`kedar@nitty-witty.com`与他联系。

前言

MySQL 是当下最流行和广泛使用的关系型数据库之一。最新发布的 MySQL 8 能够提供比以往版本更有效的高性能结果查询和更轻松的管理配置。

本书面向的读者

这本书适合的读者范围很广。对于使用过 MySQL 早期版本的 MySQL 数据库管理员和开发者，可以通过本书了解 MySQL 8 的新功能以及如何利用它们；对于那些有其他 RDBMS（如 Oracle、MSSQL、PostgreSQL 和 DB2）工作经验的读者，可以将本书作为学习 MySQL 8 的快速入门指南；对于初学者，可以将本书作为使用手册，参考书中内容找到问题的快速解决方案。

最重要的是，本书还可以让你做好应对生产环境问题的准备。读完本书后，你将有信心管理好拥有海量数据集的大型数据库服务器。

在我使用 MySQL 的 10 年中，我目睹了许多由于小错误导致的重大停机事故。本书涵盖了许多容易出错的场景，并用警告图标显著地标出。

阅读本书，初学者并不需要反复翻阅以理解概念。书中每个主题都提供了指向 MySQL 文档或其他资料的参考链接，读者可以通过参考链接来了解更多详细信息。

因为这本书也适合初学者，所以可能有一些你已经知道的方法。当你读到已熟知的内容时，可以跳过它们。

本书包括的内容

俗话说，熟能生巧。但是除了练习，你还需要一些基础知识和训练。本书可以在这些方面帮你进步。书中涵盖了大多数日常实际使用的场景。

第 1 章介绍如何在不同版本的 Linux 系统上安装 MySQL 8，如何从以前的稳定版本升级到 MySQL 8，以及如何从 MySQL 8 降级。

第 2 章介绍 MySQL 的基本用法，例如创建数据库和表；以各种方式插入、更新、删除和选择数据；将数据保存到不同的目的地；对结果进行排序和分组；多表联接；管理用户及其他数据库元素，如触发器、存储过程、函数和事件；以及获取元数据信息。

第 3 章介绍了 MySQL 8 新增的功能和内容，例如 JSON 数据类型、公用表表达式和窗口函数。

第 4 章介绍了如何配置 MySQL，及其基本配置参数。

第 5 章解释了 RDBMS 的 4 个隔离级别，以及如何将 MySQL 用于事务。

第 6 章演示了如何启用二进制日志、各种格式的二进制日志，以及如何从二进制日志中检索数据。

第 7 章介绍了各种类型的备份方法，每种方法的优缺点，以及如何根据需求选择适用的方法。

第 8 章介绍了如何从各种备份中恢复数据。

第 9 章介绍了如何设置各种复制拓扑。具体介绍了将从服务器由主从复制切换到链式复制的方法，以及将从服务器从链式复制切换到主从复制的方法，相信这些方法能够引起读者的兴趣。

第 10 章介绍了克隆表，将助你成为管理大表的行家。这一章还介绍了第三方工具的安装和使用方法。

第 11 章将讲述如何调整、创建、复制和管理表空间。

第 12 章引导读者了解错误日志、一般查询、慢查询和二进制日志。

第 13 章详细解释了查询和 schema 调优，介绍了很多方法和技巧。

第 14 章着重于安全方面，详细介绍了如何安全安装、限定网络和用户、设置和重置密码等方法。

充分利用本书

如果你具备 Linux 系统的基础知识，那么你将能更轻松地理解本书内容。

约定惯例

本书使用了许多文本约定惯例。

`CodeInText`：这种字体表示文本中的代码、数据库表名、文件夹名、文件名、文件扩展名、路径名、虚拟网址、用户输入和 Twitter 句柄。例如，MySQL 依赖于 `libaio` 库。

如果我们想要提醒你注意命令行语句中的特定部分，会将相关行或项加粗，例如：

```
shell> sudo yum repolist all | grep mysql8
mysql80-community/x86_64    MySQL 8.0 Community Server
enabled:                      16
mysql80-community-source    MySQL 8.0 Community Server
disabled
```

命令行输入或输出的写法如下：

```
mysql> ALTER TABLE table_name REMOVE PARTITIONING;
```

黑体（中文）或粗体（英文）：表示新词、重点词，或者在屏幕上显示的词。例如，菜单或对话框中的单词会使用黑体或粗体。举一个具体的例子：选择用于获取 MySQL 8.0 的**开发版本**选项卡，然后选择操作系统和版本。

这个图标表示警告或重要的注意事项。

这个图标表示技巧和诀窍。

标题

在这本书里，你会发现几个经常出现的标题。为了明确说明如何操作，各个标题的使用规则如下。

准备工作

这一部分将描述本节所包含的内容，并说明完成本节内容所需的软件或设置操作。

如何操作

这一部分包含完成本节内容需要遵循的步骤。

延伸阅读

这一部分包含与本节内容相关的其他信息，以便读者形成更完整的认知。

读者服务

轻松注册成为博文视点社区用户（www.broadview.com.cn），扫码直达本书页面。

- **提交勘误**：您对书中内容的修改意见可在 提交勘误 处提交，若被采纳，将获赠博文视点社区积分（在您购买电子书时，积分可用来抵扣相应金额）。
- **交流互动**：在页面下方 读者评论 处留下您的疑问或观点，与我们和其他读者一同学习交流。

页面入口：http://www.broadview.com.cn/35010

目录

第 1 章 安装或升级到 MySQL 8

在本章中，我们将介绍以下内容：

- 使用 YUM / APT 安装 MySQL
- 使用 RPM 或 DEB 文件安装 MySQL 8.0
- 使用通用二进制文件在 Linux 系统上安装 MySQL
- 启动或停止 MySQL 8 的运行
- 卸载 MySQL 8
- 用 systemd 管理 MySQL 服务器
- 从 MySQL 8.0 降级
- 升级到 MySQL 8.0
- 安装 MySQL 工具集

1.1 引言

在本章中你将学习安装 MySQL 8、升级到 MySQL 8.0 和从 MySQL 8.0 降级的步骤。有如下 5 种不同的安装 MySQL 8 或升级到 MySQL 8 的方式，本章介绍其中最广泛使用的 3 种安装方式。

- 软件仓库（YUM 或 APT）
- RPM 或 DEB 文件
- 通用二进制文件
- Docker（本章未涵盖）

- 源代码编译（本章未涵盖）

如果你已经安装了 MySQL 并想升级到 MySQL 8，请参考 1.9 节的升级步骤。如果安装已损坏，请参考 1.6 节中的卸载步骤。

在安装之前，请记下操作系统和 CPU 架构。接下来的操作如下面各表所示。

MySQL Linux RPM 包分发标识符

发布号	应用于
el6、el7	Red Hat Enterprise Linux、Oracle Linux、CentOS 6 或 7
fc23、fc24、fc25	Fedora 23、24 或 25
sles12	SUSE Linux Enterprise 12

MySQL Linux RPM 包 CPU 标识符

CPU 型号	可用的处理器类型或家族
i386、i586、i686	奔腾处理器或更高性能的处理器，32 位
x86_64	64 位 x86 处理器
iA64	安腾（IA-64）处理器

MySQL Debian 和 Ubuntu 7、8 安装软件包的 CPU 型号

CPU 型号	可用的处理器类型或系列
i386	奔腾处理器或更高性能的处理器，32 位
AMD64	64 位 x86 处理器

MySQL Debian 6 安装包 CPU 类型

CPU 型号	可用的处理器类型或系列
i686	奔腾处理器或更高性能的处理器，32 位
x86_64	64 位 x86 处理器

1.2 使用 YUM / APT 安装 MySQL

最常见和最简单的安装 MySQL 8 的方式是通过软件仓库将官方 Oracle MySQL 仓库添加到列表中，并通过软件包管理软件安装 MySQL。

主要有两种类型的仓库软件：

- YUM（CentOS、Red Hat、Fedora 和 Oracle Linux 系统）

- APT（Debian、Ubuntu系统）

1.2.1　如何操作

让我们看看以下几种安装 MySQL 8 的方式分别需要哪些操作步骤。

使用 YUM 仓库

1. 查看 Red Hat 或 CentOS 的版本：

```
shell> cat /etc/redhat-release
CentOS Linux release 7.3.1611 (Core)
```

2. 将 MySQL YUM 仓库添加到系统的仓库列表中。这是一次性操作，可以通过安装 MySQL 提供的 RPM 来执行。你可以从 http://dev.mysql.com/downloads/repo/yum/下载 MySQL YUM 仓库，然后根据自己的操作系统选择文件。

   使用以下命令安装下载的发行包，并将名称改为下载的 RPM 包的平台和特定版本的包名称：

```
shell> sudo yum localinstall -y mysql57-community-release-
el7-11.noarch.rpm
Loaded plugins: fastestmirror
Examining mysql57-community-release-el7-11.noarch.rpm: mysql57-
community-release-el7-11.noarch
Marking mysql57-community-release-el7-11.noarch.rpm to be installed
Resolving Dependencies
--> Running transaction check
---> Package mysql57-community-release.noarch 0:el7-11 will be
installed
--> Finished Dependency Resolution
~
  Verifying: mysql57-community-release-el7-11.noarch 1/1

Installed:
  mysql57-community-release.noarch 0:el7-11
Complete!
```

3. 或者你可以复制链接位置并使用 RPM 直接安装（安装后可以跳过下一步）：

```
shell> sudo rpm -Uvh
"https://dev.mysql.com/get/mysql57-community-release-el7-11.noarch.
rpm"
Retrieving
https://dev.mysql.com/get/mysql57-community-release-el7-11.noarch.r
pm
Preparing...
################################## [100%]
Updating / installing...
   1:mysql57-community-release-el7-11
################################## [100%]
```

4. 验证安装：

```
shell> yum repolist enabled | grep 'mysql.*-community.*'
mysql-connectors-community/x86_64     MySQL Connectors Community
42
mysql-tools-community/x86_64          MySQL Tools Community
53
mysql57-community/x86_64              MySQL 5.7 Community Server
227
```

5. 设置发布系列。在编写本书时，MySQL 8 并不是一个**通用可用**（**GA**）版本。所以 MySQL 5. 7 将被选作默认发布系列。要安装 MySQL 8，必须将发布系列设置为 8：

```
shell> sudo yum repolist all | grep mysql
mysql-cluster-7.5-community/x86_64    MySQL Cluster 7.5 Community
disabled
mysql-cluster-7.5-community-source    MySQL Cluster 7.5  Community
disabled
mysql-cluster-7.6-community/x86_64    MySQL Cluster 7.6  Community
disabled
mysql-cluster-7.6-community-source    MySQL Cluster 7.6  Community
disabled
mysql-connectors-community/x86_64     MySQL Connectors Community
enabled:   42
```

```
mysql-connectors-community-source        MySQL Connectors Community
disabled
mysql-tools-community/x86_64             MySQL Tools Community
enabled:   53
mysql-tools-community-source             MySQL Tools Community - Sou
disabled
mysql-tools-preview/x86_64               MySQL Tools Preview
disabled
mysql-tools-preview-source               MySQL Tools Preview - Sourc
disabled
mysql55-community/x86_64                 MySQL 5.5 Community Server
disabled
mysql55-community-source                 MySQL 5.5 Community Server
disabled
mysql56-community/x86_64                 MySQL 5.6 Community Server
disabled
mysql56-community-source                 MySQL 5.6 Community Server
disabled
mysql57-community/x86_64                 MySQL 5.7 Community Server
enabled:   227
mysql57-community-source                 MySQL 5.7 Community Server
disabled
mysql80-community/x86_64                 MySQL 8.0 Community Server
disabled
mysql80-community-source                 MySQL 8.0 Community Server
disabled
```

6. 禁用mysql57-community并启用mysql80-community:

```
shell> sudo yum install yum-utils.noarch -y
shell> sudo yum-config-manager --disable mysql57-community
shell> sudo yum-config-manager --enable mysql80-community
```

7. 确认mysql80-community已启用:

```
shell> sudo yum repolist all | grep mysql8
mysql80-community/x86_64     MySQL 8.0 Community Server
enabled:   16
```

```
mysql80-community-source     MySQL 8.0 Community Server
disabled
```

8. 安装 MySQL 8：

```
shell> sudo yum install -y mysql-community-server
Loaded plugins: fastestmirror
mysql-connectors-community |  2.5 kB  00:00:00
mysql-tools-community      |  2.5 kB  00:00:00
mysql80-community          |  2.5 kB 00:00:00
Loading mirror speeds from cached hostfile
 * base: mirror.web-ster.com
 * epel: mirrors.cat.pdx.edu
 * extras: mirrors.oit.uci.edu
 * updates: repos.lax.quadranet.com
Resolving Dependencies
~
Transaction test succeeded
Running transaction
  Installing : mysql-community-common-8.0.3-0.1.rc.el7.x86_64   1/4
  Installing : mysql-community-libs-8.0.3-0.1.rc.el7.x86_64     2/4
  Installing : mysql-community-client-8.0.3-0.1.rc.el7.x86_64   3/4
  Installing : mysql-community-server-8.0.3-0.1.rc.el7.x86_64   4/4
  Verifying  : mysql-community-libs-8.0.3-0.1.rc.el7.x86_64     1/4
  Verifying  : mysql-community-common-8.0.3-0.1.rc.el7.x86_64   2/4
  Verifying  : mysql-community-client-8.0.3-0.1.rc.el7.x86_64   3/4
  Verifying  : mysql-community-server-8.0.3-0.1.rc.el7.x86_64   4/4

Installed:
  mysql-community-server.x86_64 0:8.0.3-0.1.rc.el7
Dependency Installed:
  mysql-community-client.x86_64 0:8.0.3-0.1.rc.el7
  mysql-community-common.x86_64 0:8.0.3-0.1.rc.el7
  mysql-community-libs.x86_64 0:8.0.3-0.1.rc.el7

Complete!
```

9. 你可以使用以下方式检查已安装的软件包：

```
shell> rpm -qa | grep -i 'mysql.*8.*'
perl-DBD-MySQL-4.023-5.el7.x86_64
mysql-community-libs-8.0.3-0.1.rc.el7.x86_64
mysql-community-common-8.0.3-0.1.rc.el7.x86_64
mysql-community-client-8.0.3-0.1.rc.el7.x86_64
mysql-community-server-8.0.3-0.1.rc.el7.x86_64
```

使用 APT 仓库

1. 将 MySQL APT 仓库添加到系统的仓库列表中。这是一个一次性操作，可以通过安装 MySQL 提供的`.deb`文件来执行。
   你可以从`http://dev.mysql.com/downloads/repo/apt/`下载 MySQL APT 仓库，或者复制链接位置并使用`wget`直接将其下载到服务器上。你可能需要安装`wget`（`sudo apt-get install wget`）：

```
shell> wget
"https://repo.mysql.com//mysql-apt-config_0.8.9-1_all.deb"
```

2. 使用以下命令安装下载的发行包，并将名称改为下载的 APT 包的平台和特定版本名称：

```
shell> sudo dpkg -i mysql-apt-config_0.8.9-1_all.deb
(Reading database ... 131133 files and directories currently installed.)
Preparing to unpack mysql-apt-config_0.8.9-1_all.deb ...
Unpacking mysql-apt-config (0.8.9-1) over (0.8.9-1) ...
Setting up mysql-apt-config (0.8.9-1) ...
Warning: apt-key should not be used in scripts (called from
postinst maintainerscript of the package mysql-apt-config)
OK
```

3. 在安装包的过程中，你需要选择 MySQL 服务器和其他组件的版本。按 Enter 键进行选择，并按向上和向下键进行导航。
   选择 **MySQL 服务器和集群**（当前选择：mysql-5.7）。
   选择 **mysql-8.0** 预览（在本书撰写时，MySQL 8.0 不是 GA）。你可能会收到警告，诸如 **MySQL 8.0-RC 注意事项，MySQL 8. 0** 目前是 **RC**（**Release Candidate，** 备

选版本）。安装它只能预览 **MySQL** 即将发布的功能，不建议在生产环境中使用它。如果你想更改发行版本，请执行以下操作：

```
shell> sudo dpkg-reconfigure mysql-apt-config
```

4. 使用以下命令从 MySQL APT 仓库更新包信息（此步骤是必需的）：

```
shell> sudo apt-get update
```

5. 安装 MySQL。在安装过程中，你需要为 root 用户提供 MySQL 安装的密码。请务必牢记密码；如果忘记了密码，则必须重置 root 密码（请参阅 14.5 节）。这一步将安装 MySQL Server，以及客户端和数据库公共文件的软件包：

```
shell> sudo apt-get install -y mysql-community-server
~
Processing triggers for ureadahead (0.100.0-19) ...
Setting up mysql-common (8.0.3-rc-1ubuntu14.04) ...
update-alternatives: using /etc/mysql/my.cnf.fallback to provide
/etc/mysql/my.cnf (my.cnf) in auto mode
Setting up mysql-community-client-core (8.0.3-rc-1ubuntu14.04) ...
Setting up mysql-community-server-core (8.0.3-rc-1ubuntu14.04) ...
~
```

6. 验证包。`ii` 表示软件包已安装：

```
shell> dpkg -l | grep -i mysql
ii  mysql-apt-config            0.8.9-1                  all   Auto
configuration for MySQL APT Repo.
ii mysql-client                 8.0.3-rc-1ubuntu14.04 amd64 MySQL
Client meta package depending on latest version
ii mysql-common                 8.0.3-rc-1ubuntu14.04 amd64 MySQL
Common
ii  mysql-community-client      8.0.3-rc-1ubuntu14.04 amd64 MySQL
Client
ii mysql-community-client-core 8.0.3-rc-1ubuntu14.04 amd64 MySQL
Client Core Binaries
ii mysql-community-server       8.0.3-rc-1ubuntu14.04 amd64 MySQL
Server
```

```
mysql-community-server-core      8.0.3-rc-1ubuntu14.04 amd64 MySQL
Server Core Binaires
```

1.3 使用 RPM 或 DEB 文件安装 MySQL 8.0

使用仓库安装 MySQL 需要访问公共互联网。出于安全考虑，大多数生产机器都未连接到互联网。在这种情况下，你可以在系统管理中下载 RPM 或 DEB 文件，并将其复制到生产机器上。

主要有两种类型的安装文件：

- RPM（CentOS、Red Hat、Fedora 和 Oracle Linux 系统）
- DEB（Debian 和 Ubuntu 系统）

你需要安装多个软件包，下面给出了清单和简短描述。

- `mysql-community-server`：数据库服务器和相关工具。
- `mysql-community-client`：MySQL 客户端应用程序和工具。
- `mysql-community-common`：服务器和客户端库的公共文件。
- `mysql-community-devel`：开发 MySQL 数据库客户端应用程序的头文件和库，例如 Perl MySQL 模块。
- `mysql-community-libs`：某些语言和应用程序需要动态加载和使用 MySQL 的共享库（`libmysqlclient.so *`）。
- `mysql-community-libs-compat`：旧版本的共享库。如果你安装了与旧版本 MySQL 动态链接的应用程序，但希望在不破坏库依赖关系的情况下升级到当前版本，请安装此程序包。

1.3.1 如何操作

让我们看看如何使用以下类型的包来完成安装。

使用 RPM 包

1. 从 **MySQL** 下载页面（`http://dev.mysql.com/downloads/mysql/`）下载 MySQL RPM TAR 软件包，选择操作系统和 CPU 架构。在本书撰写时，MySQL 8.0

不是通用可用（GA）的。如果当你下载时它仍在开发系列中，请选择用于获取 MySQL 8.0 的 **Development Releases** 选项卡，然后选择操作系统和版本：

```
shell> wget
'https://dev.mysql.com/get/Downloads/MySQL-8.0/mysql-8.0.3-0.1.rc.e
l7.x86_64.rpm-bundle.tar'
~
Saving to: 'mysql-8.0.3-0.1.rc.el7.x86_64.rpm-bundle.tar'
~
```

2. 解压安装包：

```
shell> tar xfv mysql-8.0.3-0.1.rc.el7.x86_64.rpm-bundle.tar
```

3. 安装 MySQL：

```
shell> sudo rpm -i mysql-community-{server-8,client,common,libs}*
```

4. RPM 无法解决依赖性问题，安装过程可能会遇到问题。如果遇到了此类问题，请使用此处列出的 `yum` 命令（你应该可以访问相关的软件包）：

```
shell> sudo yum install mysql-community-
{server-8,client,common,libs}* -y
```

5. 验证安装：

```
shell> rpm -qa | grep -i mysql-community
mysql-community-common-8.0.3-0.1.rc.el7.x86_64
mysql-community-libs-compat-8.0.3-0.1.rc.el7.x86_64
mysql-community-libs-8.0.3-0.1.rc.el7.x86_64
mysql-community-server-8.0.3-0.1.rc.el7.x86_64
mysql-community-client-8.0.3-0.1.rc.el7.x86_64
```

使用 APT 包

1. 从 **MySQL** 下载页面（`http:// dev.mysql.com/downloads/mysql/`）下载 MySQL APT TAR：

```
shell> wget
"https://dev.mysql.com/get/Downloads/MySQL-8.0/mysql-server_8.0.3-r
```

```
c-1ubuntu16.04_amd64.deb-bundle.tar"
~
Saving to: 'mysql-server_8.0.3-rc-1ubuntu16.04_amd64.deb-bundle.tar'
~
```

2. 解压安装包：

```
shell> tar -xvf mysql-server_8.0.3-rc-1ubuntu16.04_amd64.deb-
bundle.tar
```

3. 安装依赖关系。如果尚未安装 `libaio1` 包，则可能需要安装它：

```
shell> sudo apt-get install -y libaio1
```

4. 将 `libstdc++6` 升级到最新版本：

```
shell> sudo add-apt-repository ppa:ubuntu-toolchain-r/test
shell> sudo apt-get update
shell> sudo apt-get upgrade -y libstdc++6
```

5. 将 `libmecab2` 升级到最新版本。如果不包含 `Universe`，则将以下面的代码添加到文件的末尾（例如 `zesty`）：

```
shell> sudo vi /etc/apt/sources.list
deb http://us.archive.ubuntu.com/ubuntu zesty main universe

shell> sudo apt-get update
shell> sudo apt-get install libmecab2
```

6. 使用以下命令预配置 MySQL 服务器软件包。它会要求你设置 root 密码：

```
shell> sudo dpkg-preconfigure mysql-community-server_*.deb
```

7. 安装数据库公共文件包、客户端软件包、客户端元数据包、服务器软件包和服务器元数据包（按此顺序安装）。你可以用下面这条命令来做到这一点：

```
shell> sudo dpkg -i mysql-{common,community-client-core,community-c
lient,client,community-server-core,community-server,server}_*.deb
```

8. 安装共享库：

```
shell> sudo dpkg -i libmysqlclient21_8.0.1-dmr-1ubuntu16.10_amd64.deb
```

9. 验证安装：

```
shell> dpkg -l | grep -i mysql
ii mysql-client                8.0.3-rc-1ubuntu14.04 amd64 MySQL
Client meta package depending on latest version
ii mysql-common                8.0.3-rc-1ubuntu14.04 amd64 MySQL
Common
ii  mysql-community-client     8.0.3-rc-1ubuntu14.04 amd64 MySQL
Client
mysql-community-client-core    8.0.3-rc-1ubuntu14.04 amd64 MySQL
Client Core Binaries
ii mysql-community-server      8.0.3-rc-1ubuntu14.04 amd64 MySQL
Server
mysql-community-server-core    8.0.3-rc-1ubuntu14.04 amd64 MySQL
Server Core Binaires
ii mysql-server                8.0.3-rc-1ubuntu16.04 amd64 MySQL
Server meta package depending on latest version
```

1.4 使用通用二进制文件在 Linux 系统上安装 MySQL

使用软件包进行安装需要先安装一些依赖项，并可能与其他软件包发生冲突。在这种情况下，可以使用下载页面上提供的通用二进制文件安装 MySQL。二进制文件使用高级编译器进行预编译，并采用最佳选项构建，以实现最佳性能。

1.4.1 如何操作

MySQL 依赖 `libaio` 库。如果未在本地安装此库，则数据目录初始化和后续的服务器启动步骤将失败。安装 `libaio` 库的代码如下。

在基于 YUM 的系统上：

```
shell> sudo yum install -y libaio
```

在基于 APT 的系统上：

```
shell> sudo apt-get install -y libaio1
```

从**MySQL**下载页面 `https://dev.mysql.com/downloads/mysql/` 下载TAR二进制文件，然后选择**Linux - Generic**作为操作系统并选择版本。可以使用 `wget` 命令直接下载到服务器上：

```
shell> cd /opt
shell> wget
"https://dev.mysql.com/get/Downloads/MySQL-8.0/mysql-8.0.3-rc-linux-glibc2.
12-x86_64.tar.gz"
```

使用以下步骤安装MySQL。

1. 添加 `mysql` 组和 `mysql` 用户。所有的文件和目录应该在 `mysql` 用户下：

    ```
    shell> sudo groupadd mysql
    shell> sudo useradd -r -g mysql -s /bin/false mysql
    ```

2. 下面是安装位置（也可以将其更改为其他位置）：

    ```
    shell> cd /usr/local
    ```

3. 解压二进制文件。将未解压的二进制文件保存在相同的位置，并将其链接到安装位置。这样可以保留多个版本，升级会变得非常容易。例如，可以下载其他版本并将其解压到其他位置，在升级时只需更改符号链接即可：

    ```
    shell> sudo tar zxvf /opt/mysql-8.0.3-rc-linux-glibc2.12-
    x86_64.tar.gz
    mysql-8.0.3-rc-linux-glibc2.12-x86_64/bin/myisam_ftdump
    mysql-8.0.3-rc-linux-glibc2.12-x86_64/bin/myisamchk
    mysql-8.0.3-rc-linux-glibc2.12-x86_64/bin/myisamlog
    mysql-8.0.3-rc-linux-glibc2.12-x86_64/bin/myisampack
    mysql-8.0.3-rc-linux-glibc2.12-x86_64/bin/mysql
    ~
    ```

4. 制作符号链接：

    ```
    shell> sudo ln -s mysql-8.0.3-rc-linux-glibc2.12-x86_64 mysql
    ```

5. 创建必要的目录并将所有权更改为 `mysql`：

    ```
    shell> cd mysql
    ```

```
shell> sudo mkdir mysql-files
shell> sudo chmod 750 mysql-files
shell> sudo chown -R mysql .
shell> sudo chgrp -R mysql .
```

6. 初始化 mysql，它会生成一个临时密码：

```
shell> sudo bin/mysqld --initialize --user=mysql
~
2017-12-02T05:55:10.822139Z 5 [Note] A temporary password is generated
for root@localhost: Aw=ee.rf(6Ua
~
```

7. 设置 SSL 的 RSA。有关 SSL 的更多详细信息，请参阅第 14 章。请注意为 root @ localhost 会生成一个临时密码 eJQdj8C*qVMq：

```
shell> sudo bin/mysql_ssl_rsa_setup
Generating a 2048 bit RSA private key
...........+++
....................................+++
writing new private key to 'ca-key.pem'
-----
Generating a 2048 bit RSA private key
...............................................................+++
...........................................+++
writing new private key to 'server-key.pem'
-----
Generating a 2048 bit RSA private key
.....+++
..........................+++
writing new private key to 'client-key.pem'
-----
```

8. 将二进制文件的所有权更改为 root，数据文件的所有权更改为 mysql：

```
shell> sudo chown -R root .
shell> sudo chown -R mysql data mysql-files
```

9. 将启动脚本复制到 init.d 中：

```
shell> sudo cp support-files/mysql.server /etc/init.d/mysql
```

10. 将 `mysql` 的二进制文件导出到 `PATH` 环境变量中：

```
shell> export PATH=$PATH:/usr/local/mysql/bin
```

11. 参考 1.5 节中的步骤来启动 MySQL 8。

安装后将在`/usr/local/mysql`中获得下表中列出的目录：

目录	内容
`bin`	`mysqld` 服务器、客户端和工具集程序
`data`	日志文件、数据库
`docs`	info 格式的 MySQL 手册
`man`	UNIX 手册页面
`include`	包含（header）文件
`lib`	库
`share`	其他支持文件，包括错误消息、示例配置文件以及用于数据库安装的 SQL 语句

1.4.2 延伸阅读

还有其他安装方法，举例如下。

1. 从源代码编译。可以从 Oracle 提供的源代码编译和生成 MySQL，可以灵活地自定义 build 参数、编译器优化选项和安装位置。强烈建议你使用 Oracle 提供的预编译二进制文件，除非需要特定的编译器选项或者正在调试 MySQL。
   本节之所以没有详述这个方法，是因为很少使用，而且它需要几个开发工具，这超出了本书的范围。要通过源代码编译进行安装，可以查看 MySQL 文档，网址为 `https://dev.mysql.com/doc/refman/8.0/en/source-installation.html`。

2. 使用 Docker。MySQL 服务器也可以使用 Docker 镜像进行安装和管理。有关安装、配置以及如何在 Docker 下使用 MySQL 的内容，请参阅 `https://hub.docker.com/r/mysql/mysql-server/`。

1.5 启动或停止 MySQL 8 的运行

安装完成后，可以使用下面的命令来启动/停止 MySQL 的运行，这些命令因不同的平台和安装方法而异。mysqld 是 mysql 服务器进程。所有的启动方法都调用 mysqld 脚本。

1.5.1 如何操作

除了启动和停止 MySQL 8 的运行，我们还将了解有关检查服务器状态的信息。让我们看看具体如何操作。

启动 MySQL 8.0 服务器

可以使用以下命令启动服务器。

1. 使用服务：

```
shell> sudo service mysql start
```

2. 使用 init.d：

```
shell> sudo /etc/init.d/mysql start
```

3. 如果你没有找到启动脚本（当执行二进制安装时），则可以从解压的位置进行复制。

```
shell> sudo cp /usr/local/mysql/support-files/mysql.server
/etc/init.d/mysql
```

4. 如果你的安装包含 systemd 支持：

```
shell> sudo systemctl start mysqld
```

5. 如果不支持 systemd，可以使用 mysqld_safe 启动 MySQL。mysqld_safe 是 mysqld 的启动脚本，用于保护 mysqld 进程。如果 mysqld 被杀死，mysqld_safe 会尝试再次启动该进程：

```
shell> sudo mysqld_safe --user=mysql &
```

启动后的状态和要做的操作如下。

1. 服务器已初始化。

2. SSL 证书和密钥文件在数据目录中生成。

3. `Validate_password` 插件已安装并启用。

4. 创建超级用户账户 `root'@'localhost`。超级用户的密码被设置并存储在 error log 文件中（不适用于二进制安装）。要显示它，请使用以下命令：

```
shell> sudo grep "temporary password" /var/log/mysqld.log
2017-12-02T07:23:20.915827Z 5 [Note] A temporary password is
generated for root@localhost: bkvotsG:h6jD
```

你可以使用该临时密码连接到 MySQL。

```
shell> mysql -u root -pbkvotsG:h6jD
mysql: [Warning] Using a password on the command line interface can
be insecure.
Welcome to the MySQL monitor. Commands end with ; or \g.
Your MySQL connection id is 7
Server version: 8.0.3-rc-log

Copyright (c) 2000, 2017, Oracle and/or its affiliates. All rights
reserved.
Oracle is a registered trademark of Oracle Corporation and/or its
affiliates. Other names may be trademarks of their respective owners.

Type 'help;' or '\h' for help. Type '\c' to clear the current input
statement.

mysql>
```

5. 使用生成的临时密码登录，并为超级用户账户设置自定义密码，尽快更改 root 密码：

```
# You will be prompted for a password, enter the one you got from
the previous step

mysql> ALTER USER 'root'@'localhost' IDENTIFIED BY 'NewPass4!';
```

```
Query OK, 0 rows affected (0.01 sec)

# password should contain at least one Upper case letter, one
lowercase letter, one digit, and one special character, and that
the total password length is at least 8 characters
```

停止 MySQL 8.0 服务器的运行

停止 MySQL 的运行并检查状态的操作与启动它的操作类似，只有一个词的差别：

1. 使用 service：

```
shell> sudo service mysqld stop
Redirecting to /bin/systemctl stopmysqld.service
```

2. 使用 init.d：

```
shell> sudo /etc/init.d/mysql stop

[ ok ] Stopping mysql (via systemctl): mysql.service.
```

3. 如果你的安装包含 systemd 支持（请参阅 1.7 节）：

```
shell> sudo systemctl stop mysqld
```

4. 使用 mysqladmin：

```
shell> mysqladmin -u root -p shutdown
```

检查 MySQL 8.0 服务器的状态

1. 使用 service：

```
shell> sudo systemctl status mysqld
● mysqld.service - MySQL Server
   Loaded: loaded (/usr/lib/systemd/system/mysqld.service; enabled;
vendor preset: disabled)
  Drop-In: /etc/systemd/system/mysqld.service.d
           └─override.conf
   Active: active (running) since Sat 2017-12-02 07:33:53 UTC; 14s
ago
```

```
       Docs: man:mysqld(8)
             http://dev.mysql.com/doc/refman/en/using-systemd.html
    Process: 10472 ExecStart=/usr/sbin/mysqld --daemonize --pid-
file=/var/run/mysqld/mysqld.pid $MYSQLD_OPTS (code=exited,
status=0/SUCCESS)
    Process: 10451 ExecStartPre=/usr/bin/mysqld_pre_systemd
(code=exited, status=0/SUCCESS)
Main PID: 10477 (mysqld)
    CGroup: /system.slice/mysqld.service
             └─10477 /usr/sbin/mysqld --daemonize --pid-
file=/var/run/mysqld/mysqld.pid --general_log=1

Dec 02 07:33:51 centos7 systemd[1]: Starting MySQL Server...
Dec 02 07:33:53 centos7 systemd[1]: Started MySQL Server.
```

2. 使用 init.d:

```
shell> sudo /etc/init.d/mysql status
● mysql.service - LSB: start and stop MySQL
   Loaded: loaded (/etc/init.d/mysql; bad; vendor preset: enabled)
   Active: inactive (dead)
     Docs: man:systemd-sysv-generator(8)

Dec 02 06:01:00 ubuntu systemd[1]: Starting LSB: start and stop
MySQL...
Dec 02 06:01:00 ubuntu mysql[20334]: Starting MySQL
Dec 02 06:01:00 ubuntu mysql[20334]:  *
Dec 02 06:01:00 ubuntu systemd[1]: Started LSB: start and stop
MySQL.
Dec 02 06:01:00 ubuntu mysql[20334]: 2017-12-02T06:01:00.969284Z
mysqld_safe A mysqld process already exists
Dec 02 06:01:55 ubuntu systemd[1]: Stopping LSB: start and stop
MySQL...
Dec 02 06:01:55 ubuntu mysql[20445]: Shutting down MySQL
Dec 02 06:01:57 ubuntu mysql[20445]: .. *
Dec 02 06:01:57 ubuntu systemd[1]: Stopped LSB: start and stop
MySQL.
Dec 02 07:26:33 ubuntu systemd[1]: Stopped LSB: start and stop
```

```
MySQL.
```

3. 如果你的安装包含 systemd 支持（请参阅 1.7 节）：

```
shell> sudo systemctl status mysqld
```

1.6 卸载 MySQL 8

如果你的安装过程出现混乱，或者你不想要 MySQL 8 版本了，可以使用以下步骤进行卸载。在卸载之前，如果需要，请确保创建备份文件（请参阅第 7 章），并停止 MySQL 的运行。

1.6.1 如何操作

在不同的系统上卸载 MySQL 8 的方式不同。我们分别看看如何操作。

在基于 YUM 的系统上

1. 检查已经安装了哪些包：

```
shell> rpm -qa | grep -i mysql-community
mysql-community-libs-8.0.3-0.1.rc.el7.x86_64
mysql-community-common-8.0.3-0.1.rc.el7.x86_64
mysql-community-client-8.0.3-0.1.rc.el7.x86_64
mysql-community-libs-compat-8.0.3-0.1.rc.el7.x86_64
mysql-community-server-8.0.3-0.1.rc.el7.x86_64
```

2. 删除包。你可能会收到有其他包依赖 MySQL 的通知，如果你计划再次安装 MySQL 则可以通过传递--nodeps 选项来忽略该警告：

```
shell> rpm -e <package-name>
```

例如：

```
shell> sudo rpm -e mysql-community-server
```

3. 删除所有包：

```
shell> sudo rpm -qa | grep -i mysql-community | xargs sudo rpm -e -
```

```
-nodeps
warning: /etc/my.cnf saved as /etc/my.cnf.rpmsave
```

在基于 APT 的系统上

1. 检查是否有任何现有的包：

```
shell> dpkg -l | grep -i mysql
```

2. 使用以下命令删除包：

```
shell> sudo apt-get remove mysql-community-server mysql-client
mysql-common mysql-community-client mysql-community-client-core
mysql-community-server mysql-community-server-core -y
Reading package lists... Done
Building dependency tree
Reading state information... Done
The following packages will be REMOVED:
  mysql-client mysql-common mysql-community-client mysql-community-
client-core mysql-community-server mysql-community-server-core
mysql-server
0 upgraded, 0 newly installed, 7 to remove and 341 not upgraded.
After this operation, 357 MB disk space will be freed.
(Reading database ... 134358 files and directories currently installed.)
Removing mysql-server (8.0.3-rc-1ubuntu16.04) ...
Removing mysql-community-server (8.0.3-rc-1ubuntu16.04) ...
update-alternatives: using /etc/mysql/my.cnf.fallback to provide
/etc/mysql/my.cnf (my.cnf) in auto mode
Removing mysql-client (8.0.3-rc-1ubuntu16.04) ...
Removing mysql-community-client (8.0.3-rc-1ubuntu16.04) ...
Removing mysql-common (8.0.3-rc-1ubuntu16.04) ...
Removing mysql-community-client-core (8.0.3-rc-1ubuntu16.04) ...
Removing mysql-community-server-core (8.0.3-rc-1ubuntu16.04) ...
Processing triggers for man-db (2.7.5-1) ...
```

或者使用以下命令删除它：

```
shell> sudo apt-get remove --purge mysql-\* -y
shell> sudo apt-get autoremove -y
```

3. 确认软件包已卸载：

```
shell> dpkg -l | grep -i mysql
ii  mysql-apt-config           0.8.9-1 all Auto
configuration for MySQL APT    Repo.
rc  mysql-common           8.0.3-rc-1ubuntu16.04 amd64
MySQL Common
rc  mysql-community-client 8.0.3-rc-1ubuntu16.04 amd64
MySQL Client
rc  mysql-community-server 8.0.3-rc-1ubuntu16.04 amd64
MySQL Server
```

rc 表示软件包已被删除（r），并且只保存配置文件（c）。

卸载二进制文件

卸载二进制文件非常简单，只需要移除符号链接。

1. 将目录更改为安装路径：

```
shell> cd /usr/local
```

2. 检查 mysql 指向的位置，该位置将显示 mysql 所引用的路径：

```
shell> sudo ls -lh mysql
```

3. 删除 mysql：

```
shell> sudo rm mysql
```

4. 删除二进制文件（可选）：

```
shell> sudo rm -f /opt/mysql-8.0.3-rc-linux-glibc2.12-x86_64.tar.gz
```

1.7 用 systemd 管理 MySQL 服务器

如果使用 RPM 或 Debian 软件包服务器安装 MySQL，则 MySQL 服务器的启动和关闭是由 systemd 管理的。在安装了关于 MySQL 的 systemd 支持的平台上，不需要再安装 mysqld_safe、mysqld_multi 或 mysqld_multi.server。MySQL 服务器的启动和关闭由 systemd 使用 systemctl 命令进行管理。你需要按如下操作配置 systemd。

基于 RPM 的系统使用 mysqld.service 文件，而基于 APT 的系统使用 mysql.server 文件。

1.7.1 如何操作

1. 创建一个本地化的 systemd 配置文件：

```
shell> sudo mkdir -pv /etc/systemd/system/mysqld.service.d
```

2. 创建或打开 conf 文件：

```
shell> sudo vi /etc/systemd/system/mysqld.service.d/override.conf
```

3. 输入以下内容：

```
[Service]
LimitNOFILE=max_open_files (ex: 102400)
PIDFile=/path/to/pid/file (ex: /var/lib/mysql/mysql.pid)
Nice=nice_level (ex: -10)
Environment="LD_PRELOAD=/path/to/malloc/library"
Environment="TZ=time_zone_setting"
```

4. 重新加载 systemd：

```
shell> sudo systemctl daemon-reload
```

5. 对于临时的更改，可以在不编辑 conf 文件的情况下重新加载：

```
shell> sudo systemctl set-environment MYSQLD_OPTS="--general_log=1"
or unset using
shell> sudo systemctl unset-environment MYSQLD_OPTS
```

6. 修改 systemd 环境后，重新启动服务器使更改生效。
启用 mysql.serviceshell> sudo systemctl，并启用 mysql.service：

```
shell> sudo systemctl unmask mysql.service
```

7. 重新启动 mysql。

   在 RPM 平台上：

```
shell> sudo systemctl restart mysqld
```

   在 Debian 平台上：

```
shell> sudo systemctl restart mysql
```

1.8 从 MySQL 8.0 降级

如果应用程序未按预期执行，可以降级到以前的通用可用（GA）版本（MySQL 5.7）。在降级之前，建议采取逻辑备份（请参阅第 7 章）。请注意，每次只能降级一个版本。假设你想从 MySQL 8.0 降级到 MySQL 5.6，你必须先降级到 MySQL 5.7，然后再从 MySQL 5.7 降级到 MySQL 5.6。

可以通过两种方式做到这一点：

- 就地降级（在 MySQL 8 中降级）
- 逻辑降级

1.8.1 如何操作

在接下来的小节中，我们将学习如何使用各种仓库、软件包等来完成安装、卸载、升级和降级。

就地降级

对于 MySQL 8.0 中 GA 状态版本之间的降级（请注意，不能使用此方法降级到 MySQL 5.7）：

1. 关闭旧的 MySQL 版本。
2. 替换 MySQL 8.0 二进制文件或较旧的二进制文件。
3. 在现有数据目录上重新启动 MySQL。
4. 运行 mysql_upgrade 实用程序。

使用YUM库

1. 设定MySQL慢关闭，这样能确保UNDO日志为空，并且能在不同发行版之间存在文件格式差异的情况下完全准备好数据文件：

```
mysql> SET GLOBAL innodb_fast_shutdown = 0;
```

2. 按照1.5.1节所述关闭mysql服务器：

```
shell> sudo systemctl stop mysqld
```

3. 从数据目录中删除InnoDB REDO日志文件（ib_logfile *文件），以避免在不同发行版之间发生与重做日志文件格式更改相关的降级问题。

```
shell> sudo rm -rf /var/lib/mysql/ib_logfile*
```

4. 降级MySQL。要降级服务器，需要卸载MySQL 8.0，参见1.6节。配置文件会自动存储为备份。

   列出可用的版本：

```
shell> sudo yum list mysql-community-server
```

   降级相对棘手，在降级之前最好先删除现有的软件包：

```
shell> sudo rpm -qa | grep -i mysql-community | xargs sudo rpm -e --nodeps
warning: /etc/my.cnf saved as /etc/my.cnf.rpmsave
```

   安装旧版本：

```
shell> sudo yum install -y mysql-community-server-<version>
```

使用APT库

1. 重新配置MySQL并选择较旧的版本：

```
shell> sudo dpkg-reconfigure mysql-apt-config
```

2. 运行apt-get update：

```
shell> sudo apt-get update
```

3. 删除当前版本：

```
shell> sudo apt-get remove mysql-community-server mysql-client
mysql-common mysql-community-client mysql-community-client-core
mysql-community-server mysql-community-server-core -y
shell> sudo apt-get autoremove
```

4. 安装旧版本（自重新配置后自动选择）:

```
shell> sudo apt-get install -y mysql-server
```

使用 RPM 或 APT 软件包

卸载现有软件包（请参阅 1.6 节）并安装新软件包，这些软件包可从 **MySQL** 下载页面（http://dev.mysql.com/downloads/mysql/）下载（请参阅 1.3 节）。

使用通用二进制文件

如果你已经通过二进制文件安装了 MySQL，则必须删除旧版本的符号链接（请参阅 1.6 节），并执行全新安装（请参阅 1.4 节）:

1. 按照 1.5 节所述来启动服务器。请注意，所有版本的启动过程都是相同的。
2. 运行 `mysql_upgrade` 实用程序，如下。

```
shell> sudo mysql_upgrade -u root -p
```

3. 重新启动 MySQL 服务器以确保对系统表所做的任何更改都能生效，如下。

```
shell> sudo systemctl restart mysqld
```

逻辑降级

以下是概要步骤。

1. 使用逻辑备份从 MySQL 8.0 版本导出现有数据（请参阅第 7 章来快速了解逻辑备份方法）。
2. 安装 MySQL 5.7。
3. 将转储文件加载到 MySQL 5.7 版本中（请参阅第 8 章来了解恢复方法）。

4. 运行 mysql_upgrade 实用程序。

以下是详细的步骤。

1. 你需要对数据库进行逻辑备份。（请参阅第 7 章了解一种更快速的备份方式 mydumper）：

```
shell> mysqldump -u root -p --add-drop-table --routines --events --all-databases --force > mysql80.sql
```

2. 按照 1.5 节中的描述关闭 MySQL 服务器。

3. 移动数据目录。如果你想保留 MySQL 8，那么就不需要恢复 SQL 备份（即步骤 1 中的操作），而需要移回数据目录：

```
shell> sudo mv /var/lib/mysql /var/lib/mysql80
```

4. 降级 MySQL。要降级服务器，我们需要卸载 MySQL 8，配置文件会自动备份。

使用 YUM 库

卸载后，安装旧版本。

1. 切换仓库：

```
shell> sudo yum-config-manager --disable mysql80-community
shell> sudo yum-config-manager --enable mysql57-community
```

2. 确认 mysql57-community 已启用：

```
shell> yum repolist enabled | grep    "mysql.*-community.*"
!mysql-connectors-community/x86_64    MySQL Connectors Community
42
!mysql-tools-community/x86_64         MySQL Tools Community
53
!mysql57-community/x86_64             MySQL 5.7 Community Server
227
```

3. 降级相对棘手，在降级之前最好删除现有的软件包：

```
shell> sudo rpm -qa | grep -i mysql-community | xargs sudo rpm -e --nodeps
```

```
warning: /etc/my.cnf saved as /etc/my.cnf.rpmsave
```

4. 列出可用版本：

```
shell> sudo yum list mysql-community-server
Loaded plugins: fastestmirror
Loading mirror speeds from cached hostfile
* base: mirror.rackspace.com
* epel: mirrors.develooper.com
* extras: centos.s.uw.edu
* updates: mirrors.syringanetworks.net
Available Packages
mysql-community-server.x86_64 5.7.20-1.el7
mysql57-community
```

5. 安装 MySQL 5.7：

```
shell> sudo yum install -y mysql-community-server
```

使用 APT 库

1. 重新配置 `apt`，以切换到 MySQL 5.7：

```
shell> sudo dpkg-reconfigure mysql-apt-config
```

2. 运行 `apt-get update`：

```
shell> sudo apt-get update
```

3. 删除当前版本：

```
shell> sudo apt-get remove mysql-community-server mysql-client
mysql-common mysql-community-client mysql-community-client-core
mysql-community-server mysql-community-server-core -y shell> sudo
apt-get autoremove
```

4. 安装 MySQL 5.7：

```
shell> sudo apt-get install -y mysql-server
```

使用 RPM 或 APT 包

卸载现有软件包（请参阅 1.6 节）并安装新软件包，这些软件包可从 **MySQL** 下载页面（请参阅 1.3 节）进行下载。

使用通用二进制文件

如果你已经通过二进制文件安装了 MySQL，则必须删除旧版本的符号链接（请参阅 1.6 节），并执行全新安装（请参阅 1.4 节）。

一旦降级 MySQL，就必须恢复备份并运行 `mysql_upgrade` 实用程序。

1. 启动 MySQL（请参阅 1.5 节）。你需要重新设置密码。

2. 恢复备份（这可能需要很长时间，具体取决于备份的大小）。有关 `myloader` 的快速恢复方法，请参阅第 8 章。

   ```
   shell> mysql -u root -p < mysql80.sql
   ```

3. 运行 `mysql_upgrade`：

   ```
   shell> mysql_upgrade -u root -p
   ```

4. 重新启动 MySQL 服务器，以确保对系统表所做的任何更改都能生效（请参阅 1.5 节）：

   ```
   shell> sudo /etc/init.d/mysql restart
   ```

1.9 升级到 MySQL 8.0

MySQL 8 使用全局数据字典，其中包含有关事务表中数据库对象的信息。在以前的版本中，字典数据存储在元数据文件和非事务性系统表中。你需要将数据目录从基于文件的结构升级到数据字典结构。

就像降级一样，可以使用两种方法升级到 MySQL 8.0：

- 就地升级
- 逻辑升级

在升级之前，还应该检查一些先决条件。

1.9.1 准备工作

1. 检查不再使用的数据类型或触发器是否有缺失或空的定义符，或者无效的创建上下文（creation context）：

```
shell> sudo mysqlcheck -u root -p --all-databases --check-upgrade
```

2. 绝对不能使用不具备本机分区支持的存储引擎的分区表。要识别这些表，请执行以下查询：

```
shell> SELECT TABLE_SCHEMA, TABLE_NAME FROM
INFORMATION_SCHEMA.TABLES WHERE ENGINE NOT IN ('innodb',
'ndbcluster') AND CREATE_OPTIONS LIKE '%partitioned%';
```

如果有这些表中的任何一个，请将它们更改为 `InnoDB`：

```
mysql> ALTER TABLE table_name ENGINE = INNODB;
```

或者删除分区：

```
mysql> ALTER TABLE table_name REMOVE PARTITIONING;
```

3. 在 MySQL 5.7 的 `mysql` 系统数据库中，不能有 MySQL 8.0 的数据字典使用过的表名称。要识别具有这些名称的表，请执行以下查询：

```
mysql> SELECT TABLE_SCHEMA, TABLE_NAME FROM
INFORMATION_SCHEMA.TABLES WHERE LOWER(TABLE_SCHEMA) = 'mysql' and
LOWER(TABLE_NAME) IN ('catalogs', 'character_sets', 'collations',
'column_type_elements', 'columns', 'events',
'foreign_key_column_usage', 'foreign_keys', 'index_column_usage',
'index_partitions', 'index_stats', 'indexes',
'parameter_type_elements', 'parameters', 'routines', 'schemata',
'st_spatial_reference_systems', 'table_partition_values',
'table_partitions', 'table_stats', 'tables', 'tablespace_files',
'tablespaces', 'triggers', 'version', 'view_routine_usage',
'view_table_usage');
```

4. 不能存在外键约束名称超过 64 个字符的表。要识别约束名称太长的表，请执行以下查询：

```
mysql> SELECT CONSTRAINT_SCHEMA, TABLE_NAME, CONSTRAINT_NAME FROM
INFORMATION_SCHEMA.REFERENTIAL_CONSTRAINTS WHERE
LENGTH(CONSTRAINT_NAME) > 64;
```

5. MySQL 8.0 不支持的表（如 ndb）应移至 InnoDB 中：

```
mysql> ALTER TABLE tablename ENGINE=InnoDB;
```

1.9.2　如何操作

和之前一样，接下来将详细介绍各种系统、包等。

就地升级

以下是概要步骤。

1. 关闭旧的 MySQL 版本。
2. 将旧的 MySQL 二进制文件或包替换为新的文件（包含不同类型的安装方法的详细步骤）。
3. 在现有数据目录上重新启动 MySQL。
4. 运行 mysql_upgrade 实用程序。
5. 在 MySQL 5.7 服务器中，如果存在已加密的 InnoDB 表空间，则通过执行以下语句来更换 keyring 主密钥。

```
mysql> ALTER INSTANCE ROTATE INNODB MASTER KEY;
```

以下是详细步骤。

1. 配置 MySQL 5.7 服务器以执行慢关闭。在慢关闭的情况下，InnoDB 在关闭之前执行完整的清除和更改缓冲区合并操作，以确保 UNDO 日志为空，并且确保当发行版对应的文件格式不同时，数据文件也能完全准备好。
   这一步是最重要的，如果不做这一步，最终会出现以下错误：

```
[ERROR] InnoDB: Upgrade after a crash is not supported.
```

这个 REDO 日志是用 MySQL 5.7.18 创建的。请按照 `http://dev.mysql.com/doc/refman/8.0/en/upgrading.html` 上的说明操作。

```
mysql> SET GLOBAL innodb_fast_shutdown = 0;
```

2. 按照 1.5 节中的描述关闭 MySQL 服务器。

升级 MySQL 二进制文件或软件包。

基于 YUM 的系统

1. 切换仓库：

```
shell> sudo yum-config-manager --disable mysql57-community
shell> sudo yum-config-manager --enable mysql80-community
```

2. 确认 `mysql80-community` 已启用：

```
shell> sudo yum repolist all | grep mysql8
mysql80-community/x86_64        MySQL 8.0   Community Server
enabled:   16
mysql80-community-source        MySQL 8.0   Community Server
disabled
```

3. 运行 yum 更新：

```
shell> sudo yum update mysql-server
```

基于 APT 的系统

1. 重新配置 `apt`，以切换到 MySQL 8.0：

```
shell> sudo dpkg-reconfigure mysql-apt-config
```

2. 运行 `apt-get update`：

```
shell> sudo apt-get update
```

3. 删除当前版本：

```
shell> sudo apt-get remove mysql-community-server mysql-client
```

```
mysql-common mysql-community-client mysql-community-client-core
mysql-community-server mysql-community-server-core -y
shell> sudo apt-get autoremove
```

4. 安装 MySQL 8：

```
shell> sudo apt-get update
shell> sudo apt-get install mysql-server
shell> sudo apt-get install libmysqlclient21
```

使用 RPM 或 APT 包

卸载现有软件包（请参阅 1.6 节）并安装新软件包，这些软件包可以从 **MySQL 下载页面**（请参阅 1.3 节）进行下载。

使用通用二进制文件

如果你已经通过二进制文件安装了 MySQL，则必须删除旧版本的符号链接（参考 1.6 节），并执行全新安装（请参阅 1.4 节）：

启动 MySQL 8.0 服务器（请参阅 1.5 节）。如果存在加密的 `InnoDB` 表空间，请使用 `--early-plugin-load` 选项加载 `keyring` 插件。

服务器会自动检测数据字典表是否存在。如果不存在，服务器则会在数据目录中创建数据字典并用元数据填充之，然后继续正常的启动顺序。在此过程中，服务器为所有数据库对象——包括数据库、表空间、系统和用户表、视图以及存储程序（存储过程、函数、触发器、事件调度程序事件）——升级元数据。服务器还会删除先前用于元数据存储的文件。例如，升级后，你会注意到你的表中不再有.frm 文件了。

服务器创建一个名为 `backup_metadata_57` 的目录，并将 MySQL 5.7 使用的文件移入其中。服务器将 `event` 和 `proc` 表重命名为 `event_backup_57` 和 `proc_backup_57`。如果此升级失败，服务器将恢复对数据目录的所有更改。在这种情况下，你应该删除所有 REDO 日志文件，在同一数据目录中启动你的 MySQL 5.7 服务器，并修复所有错误。然后，执行 MySQL 5.7 服务器的另一个慢关闭，并启动 MySQL 8.0 服务器以再次尝试。

运行 `mysql_upgrade` 实用程序：

```
shell> sudo mysql_upgrade -u root -p
```

Mysql_upgrade 检查所有数据库中的所有表与当前版本的 MySQL 是否存在不兼容的现象。它会检查 MySQL 5.7 和 MySQL 8.0 的 mysql 系统数据库的不兼容情况，并进行所有需要的更改，以便你利用新的特性或功能。mysql_upgrade 还使 MySQL 8.0 的 performance schema、INFORMATION_SCHEMA 和 sys schema 对象保持最新。

重新启动 MySQL 服务器（请参阅 1.5 节）。

逻辑升级

以下是概要步骤。

1. 使用 mysqldump 从旧的 MySQL 版本导出现有数据。
2. 安装新的 MySQL 版本。
3. 将转储文件加载到新的 MySQL 版本中。
4. 运行 mysql_upgrade 实用程序。

以下是详细步骤。

1. 需要对数据库进行逻辑备份（请参阅第 7 章来了解一种更快的备份方式 mydumper）：

```
shell> mysqldump -u root -p --add-drop-table --routines --events -
all-databases --ignore-table=mysql.innodb_table_stats --ignore-
table=mysql.innodb_index_stats --force > data-for-upgrade.sql
```

2. 关闭 MySQL 服务器（请参阅 1.5 节）。

3. 安装新的 MySQL 版本（请参阅 1.9.2 节中“就地升级”部分提到的方法）。

4. 启动 MySQL 服务器（请参阅 1.5 节）。

5. 重置临时 root 密码：

```
shell> mysql -u root -p
Enter password: **** (enter temporary root password from error log)
```

```
mysql> ALTER USER USER() IDENTIFIED BY 'your new password';
```

6. 恢复备份（这可能需要很长时间，具体取决于备份的大小）。要了解有关 myloader 的快速恢复方法，请参阅第 8 章。

```
shell> mysql -u root -p --force < data-for-upgrade.sql
```

7. 运行 mysql_upgrade 实用程序：

```
shell> sudo mysql_upgrade -u root -p
```

8. 重新启动 MySQL 服务器（请参阅 1.5 节）。

1.10　安装 MySQL 工具集

MySQL 工具集为我们提供了非常方便的工具，可以在不需要太多人工操作的情况下顺利执行日常操作。

1.10.1　如何操作

可以按照下面的方式将 MySQL 工具集安装在基于 YUM 和 APT 的系统上。让我们分别来看看。

基于 YUM 的系统

通过选择 Red Hat Enterprise Linux / Oracle Linux 从 **MySQL** 下载页面（https://dev.mysql.com/downloads/utilities/）下载文件，或者直接使用 wget 从此链接开始下载：

```
shell> wget
https://cdn.mysql.com//Downloads/MySQLGUITools/mysql-utilities-1.6.5-1.el7.
noarch.rpm

shell> sudo yum localinstall -y mysql-utilities-1.6.5-1.el7.noarch.rpm
```

基于 APT 的系统

通过选择 Ubuntu Linux，从 **MySQL** 下载页面（https://dev.mysql.com/

downloads/utilities/）下载文件，或直接使用 wget 从此链接下载：

```
shell> wget
"https://cdn.mysql.com//Downloads/MySQLGUITools/mysql-utilities_1.6.5-1ubun
tu16.10_all.deb"
shell> sudo dpkg -i mysql-utilities_1.6.5-1ubuntu16.10_all.deb
shell> sudo apt-get install -f
```

第2章 使用 MySQL

在本章中，我们将介绍以下内容：

- 使用命令行客户端连接到 MySQL
- 创建数据库
- 创建表
- 插入、更新和删除行
- 加载示例数据
- 查询数据
- 对结果排序
- 对结果分组（聚合函数）
- 创建用户
- 授予和撤销用户的访问权限
- 查询数据并保存到文件和表中
- 将数据加载到表中
- 表关联
- 存储过程
- 函数
- 触发器
- 视图
- 事件
- 获取有关数据库和表的信息

2.1 引言

我们将在本章学习很多内容，让我们逐个仔细看看。

2.2 使用命令行客户端连接到 MySQL

到目前为止，我们已经学会了如何在各种平台上安装 MySQL 8.0。安装完成后，我们将获得名为 `mysql` 的命令行客户端工具，可以用它来连接到任何 MySQL 服务器。

2.2.1 准备工作

首先，你需要知道应该连接哪台服务器。如果你在一台主机上安装了 MySQL 服务器，却尝试从另一台主机（通常称为客户机）连接到服务器，则应指定主机名或服务器 IP 地址，`mysql-client` 软件包应安装在客户端上。在第 1 章中，我们安装了 MySQL 服务器和客户端软件包。如果你已经连接到服务器（通过 SSH），则可以指定 `localhost`、`127.0.0.1` 或：`: 1`。

其次，由于已连接到服务器，因此你需要指定的下一个参数是要在服务器上连接的端口。默认情况下，MySQL 在端口 `3306` 上运行，因此应指定 `3306`。

知道连接到哪里之后，接下来就是要知道登录服务器的用户名和密码。目前我们还没有创建任何用户，所以使用 root 用户进行连接，密码则是你在安装过程中设置的密码。如果已更改密码，请使用新密码。

2.2.2 如何操作

可以使用以下命令中的任一条来连接到 MySQL 客户端：

```
shell> mysql-h localhost -P 3306 -u <username> -p<password>
shell> mysql--host=localhost--port=3306 --user=root --password=<password>
shell> mysql--host localhost--port 3306 --user root --password=<password>
```

强烈建议不要在命令行中输入密码，可以将系统提示你输入密码的字段保留为空白：

```
shell> mysql --host=localhost --port=3306 --user=root --password
Enter Password:
```

1. -P 参数（大写形式）用于指定端口。
2. -p 参数（小写形式）用于指定密码。
3. -p 参数后没有空格。
4. 对于密码，=后没有空格。

默认情况下，主机为 localhost，端口为 3306，用户为当前 shell 用户。

1. 查看当前用户：

```
shell> whoami
```

2. 要断开连接，请按 Ctrl + D 组合键或键入 exit：

```
mysql> ^DBye
shell>
```

   或使用：

```
mysql> exit;
Bye
shell>
```

3. 连接到 mysql 提示符后，可以执行以分隔符结尾的命令。默认分隔符是分号（；）：

```
mysql> SELECT 1;
+---+
| 1 |
+---+
| 1 |
+---+
1 row in set (0.00 sec)
```

4. 要撤销命令，请按 Ctrl + C 组合键或键入\c：

```
mysql> SELECT ^C
mysql> SELECT \c
```

使用 root 用户连接到 MySQL 并不是推荐做法。可以通过授予适当的权限来创建用户和设置用户的访问权限，这些权限将在 2.10 和 2.11 两节中讨论。在那之前，你可以暂时使用 root 用户连接到 MySQL。

2.2.3 延伸阅读

连接之后你可能会注意到一条警告：

```
Warning:Using a password on the command line interface can be insecure.
```

要了解连接的安全方式，请参阅第 14 章。

一旦连接到命令行提示符后，就可以执行 SQL 语句，该语句可以以;、\g 或\G 结尾。
;或\g 对应的输出水平显示，\G 对应的输出垂直显示。

2.3 创建数据库

我们已经安装了 MySQL 8.0 并连接到了 MySQL 服务器。现在是时候存储一些数据了，这才是数据库的主要用途。在任何**关系数据库管理系统**（**RDBMS**）中，数据都存储在“行”中，这是数据库的基本构建单元。行包含可以存储多组值的列。

例如，假设你想将有关你的客户（customer）的信息存储在数据库中。

数据集如下：

```
customer id=1, first_name=Mike, last_name=Christensen, country=USA
customer id=2, first_name=Andy, last_name=Hollands, country=Australia
customer id=3, first_name=Ravi, last_name=Vedantam, country=India
customer id=4, first_name= Rajiv, last_name=Perera, country=Sri Lanka
```

你可以把这些数据保存到行中，例如：(1,'Mike','Christensen','USA')、(2,'Andy','Hollands','Australia')、(3,'Ravi','Vedantam','India')、(4,'Rajiv','Perera','Sri Lanka')。这个数据集有 4 行，由 4 列(id、first_name、last_nameand 和 country) 描述，存储在一个表中。表中可以容纳的列数必须在创建表的时候定义，这也是 RDBMS 的主要局限。虽然我们可以随时更改表的定义，但更改的同时整个表都需要重新构建。在某些情况下，表在被修改时将不可用。更改表结构将在第 9 章中详细讨论。

数据库是许多表的集合，而数据库服务器可以容纳许多这样的数据库。逻辑关系如下：

数据库服务器→数据库→表（由列定义）→行

数据库和表称为数据库对象。任何操作（如创建、修改或删除数据库对象）都称为**数据定义语言（DDL）**操作。

数据按某种蓝图组织构建数据库（分为数据库和表），这种数据的组织形式被称为 **schema**。

2.3.1 如何操作

连接到 MySQL 服务器：

```
shell> mysql -u root -p
Enter Password:
mysql> CREATE DATABASE company;
mysql> CREATE DATABASE `my.contacts`;
```

反标记字符（ ` ）用于引用标识符，如数据库和表名。当数据库名称包含特殊字符，如句点（ . ）时，需要使用反标记字符。

你可以在不同数据库之间切换：

```
mysql> USE company
mysql> USE `my.contacts`
```

通过在命令行中指定数据库，可以直接连接到你想要连接的数据库，无须切换：

```
shell> mysql -u root -p company
```

要查找连接到了哪个数据库，请使用以下命令：

```
mysql> SELECT DATABASE();
+------------+
| DATABASE() |
+------------+
| company    |
+------------+
1 row in set (0.00 sec)
```

要查找你有权访问的所有数据库，请使用：

```
mysql> SHOW DATABASES;
+--------------------+
| Database           |
+--------------------+
| company            |
| my.contacts        |
| information_schema |
| mysql              |
| performance_schema |
| sys                |
+--------------------+
6 rows in set (0.00 sec)
```

数据库被创建为数据目录中的一个目录。基于仓库安装的默认数据目录是/var/lib/mysql,如果是通过二进制文件安装的,数据目录则是/usr/local/mysql/data/。可以通过执行下列操作获知当前的数据目录：

```
mysql> SHOW VARIABLES LIKE 'datadir';
+---------------+------------------------+
| Variable_name | Value                  |
+---------------+------------------------+
| datadir       | /usr/local/mysql/data/ |
+---------------+------------------------+
1 row in set (0.00 sec)
```

检查数据目录内的文件：

```
shell> sudo ls -lhtr /usr/local/mysql/data/
total 185M
-rw-r----- 1   mysql mysql  56     Jun  2  16:57   auto.cnf
-rw-r----- 1   mysql mysql  257    Jun  2  16:57   performance_sche_3.SDI
drwxr-x --- 2  mysql mysql 4.0K    Jun  2  16:57   performance_schema
drwxr-x --- 2  mysql mysql 4.0K    Jun  2  16:57   mysql
-rw-r----- 1   mysql mysql  242    Jun  2  16:57   sys_4.SDI
drwxr-x --- 2  mysql mysql 4.0K    Jun  2  16:57   sys
-rw------- 1   mysql root   1.7K   Jun  2  16:58   ca-key.pem
-rw-r-- r-- 1  mysql root   1.1K   Jun  2  16:58   ca.pem
-rw------- 1   mysql root   1.7K   Jun  2  16:58   server-key.pem
-rw-r-- r-- 1  mysql root   1.1K   Jun  2  16:58   server-cert.pem
-rw------- 1   mysql root   1.7K   Jun  2  16:58   client-key.pem
-rw-r-- r-- 1  mysql root   1.1K   Jun  2  16:58   client-cert.pem
-rw------- 1   mysql root   1.7K   Jun  2  16:58   private_key.pem
-rw-r-- r-- 1  mysql root    451   Jun  2  16:58   public_key.pem
-rw-r----- 1   mysql mysql 1.4K    Jun  2  17:46   ib_buffer_pool
-rw-r----- 1   mysql mysql    5    Jun  2  17:46   server1.pid
-rw-r----- 1   mysql mysql  247    Jun  3  13:55   company_5.SDI
drwxr-x --- 2  mysql mysql  4.0K   Jun  4  08:13   company
-rw-r----- 1   mysql mysql  12K    Jun  4  18:58   server1.err
-rw-r----- 1   mysql mysql  249    Jun  5  16:17   employees_8.SDI
drwxr-x --- 2  mysql mysql  4.0K   Jun  5  16:17   employees
-rw-r----- 1   mysql mysql  76M    Jun  5  16:18   ibdata1
-rw-r----- 1   mysql mysql  48M    Jun  5  16:18   ib_logfile1
-rw-r----- 1   mysql mysql  48M    Jun  5  16:18   ib_logfile0
-rw-r----- 1   mysql mysql 12M     Jun 10  10:29   ibtmp1
```

2.3.2　延伸阅读

你可能想了解其他文件和目录，例如 information_schema 和 performance_schema，这些你尚未创建。information_schema 的信息将在 2.20 节中讨论，而 performance_schema 将在 13.15 节中讨论。

2.4 创建表

在表中定义列时，应该指定列的名称、数据类型（整型、浮点型、字符串等）和默认值（如果有的话）。MySQL 支持各种数据类型。更多有关信息请参阅 MySQL 文档（https://dev.mysql.com/doc/refman/8.0/en/data-types.html）。下面是所有数据类型的概述，其中 JSON 数据类型是一个新的扩展类型，将在第 3 章中讨论。

1. 数字：TINYINT、SMALLINT、MEDIUMINT、INT、BIGINT 和 BIT。
2. 浮点数：DECIMAL、FLOAT 和 DOUBLE。
3. 字符串：CHAR、VARCHAR、BINARY、VARBINARY、BLOB、TEXT、ENUM 和 SET。
4. Spatial 数据类型，更多详细信息请参阅 https://dev.mysql.com/doc/refman/8.0/en/spatial-extensions.html。
5. JSON 数据类型，将在第 3 章中详细讨论。

你可以在一个数据库中创建多张表。

2.4.1 如何操作

这些表包含列定义：

```
mysql> CREATE TABLE IF NOT EXISTS `company`.`customers` (
`id` int unsigned AUTO_INCREMENT PRIMARY KEY,
`first_name` varchar(20),
`last_name` varchar(20),
`country` varchar(20)
) ENGINE=InnoDB;
```

其中的选项解释如下。

- **句点符号**：表可以使用 database.table 引用。如果已经连接到数据库，则可以简单地使用 customers 而不是 company.customers。
- IF NOT EXISTS：如果存在一个具有相同名字的表，并且你指定了这个子句，MySQL 只会抛出一个警告，告知表已经存在。否则，MySQL 将抛出一个错误。
- id：它被声明为一个整型数，因为它只包含整型数。除此之外，还有两个关键字，AUTO_INCREMENT 和 PRIMARY KEY。

- AUTO_INCREMENT：自动生成线性递增序列，因此不必担心为每一行的 id 分配值。
- PRIMARY KEY：每行都由一个非空的 UNIQUE 列标识。只有一列应该在表中定义。如果一个表包含 AUTO_INCREMENT 列，则它会被视为 PRIMARY KEY。
- first_name、last_name 和 country：它们包含字符串，因此它们被定义为 varchar。
- **Engine**：与列定义一起，还应该指定存储引擎。一些类型的存储引擎包括 InnoDB、MyISAM、FEDERATED、BLACKHOLE、CSV 和 MEMORY。在所有引擎中，InnoDB 是唯一的事务引擎，也是默认引擎。要了解更多关于事务的信息，请参阅第 5 章。

要列出所有存储引擎，请执行以下操作：

```
mysql> SHOW ENGINES\G
*************************** 1.row ***************************
      Engine: MRG_MYISAM
     Support: YES
     Comment: Collection of identical MyISAM tables
Transactions: NO
          XA: NO
  Savepoints: NO
*************************** 2.row ***************************
      Engine: FEDERATED
     Support: NO
     Comment: Federated MySQL storage engine
Transactions: NULL
          XA: NULL
  Savepoints: NULL
*************************** 3.row ***************************
      Engine: InnoDB
     Support: DEFAULT
     Comment: Supports transactions, row-level locking, and foreign keys
Transactions: YES
          XA: YES
  Savepoints: YES
*************************** 4.row ***************************
      Engine: BLACKHOLE
```

```
     Support: YES
     Comment: /dev/null storage engine (anything you write to it disappears)
Transactions: NO
          XA: NO
  Savepoints: NO
*************************** 5.row ***************************
      Engine: CSV
     Support: YES
     Comment: CSV storage engine
Transactions: NO
          XA: NO
  Savepoints: NO
*************************** 6.row ***************************
      Engine: MEMORY
     Support: YES
     Comment: Hash based, stored in memory, useful for temporary tables
Transactions: NO
          XA: NO
  Savepoints: NO
*************************** 7.row ***************************
      Engine: PERFORMANCE_SCHEMA
     Support: YES
     Comment: Performance Schema
Transactions: NO
          XA: NO
  Savepoints: NO
*************************** 8.row ***************************
      Engine: ARCHIVE
     Support: YES
     Comment: Archive storage engine
Transactions: NO
          XA: NO
  Savepoints: NO
*************************** 9.row ***************************
      Engine: MyISAM
     Support: YES
     Comment: MyISAM storage engine
```

```
Transactions: NO
          XA: NO
  Savepoints: NO
9 rows in set (0.00 sec)
```

你可以在一个数据库中创建多张表。

创建另一张表来跟踪付款进度：

```
mysql> CREATE TABLE `company`.`payments`(
`customer_name` varchar(20) PRIMARY KEY,
`payment` float
);
```

要列出所有表，请使用：

```
mysql> SHOW TABLES;
+   -------------------+
| Tables_in_company    |
+   -------------------+
| customers            |
| payments             |
+   -------------------+
2   rows in set (0.00   sec)
```

要查看表结构，请执行以下操作：

```
mysql> SHOW CREATE TABLE customers\G
*************************** 1.row ***************************
 Table:customers
Create Table:CREATE TABLE `customers` (
 `id` int(10) unsigned NOT NULL AUTO_INCREMENT,
 `first_name` varchar(20) DEFAULT NULL,
 `last_name` varchar(20) DEFAULT NULL,
 `country` varchar(20) DEFAULT NULL,
 PRIMARY KEY (`id`)
)ENGINE=InnoDB AUTO_INCREMENT=9 DEFAULT CHARSET=utf8mb4
1 row in set (0.00 sec)
```

或者使用下面的语句：

```
mysql> DESC customers;
+------------+-----------------+------+-----+---------+----------------+
| Field      | Type            | Null | Key | Default | Extra          |
+------------+-----------------+------+-----+---------+----------------+
| id         | int(10) unsigned| NO   | PRI | NULL    | auto_increment |
| first_name | varchar(20)     | YES  |     | NULL    |                |
| last_name  | varchar(20)     | YES  |     | NULL    |                |
| country    | varchar(20)     | YES  |     | NULL    |                |
+------------+-----------------+------+-----+---------+----------------+
4 rows in set (0.01 sec)
```

MySQL 会在数据目录内创建.ibd 文件：

```
shell> sudo ls -lhtr /usr/local/mysql/data/company
total 256K
-rw-r----- 1 mysql mysql 128K Jun 4 07:36 customers.ibd
-rw-r----- 1 mysql mysql 128K Jun 4 08:24 payments.ibd
```

克隆表结构

你可以将一个表的结构克隆到新表中：

```
mysql> CREATE TABLE new_customers LIKE customers;
Query OK, 0 rows affected (0.05 sec)
```

可以验证新表的结构：

```
mysql> SHOW CREATE TABLE new_customers\G
*************************** 1.row ***************************
       Table:new_customers
Create Table:CREATE TABLE `new_customers` (
  `id` int(10) unsigned NOT NULL AUTO_INCREMENT,
  `first_name` varchar(20) DEFAULT NULL,
  `last_name` varchar(20) DEFAULT NULL,
  `country` varchar(20) DEFAULT NULL,
  PRIMARY KEY (`id`)
)ENGINE=InnoDB DEFAULT CHARSET=utf8mb4
1 row in set (0.00 sec)
```

2.4.2 延伸阅读

要了解创建表的更多其他选项，请参阅 https://dev.mysql.com/doc/refman/8.0/en/create-table.html。分区表和压缩表将分别在第 10 章和第 11 章中讨论。

2.5 插入、更新和删除行

INSERT、UPDATE、DELETE 和 SELECT 操作称为**数据操作语言**（**DML**）语句。INSERT、UPDATE 和 DELETE 也称为写操作，或者简称为写（**write**）。SELECT 是一个读操作，简称为读（**read**）。

2.5.1 如何操作

让我们来逐个仔细看看。建议稍后自己尝试一下，相信你会喜欢学习这个过程。在本节结束后，我们也会掌握 TRUNCATE TABLE（截断表）。

插入

INSERT 语句用于在表中创建新记录：

```
mysql> INSERT IGNORE INTO `company`.`customers`(first_name,
last_name,country)
VALUES
('Mike', 'Christensen', 'USA'),
('Andy', 'Hollands', 'Australia'),
('Ravi', 'Vedantam', 'India'),
('Rajiv', 'Perera', 'Sri Lanka');
```

或者可以明确地写出 id 列，如果你想插入特定的 id：

```
mysql> INSERT IGNORE INTO `company`.`customers`(id, first_name,
last_name,country)
VALUES
(1, 'Mike', 'Christensen', 'USA'),
(2, 'Andy', 'Hollands', 'Australia'),
(3, 'Ravi', 'Vedantam', 'India'),
(4, 'Rajiv', 'Perera', 'Sri Lanka');
```

```
Query OK, 0 rows affected, 4 warnings (0.00 sec)
Records:4 Duplicates:4 Warnings:4
```

IGNORE：如果该行已经存在，并给出了 IGNORE 子句，则新数据将被忽略，INSERT 语句仍然会执行成功，同时生成一个警告和重复数据的数目。反之，如果未给出 IGNORE 子句，则 INSERT 语句会生成一条错误信息。行的唯一性由主键标识：

```
mysql> SHOW WARNINGS;
+---------+------+---------------------------------------+
| Level   | Code | Message                               |
+---------+------+---------------------------------------+
| Warning | 1062 | Duplicate entry '1' for key 'PRIMARY' |
| Warning | 1062 | Duplicate entry '2' for key 'PRIMARY' |
| Warning | 1062 | Duplicate entry '3' for key 'PRIMARY' |
| Warning | 1062 | Duplicate entry '4' for key 'PRIMARY' |
+---------+------+---------------------------------------+
4 rows in set (0.00 sec)
```

更新

UPDATE 语句用于修改表中的现有记录：

```
mysql> UPDATE customers SET first_name='Rajiv', country='UK' WHERE id=4;
Query OK, 1 row affected (0.00 sec)
Rows matched:1  Changed:1   Warnings:0
```

WHERE：这是用于过滤的子句。在 WHERE 子句后指定的任何条件都会用于过滤，被筛选出来的行都会被更新。

WHERE 子句是强制性的。如果没有给出它，UPDATE 会更新整个表。建议在事务中修改数据，以便在发现任何错误时轻松地回滚这些更改。可以参考第 5 章了解有关事务的更多信息。

删除

删除记录可按如下方式完成：

```
mysql> DELETE FROM customers WHERE id=4 AND first_name='Rajiv';
Query OK, 1 row affected (0.03 sec)
```

WHERE 子句是强制性的。如果没有给出它，DELETE 将删除表中的所有行。
建议在事务中修改数据，以便在发现任何错误时轻松地回滚更改。

REPLACE、INSERT、ON DUPLICATE KEY UPDATE

在很多情况下，我们需要处理重复项。行的唯一性由主键标识。如果行已经存在，则 REPLACE 会简单地删除行并插入新行；如果行不存在，则 REPLACE 等同于 INSERT。

如果你想在行已经存在的情况下处理重复项，则需要使用 ON DUPLICATE KEY UPDATE。如果指定了 ON DUPLICATE KEY UPDATE 选项，并且 INSERT 语句在 PRIMARY KEY 中引发了重复值，则 MySQL 会用新值更新已有行。

假设你希望每次从同一客户那里收到付款后更新之前的金额，并且在客户首次付款时插入新记录，那么你需要定义一个金额栏，并在每次收到新付款时进行更新：

```
mysql> REPLACE INTO customers VALUES (1,'Mike','Christensen','America');
Query OK, 2 rows affected (0.03 sec)
```

可以看到有两行受到影响，一个重复行被删除，一个新行被插入：

```
mysql> INSERT INTO payments VALUES('Mike Christensen', 200) ON DUPLICATE KEY
UPDATE payment=payment+VALUES(payment);
Query OK, 1 row affected (0.00 sec)

mysql> INSERT INTO payments VALUES('Ravi Vedantam',500) ON DUPLICATE KEY
UPDATE payment=payment+VALUES(payment);
Query OK, 1 row affected (0.01 sec)
```

当 Mike Christensen 下次支付 300 美元时，将更新该行并将此付款金额添加到以前的金额中：

```
mysql> INSERT INTO payments VALUES('Mike Christensen', 300) ON DUPLICATE KEY
UPDATE payment=payment+VALUES(payment);
```

```
Query OK, 2 rows affected (0.00 sec)
```

VALUES（payment）：指 INSERT 语句中给出的值，payment 指的是表中的列。

TRUNCATING TABLE

删除整个表需要很长时间，因为 MySQL 需要逐行执行操作。删除表的所有行（保留表结构）的最快方法是使用 TRUNCATE TABLE 语句。

TRUNCATING TABLE 是 MySQL 中的 DDL 操作，也就是说一旦数据被清空，就不能被回滚：

```
mysql> TRUNCATE TABLE customers;
Query OK, 0 rows affected (0.03 sec)
```

2.6 加载示例数据

你已经创建了 schema（数据库和表）以及一些数据（通过 INSERT、UPDATE 和 DELETE）。为了解释后面的各节，我们需要更多的数据。MySQL 提供了一个示例 employee 数据库和大量数据供我们学习使用。在本章中，我们将讨论如何获取这些数据并将其存储在我们的数据库中。

2.6.1 如何操作

1. 下载压缩文件：

```
shell> wget
'https://codeload.github.com/datacharmer/test_db/zip/master' -O
master.zip
```

2. 解压缩文件：

```
shell> unzip master.zip
```

3. 加载数据：

```
shell> cd test_db-master
```

```
shell> mysql -u root -p < employees.sql
mysql:[Warning] Using a password on the command line interface can be
insecure.
INFO
CREATING DATABASE STRUCTURE
INFO
storage engine:InnoDB
INFO
LOADING departments
INFO
LOADING employees
INFO
LOADING dept_emp
INFO
LOADING dept_manager
INFO
LOADING titles
INFO
LOADING salaries
data_load_time_diff
NULL
```

4. 验证数据：

```
shell> mysql -u root -p   employees -A
mysql:[Warning] Using a password on the command line interface can be
insecure.
Welcome to the MySQL monitor. Commands end with ;or \g.
Your MySQL connection id is 35
Server version:8.0.3-rc-log MySQL Community Server (GPL) Copyright (c)
2000, 2017, Oracle and/or its affiliates.All rights reserved.
Oracle is a registered trademark of Oracle Corporation and/or its
affiliates.Other names may be trademarks of their respective owners.

Type 'help;'or '\h' for help.Type '\c' to clear the current input
statement.

mysql> SHOW TABLES;
```

```
+-------------------------+
| Tables_in_employees     |
+-------------------------+
| current_dept_emp        |
| departments             |
| dept_emp                |
| dept_emp_latest_date    |
| dept_manager            |
| employees               |
| salaries                |
| titles                  |
+-------------------------+
8 rows in set (0.00 sec)

mysql> DESC employees\G
*************************** 1.row ***************************
  Field:emp_no
   Type:int(11)
   Null:NO
    Key:PRI
Default:NULL
  Extra:
*************************** 2.row ***************************
  Field:birth_date
   Type:date
   Null:NO
    Key:
Default:NULL
  Extra:
*************************** 3.row ***************************
  Field:first_name
   Type:varchar(14)
   Null:NO
    Key:
Default:NULL
  Extra:
*************************** 4.row ***************************
```

```
  Field:last_name
   Type:varchar(16)
   Null:NO
    Key:
Default:NULL
  Extra:
*************************** 5.row ***************************
  Field:gender
   Type:enum('M','F')
   Null:NO
    Key:
Default:NULL
  Extra:
*************************** 6.row ***************************
  Field:hire_date
   Type:date
   Null:NO
    Key:
Default:NULL
  Extra:
6 rows in set (0.00 sec)
```

2.7　查询数据

我们已在表中插入并更新数据，现在来学习如何从数据库中检索信息。在本节中，我们将讨论如何从创建的示例 employee 数据库中检索数据。

使用 SELECT 可以做很多事情。本节将讨论最常见的用例。有关语法和其他用例的更多详细信息，请参阅 https://dev.mysql.com/doc/refman/8.0/en/select.html。

2.7.1　如何操作

从 employee 数据库的 departments 表中选择所有数据。可以使用星号（*）从表中选择所有列。如果只选择你需要的数据，则不建议使用星号（*）：

```
mysql> SELECT * FROM departments;
```

```
+---------+--------------------+
| dept_no | dept_name          |
+---------+--------------------+
| d009    | Customer Service   |
| d005    | Development        |
| d002    | Finance            |
| d003    | Human Resources    |
| d001    | Marketing          |
| d004    | Production         |
| d006    | Quality Management |
| d008    | Research           |
| d007    | Sales              |
+---------+--------------------+
9 rows in set (0.00 sec)
```

选择列

假设你需要 dept_manager 的 emp_no 和 dept_no 列：

```
mysql> SELECT emp_no, dept_no FROM dept_manager;
+--------+---------+
| emp_no | dept_no |
+--------+---------+
| 110022 | d001    |
| 110039 | d001    |
| 110085 | d002    |
| 110114 | d002    |
| 110183 | d003    |
| 110228 | d003    |
| 110303 | d004    |
| 110344 | d004    |
| 110386 | d004    |
| 110420 | d004    |
| 110511 | d005    |
| 110567 | d005    |
| 110725 | d006    |
| 110765 | d006    |
| 110800 | d006    |
```

```
| 110854   |    d006        |
| 111035   |    d007        |
| 111133   |    d007        |
| 111400   |    d008        |
| 111534   |    d008        |
| 111692   |    d009        |
| 111784   |    d009        |
| 111877   |    d009        |
| 111939   |    d009        |
+---------+-------------+
24 rows in set (0.00 sec)
```

计数

从 employees 表中查找员工的数量：

```
mysql> SELECT COUNT(*) FROM employees;
+----------+
| COUNT(*) |
+----------+
|   300024 |
+----------+
1 row in set (0.03 sec)
```

条件过滤

找到 first_name 为 Georgi 且 last_name 为 Facello 的员工的 emp_no：

```
mysql> SELECT emp_no FROM employees WHERE first_name='Georgi' AND
last_name='Facello';
+----------+
| emp_no   |
+----------+
|  10001   |
|  55649   |
+----------+
2 rows in set (0.08 sec)
```

所有的过滤条件都是通过 WHERE 子句给出的，除整型数和浮点数之外，其他所有内容都应放在引号内。

操作符

MySQL 支持使用许多操作符来筛选结果，可以在 https://dev.mysql.com/doc/refman/8.0/en/comparison-operators.html 上获取所有操作符的列表。我们将在这里讨论其中一些操作符。在下面的例子中将详细解释 LIKE 和 RLIKE 操作符。

- **equality**：参考前面使用=进行过滤的例子。
- IN：检查一个值是否在一组值中。

  例如，找出姓氏为 Christ、Lamba 或 Baba 的所有员工的人数：

```
mysql> SELECT COUNT(*) FROM employees WHERE last_name IN ('Christ',
'Lamba', 'Baba');
+ ----------+
| COUNT(*)  |
+ ----------+
|      626  |
+ ----------+
1 row in set (0.08 sec)
```

- BETWEEN...AND：检查一个值是否在一个范围内。

  例如，找出 1986 年 12 月入职的员工人数：

```
mysql> SELECT COUNT(*) FROM employees WHERE hire_date BETWEEN
'1986-12-01' AND '1986-12-31';
+ ----------+
| COUNT(*)  |
+ ----------+
|     3081  |
+ ----------+
1 row in set (0.06 sec)
```

- NOT：你可以简单地用 NOT 运算符来否定结果。

  例如，找出不是在 1986 年 12 月入职的员工的人数：

```
mysql> SELECT COUNT(*) FROM employees WHERE hire_date NOT BETWEEN
'1986-12-01' AND '1986-12-31';
+ ----------+
| COUNT(*)  |
+ ----------+
|   296943  |
+ ----------+
1 row in set (0.08 sec)
```

简单模式匹配

可以使用 LIKE 运算符来实现简单模式匹配。使用下画线（_）来精准匹配一个字符，使用（%）来匹配任意数量的字符。

- 找出名字以 Christ 开头的所有员工的人数：

```
mysql> SELECT COUNT(*) FROM employees WHERE first_name LIKE 'christ%';
+ ----------+
| COUNT(*)  |
+ ----------+
|     1157  |
+ ----------+
1 row in set (0.06 sec)
```

- 找出名字以 Christ 开头并以 ed 结尾的所有员工的人数：

```
mysql> SELECT COUNT(*) FROM employees WHERE first_name LIKE 'christ%ed';
+ ----------+
| COUNT(*)  |
+ ----------+
|      228  |
+ ----------+
1 row in set (0.06 sec)
```

- 找出名字中包含 sri 的所有员工的人数：

```
mysql> SELECT COUNT(*) FROM employees WHERE first_name LIKE '%sri%';
+ ----------+
| COUNT(*)  |
```

```
+ ----------+
|      253  |
+ ----------+
1 row in set (0.08 sec)
```

- 找到名字以 er 结尾的所有员工的人数：

```
mysql> SELECT COUNT(*) FROM employees WHERE first_name LIKE '%er';
+ ----------+
| COUNT(*)  |
+ ----------+
|    5388   |
+ ----------+
1 row in set (0.08 sec)
```

- 找出名字以任意两个字符开头、后面跟随 ka、再后面跟随任意数量字符的所有员工的人数：

```
mysql> SELECT COUNT(*) FROM employees WHERE first_name LIKE '__ka%';
+ ----------+
| COUNT(*)  |
+ ----------+
|     1918  |
+ ----------+
1 row in set (0.06 sec)
```

正则表达式

你可以利用 RLIKE 或 REGEXP 运算符在 WHERE 子句中使用正则表达式。使用 REGEXP 的方法有多种（见下表），更多示例请参阅 https://dev.mysql.com/doc/refman/8.0/en/regexp.html。

表达式	描述
*	零次或多次重复
+	一个或多个重复
?	可选字符
.	任何字符
\.	区间

续表

表达式	描述
^	以……开始
$	以……结束
[abc]	只有 *a*、*b* 或 *c*
[^abc]	非 *a*，非 *b*，亦非 *c*
[a-z]	字符 *a* 到 *z*
[0-9]	数字 0 到 9
^...$	开始和结束
\d	任何数字
\D	任何非数字字符
\s	任何空格
\S	任何非空白字符
\w	任何字母数字字符
\W	任何非字母数字字符
{m}	*m* 次重复
{m,n}	*m* 到 *n* 次重复

- 找出名字以 Christ 开头的所有员工的人数：

```
mysql> SELECT COUNT(*) FROM employees WHERE first_name RLIKE '^christ';
+ ----------+
| COUNT(*)  |
+ ----------+
|     1157  |
+ ----------+
1 row in set (0.18 sec)
```

- 找出姓氏以 ba 结尾的所有员工的人数：

```
mysql> SELECT COUNT(*) FROM employees WHERE last_name REGEXP 'ba$';
+ ----------+
| COUNT(*)  |
+ ----------+
|     1008  |
+ ----------+
1 row in set (0.15 sec)
```

- 查找姓氏不包含元音（a、e、i、o和u）的所有员工的人数：

```
mysql> SELECT COUNT(*) FROM employees WHERE last_name NOT REGEXP
'[aeiou]';
+ ----------+
| COUNT(*)  |
+ ----------+
|      148  |
+ ----------+
1 row in set (0.11 sec)
```

限定结果

查询 hire_date 在1986年之前的任何10名员工的姓名。可以在查询语句末尾使用 LIMIT 子句来实现此查询：

```
mysql> SELECT first_name, last_name FROM employees WHERE hire_date <
'1986-01-01' LIMIT 10;
+------------+------------+
| first_name | last_name  |
+------------+------------+
| Bezalel    | Simmel     |
| Sumant     | Peac       |
| Eberhardt  | Terkki     |
| Otmar      | Herbst     |
| Florian    | Syrotiuk   |
| Tse        | Herber     |
| Udi        | Jansch     |
| Reuven     | Garigliano |
| Erez       | Ritzmann   |
| Premal     | Baek       |
+------------+------------+
10 rows in set (0.00 sec)
```

使用表别名

默认情况下，SELECT 子句中给出的任何列都将显示在结果中。在前面的示例中，我们已经得出了统计数值，但它显示为 COUNT(*)。你可以使用 AS 别名来更改 COUNT(*)：

```
mysql> SELECT COUNT(*) AS count FROM employees WHERE hire_date BETWEEN
'1986-12-01' AND '1986-12-31';
+--------+
| count |
+--------+
| 3081  |
+--------+
1 row in set (0.06 sec)
```

2.8 对结果排序

可以根据列或别名列对结果进行排序，也可以用 DESC 指定按降序或用 ASC 指定按升序来排序。默认情况下，排序将按照升序进行。你可以将 LIMIT 子句与 ORDER BY 结合使用以限定结果集。

2.8.1 如何操作

查找薪水最高的前 5 名员工的员工编号。

```
mysql> SELECT emp_no,salary FROM salaries ORDER BY salary DESC LIMIT 5;
+--------+--------+
| emp_no | salary |
+--------+--------+
| 43624  | 158220 |
| 43624  | 157821 |
| 254466 | 156286 |
| 47978  | 155709 |
| 253939 | 155513 |
+--------+--------+
5 rows in set (0.74 sec)
```

你可以在 SELECT 语句中提及列的位置，而不是指定列名称。例如，你想对位于第二列的工资进行排序，那么可以指定 ORDER BY 2：

```
mysql> SELECT emp_no,salary FROM salaries ORDER BY 2 DESC LIMIT 5;
+--------+--------+
| emp_no | salary |
```

```
+--------+--------+
| 43624  | 158220 |
| 43624  | 157821 |
| 254466 | 156286 |
| 47978  | 155709 |
| 253939 | 155513 |
+--------+--------+
5 rows in set (0.78 sec)
```

2.9 对结果分组（聚合函数）

你可以在列上使用 GROUP BY 子句对结果进行分组，然后使用 AGGREGATE（聚合）函数，例如 COUNT、MAX、MIN 和 AVERAGE。还可以在 group by 子句中的列上使用函数。请参阅下面的 SUM 示例，其中用到了 YEAR() 函数。

2.9.1 如何操作

前面提到的每个聚合函数都会在本节详细介绍。

COUNT

1. 分别找出男性和女性员工的人数：

```
mysql> SELECT gender, COUNT(*) AS count FROM employees GROUP BY
gender;
+--------+--------+
| gender | count  |
+--------+--------+
| M      | 179973 |
| F      | 120051 |
+--------+--------+
2 rows in set (0.14 sec)
```

2. 如果你希望查找员工名字中最常见的 10 个名字，可以使用 GROUP BY first_name 对所有名字分组，然后使用 COUNT(first_name)在各组内计数，最后使用 ORDER BY 计数对结果进行排序，并将返回结果行数限制为前 10 行：

```
mysql> SELECT first_name, COUNT(first_name) AS count FROM employees
GROUP BY first_name ORDER BY count DESC LIMIT 10;
+-------------+-------+
| first_name  | count |
+-------------+-------+
| Shahab      |   295 |
| Tetsushi    |   291 |
| Elgin       |   279 |
| Anyuan      |   278 |
| Huican      |   276 |
| Make        |   275 |
| Panayotis   |   272 |
| Sreekrishna |   272 |
| Hatem       |   271 |
| Giri        |   270 |
+-------------+-------+
10 rows in set (0.21 sec)
```

SUM

查找每年给予员工的薪水总额，并按薪水高低对结果进行排序。YEAR() 函数将返回给定日期所在的年份：

```
mysql> SELECT '2017-06-12', YEAR('2017-06-12');
+------------+--------------------+
| 2017-06-12 | YEAR('2017-06-12') |
+------------+--------------------+
| 2017-06-12 |               2017 |
+------------+--------------------+
1 row in set (0.00 sec)
mysql> SELECT YEAR(from_date), SUM(salary) AS sum FROM salaries GROUP BY
YEAR(from_date) ORDER BY sum DESC;
+-----------------+-------------+
| YEAR(from_date) | sum         |
+-----------------+-------------+
|            2000 | 17535667603 |
|            2001 | 17507737308 |
```

```
|            1999 | 17360258862 |
|            1998 | 16220495471 |
|            1997 | 15056011781 |
|            1996 | 13888587737 |
|            1995 | 12638817464 |
|            1994 | 11429450113 |
|            2002 | 10243347616 |
|            1993 | 10215059054 |
|            1992 |  9027872610 |
|            1991 |  7798804412 |
|            1990 |  6626146391 |
|            1989 |  5454260439 |
|            1988 |  4295598688 |
|            1987 |  3156881054 |
|            1986 |  2052895941 |
|            1985 |   972864875 |
+-----------------+-------------+
18 rows in set (1.47 sec)
```

AVERAGE

查找平均工资最高的 10 名员工：

```
mysql> SELECT emp_no, AVG(salary) AS avg FROM salaries GROUP BY emp_no
ORDER BY avg DESC LIMIT 10;
+---------- +------------- +
| emp_no    | avg          |
+---------- +------------- +
| 109334    | 141835.3333  |
| 205000    | 141064.6364  |
| 43624     | 138492.9444  |
| 493158    | 138312.8750  |
| 37558     | 138215.8571  |
| 276633    | 136711.7333  |
| 238117    | 136026.2000  |
| 46439     | 135747.7333  |
| 254466    | 135541.0625  |
| 253939    | 135042.2500  |
```

```
+---------- +-------------- +
10 rows in set (0.91 sec)
```

DISTINCT

可以使用DISTINCT子句过滤出表中的不同条目：

```
mysql> SELECT DISTINCT title FROM titles;
+--------------------+
| title              |
+--------------------+
| Senior Engineer    |
| Staff              |
| Engineer           |
| Senior Staff       |
| Assistant Engineer |
| Technique Leader   |
| Manager            |
+--------------------+
7 rows in set (0.30 sec)
```

使用HAVING过滤

可以通过添加HAVING子句来过滤GROUP BY子句的结果。

例如，找到平均工资超过140,000美元的员工：

```
mysql> SELECT emp_no, AVG(salary) AS avg FROM salaries GROUP BY emp_no
HAVING avg > 140000 ORDER BY avg DESC;
+---------- +-------------- +
| emp_no    |  avg          |
+---------- +-------------- +
| 109334    | 141835.3333   |
| 205000    | 141064.6364   |
+---------- +-------------- +
2 rows in set (0.80 sec)
```

2.9.2 延伸阅读

还有许多其他聚合函数，请参阅 https://dev.mysql.com/doc/refman/8.0/en/group-by-functions.html 以获取更多信息。

2.10 创建用户

到目前为止，我们一直在使用 root 用户连接到 MySQL 并执行语句。但其实不应该在访问 MySQL 时使用 root 用户，除非是 localhost 的管理任务。你应该创建用户、限制访问、限制资源使用，等等。为了创建新用户，你需要拥有 CREATE USER 权限，这将在 2.10.1 节讨论。在初始设置过程中，你可以使用 root 用户创建其他用户。

2.10.1 如何操作

使用 root 用户连接到 mysql 并执行 CREATE USER 命令来创建新用户。

```
mysql> CREATE USER IF NOT EXISTS 'company_read_only'@'localhost'
IDENTIFIED WITH mysql_native_password
BY 'company_pass'
WITH MAX_QUERIES_PER_HOUR 500
MAX_UPDATES_PER_HOUR 100;
```

如果密码的安全性不够强，可能会出现以下错误。

```
ERROR 1819 (HY000):Your password does not satisfy the current policy
requirements
```

上述声明将为用户创建以下内容。

- 用户名：company_read_only。
- 仅从 localhost 访问。
- 可以限制对 IP 范围的访问，例如 10.148.%.%。通过给出%，用户可以从任何主机访问。
- 密码：company_pass。
- 使用 mysql_native_password（默认）身份验证。
- 还可以指定任何可选的身份验证，例如 sha256_password、LDAP 或 Kerberos。

- 用户可以在一小时内执行的最大查询数为 500。
- 用户可以在一小时内执行的最大更新次数为 100 次。

当客户端连接到 MySQL 服务器时，它会经历两个访问控制阶段：

1. 连接验证
2. 请求验证

在连接验证过程中，服务器通过用户名和连接的主机名来识别连接。服务器会调用用户认证插件并验证密码。服务器还会检查用户是否被锁定。

在请求验证阶段，服务器会检查用户是否有足够的权限执行每项操作。

在前面的语句中，必须以明文形式输入密码，这些密码可以记录在命令历史记录文件 `$ HOME/.mysql_history` 中。为了避免这种情况（以明文形式输入密码），你可以在本地服务器上计算 hash 值并直接指定 hash 字符串。语法与之前几乎相同，除了需要把 `mysql_native_password BY 'company_pass'` 更改为 `mysql_native_password AS 'hashed_string'`：

```
mysql> SELECT PASSWORD('company_pass');
+-------------------------------------------+
|PASSWORD('company_pass')                   |
+-------------------------------------------+
| *EBD9E3BFD1489CA1EB0D2B4F29F6665F321E8C18 |
+-------------------------------------------+
1 row in set, 1 warning (0.00 sec)
mysql> CREATE USER IF NOT EXISTS 'company_read_only'@'localhost' IDENTIFIED
WITH mysql_native_password
AS '*EBD9E3BFD1489CA1EB0D2B4F29F6665F321E8C18'
WITH MAX_QUERIES_PER_HOUR 500
MAX_UPDATES_PER_HOUR 100;
```

你可以通过授予权限的方式直接创建用户，请参阅 2.11 节。但是，MySQL 将在下一版本中弃用此功能。

2.10.2 延伸阅读

有关创建用户的更多选择，请参阅 https://dev.mysql.com/doc/refman/8.0/en/create-user.html。其他安全的认证方法（如 SSL）将在第 14 章中讨论。

2.11 授予和撤销用户的访问权限

你可以限制用户访问特定数据库或表，或限制特定操作，如 SELECT、INSERT 和 UPDATE。你需要拥有 GRANT 权限，才能为其他用户授予权限。

2.11.1 如何操作

在初始设置期间，可以使用 root 用户授予权限，还可以创建管理员账户来管理用户。

授予权限

- 将 READ ONLY（SELECT）权限授予 company_read_only 用户：

```
mysql> GRANT SELECT ON company.* TO
'company_read_only'@'localhost';
Query OK, 0 rows affected (0.06 sec)
```

  星号（*）表示数据库内的所有表。

- 将 INSERT 权限授予新的 company_insert_only 用户：

```
mysql> GRANT INSERT ON company.* TO
'company_insert_only'@'localhost' IDENTIFIED BY 'xxxx';
Query OK, 0 rows affected, 1 warning (0.05 sec)

mysql> SHOW WARNINGS\G
*************************** 1.row ***************************
  Level: Warning
   Code: 1287
Message: Using GRANT for creating new user is deprecated and will
be removed in future release. Create new user with CREATE USER
statement.
```

```
1 row in set (0.00 sec)
```

- 将WRITE权限授予新的company_write用户：

```
mysql> GRANT INSERT, DELETE, UPDATE ON company.* TO
'company_write'@'%' IDENTIFIED WITH mysql_native_password AS
'*EBD9E3BFD1489CA1EB0D2B4F29F6665F321E8C18';
Query OK, 0 rows affected, 1 warning (0.04 sec)
```

- 限制查询指定的表。将employees_read_only用户限制为仅能查询employees表：

```
mysql> GRANT SELECT ON employees.employees TO
'employees_read_only'@'%' IDENTIFIED WITH mysql_native_password AS
'*EBD9E3BFD1489CA1EB0D2B4F29F6665F321E8C18';
Query OK, 0 rows affected, 1 warning (0.03 sec)
```

- 可以进一步将访问权限限制为仅能查询指定列。限制employees_ro用户仅能访问employees表的first_name列和last_name列：

```
mysql> GRANT SELECT(first_name,last_name) ON employees.employees TO
'employees_ro'@'%' IDENTIFIED WITH mysql_native_password AS
'*EBD9E3BFD1489CA1EB0D2B4F29F6665F321E8C18';
Query OK, 0 rows affected, 1 warning (0.06 sec)
```

- 扩展授权。可以通过执行新授权来扩展授权。将权限扩展到employees_col_ro用户，以访问薪资（salaries）表中的薪水：

```
mysql> GRANT SELECT(salary) ON employees.salaries TO
'employees_ro'@'%';
Query OK, 0 rows affected (0.00 sec)
```

- 创建SUPER用户。需要一个管理员账户来管理该服务器。ALL表示除GRANT权限之外的所有权限。

```
mysql> CREATE USER 'dbadmin'@'%' IDENTIFIED WITH
mysql_native_password BY 'DB@dm1n';
Query OK, 0 rows affected (0.01 sec)

mysql> GRANT ALL ON *.* TO 'dbadmin'@'%';
Query OK, 0 rows affected (0.01 sec)
```

- 授予 GRANT 特权。用户拥有 GRANT OPTION 权限才能授予其他用户权限。可以将 GRANT 特权扩展到 dbadmin 超级用户：

```
mysql> GRANT GRANT OPTION ON *.* TO 'dbadmin'@'%';
Query OK, 0 rows affected (0.03 sec)
```

更多有关的权限类型请参阅 https://dev.mysql.com/doc/refman/8.0/en/grant.html。

检查授权

你可以检查所有用户的授权。检查 employee_col_ro 用户的授权：

```
mysql> SHOW GRANTS FOR 'employees_ro'@'%'\\G
*************************** 1. row ***************************
Grants for employees_ro@%: GRANT USAGE ON *.* TO `employees_ro`@`%`
*************************** 2. row ***************************
Grants for employees_ro@%: GRANT SELECT (`first_name`, `last_name`) ON
`employees`.`employees` TO `employees_ro`@`%`
*************************** 3. row ***************************
Grants for employees_ro@%: GRANT SELECT (`salary`) ON
`employees`.`salaries` TO `employees_ro`@`%`
```

检查 dbadmin 用户的授权。可以看到 dbadmin 用户拥有的所有授权：

```
mysql> SHOW GRANTS FOR 'dbadmin'@'%'\G
*************************** 1.row ***************************
Grants for dbadmin@%:GRANT SELECT, INSERT, UPDATE, DELETE, CREATE, DROP,
RELOAD, SHUTDOWN, PROCESS, FILE, REFERENCES, INDEX, ALTER, SHOW DATABASES,
SUPER, CREATE TEMPORARY TABLES, LOCK TABLES, EXECUTE, REPLICATION SLAVE,
REPLICATION CLIENT, CREATE VIEW, SHOW VIEW, CREATE ROUTINE, ALTER ROUTINE,
CREATE USER, EVENT, TRIGGER, CREATE TABLESPACE, CREATE ROLE, DROP ROLE ON
*.* TO `dbadmin`@`%` WITH GRANT OPTION
*************************** 2.row ***************************
Grants for dbadmin@%:GRANT
BINLOG_ADMIN,CONNECTION_ADMIN,ENCRYPTION_KEY_ADMIN,GROUP_REPLICATION_ADMIN,
REPLICATION_SLAVE_ADMIN,ROLE_ADMIN,SET_USER_ID,SYSTEM_VARIABLES_ADMIN ON
*.* TO `dbadmin`@`%`
```

```
2 rows in set (0.00 sec)
```

撤销权限

撤销权限与创建权限的语法相同。向用户授予权限用 TO，撤销用户的权限用 FROM。

- 撤销'company_write'@'%'用户的 DELETE 访问权限：

```
mysql> REVOKE DELETE ON company.* FROM 'company_write'@'%';
Query OK, 0 rows affected (0.04 sec)
```

- 撤销 employee_ro 用户对薪水列的访问权限：

```
mysql> REVOKE SELECT(salary) ON employees.salaries FROM
'employees_ro'@'%';
Query OK, 0 rows affected (0.03 sec)
```

修改 mysql.user 表

所有用户信息及权限都存储在 mysql.user 表中。如果你有权访问 mysql.user 表，则可以直接通过修改 mysql.user 表来创建用户并授予权限。

如果你使用 GRANT、REVOKE、SET PASSWORD 或 RENAME USER 等账户管理语句间接修改授权表，则服务器会通知这些更改，并立即再次将授权表加载到内存中。

如果使用 INSERT、UPDATE 或 DELETE 等语句直接修改授权表，则更改不会影响权限检查，除非你重新启动服务器或指示其重新加载表。如果直接更改授权表，但忘记了重新加载表，那么在重新启动服务器之前，这些更改无效。

可以通过执行 FLUSH PRIVILEGES 语句来完成 GRANT 表的重新加载。

查询 mysql.user 表以找出 dbadmin 用户的所有条目：

```
mysql> SELECT * FROM mysql.user WHERE user='dbadmin'\G
*************************** 1.row ***************************
                  Host:%
                  User:dbadmin
           Select_priv:Y
           Insert_priv:Y
           Update_priv:Y
```

```
              Delete_priv:Y
              Create_priv:Y
                Drop_priv:Y
              Reload_priv:Y
            Shutdown_priv:Y
             Process_priv:Y
                File_priv:Y
               Grant_priv:Y
          References_priv:Y
               Index_priv:Y
               Alter_priv:Y
             Show_db_priv:Y
               Super_priv:Y
    Create_tmp_table_priv:Y
         Lock_tables_priv:Y
             Execute_priv:Y
          Repl_slave_priv:Y
         Repl_client_priv:Y
         Create_view_priv:Y
           Show_view_priv:Y
      Create_routine_priv:Y
       Alter_routine_priv:Y
         Create_user_priv:Y
               Event_priv:Y
             Trigger_priv:Y
   Create_tablespace_priv:Y
                 ssl_type:
               ssl_cipher:
              x509_issuer:
             x509_subject:
            max_questions:0
              max_updates:0
          max_connections:0
     max_user_connections:0
                   plugin:mysql_native_password
    authentication_string:*AB7018ADD9CB4EDBEB680BB3F820479E4CE815D2
         password_expired:N
```

```
 password_last_changed:2017-06-10 16:24:03
     password_lifetime:NULL
        account_locked:N
      Create_role_priv:Y
        Drop_role_priv:Y
1 row in set (0.00 sec)
```

你可以看到 dbadmin 用户能从任意主机（%）访问数据库，只需更新 mysql.user 表并重新加载授权表，即可将它们限制为从 localhost 访问数据库：

```
mysql> UPDATE mysql.user SET host='localhost' WHERE user='dbadmin';
Query OK, 1 row affected (0.02 sec)
Rows matched:1  Changed:1   Warnings:<m i='20</m>15'>0

mysql> FLUSH PRIVILEGES;
Query OK, 0 rows affected (0.00 sec)
```

设置用户密码有效期

可以设置一段时间作为用户密码的有效期，过期之后用户则需要更改密码。

当应用程序开发人员要求访问数据库时，可以使用默认密码创建该账户，并将其设置为过期状态。然后与开发人员分享此密码，他们则必须更改密码才能继续使用 MySQL。

创建所有账户并设置其密码过期日期等于 default_password_lifetime 变量的值，默认情况下用户被禁用。

- 创建一个具有过期密码的用户。当开发人员第一次登录并尝试执行任何语句时，错误 1820(HY000)：将被抛出。在执行此语句之前，必须使用 ALTER USER 语句重置密码：

  ```
  mysql> CREATE USER 'developer'@'%' IDENTIFIED WITH
  mysql_native_password AS
  '*EBD9E3BFD1489CA1EB0D2B4F29F6665F321E8C18' PASSWORD EXPIRE;
  Query OK, 0 rows affected (0.04 sec

  shell> mysql -u developer -pcompany_pass
  ```

```
mysql:[Warning] Using a password on the command line interface can be
insecure.
Welcome to the MySQL monitor. Commands end with ;or \g.
Your MySQL connection id is 31
Server version:8.0.3-rc-log
Copyright (c) 2000, 2017, Oracle and/or its affiliates.All rights
reserved.

Oracle is a registered trademark of Oracle Corporation and/or its
affiliates.Other names may be trademarks of their respective owners.

Type 'help;'or '\h' for help.Type '\c' to clear the current input
statement.

mysql> SHOW DATABASES;
ERROR 1820 (HY000):You must reset your password using ALTER USER
statement before executing this statement.
```

开发人员必须使用以下命令更改密码：

```
mysql> ALTER USER 'developer'@'%' IDENTIFIED WITH mysql_native_password BY
'new_company_pass';
Query OK, 0 rows affected (0.03 sec)
```

- 手动设置过期用户：

```
mysql> ALTER USER 'developer'@'%' PASSWORD EXPIRE;
Query OK, 0 rows affected (0.06 sec)
```

- 要求用户每隔 90 天更改一次密码：

```
mysql> ALTER USER 'developer'@'%' PASSWORD EXPIRE INTERVAL 90 DAY;
Query OK, 0 rows affected (0.04 sec)
```

锁定用户

如果发现账户有任何问题，可以将其锁定。MySQL 支持使用 CREATE USER 或 ALTER USER 锁定用户。

通过将 ACCOUNT LOCK 子句添加到 ALTER USER 语句来锁定账户：

```
mysql> ALTER USER 'developer'@'%' ACCOUNT LOCK;
Query OK, 0 rows affected (0.05 sec)
```

开发者会收到该账户已被锁定的报错消息：

```
shell> mysql -u developer -pnew_company_pass
mysql:[Warning] Using a password on the command line interface can be
insecure.
ERROR 3118 (HY000):Access denied for user 'developer'@'localhost'.Account
is locked.
```

确认后可以解锁该账户：

```
mysql> ALTER USER 'developer'@'%' ACCOUNT UNLOCK;
Query OK, 0 rows affected (0.00 sec)
```

为用户创建角色

MySQL 的角色是一个权限的集合。与用户账户一样，角色的权限可以被授予和撤销。用户账户被授予角色后，该角色就会将其拥有的权限授予该账户。之前，我们为不同的用户创建了读取、写入和管理权限。对于写入权限，我们已授予用户 INSERT、DELETE 和 UPDATE 权限。现在你可以将这些权限授予某个角色，然后为用户分配该角色。通过这种方式，可以避免为许多用户账户单独授予权限的麻烦。

- 创建角色：

  ```
  mysql> CREATE ROLE 'app_read_only', 'app_writes', 'app_developer';
  Query OK, 0 rows affected (0.01 sec)
  ```

- 使用 GRANT 语句为角色分配权限：

  ```
  mysql> GRANT SELECT ON employees.* TO 'app_read_only';
  Query OK, 0 rows affected (0.00 sec)

  mysql> GRANT INSERT, UPDATE, DELETE ON employees.* TO 'app_writes';
  Query OK, 0 rows affected (0.00 sec)
  ```

```
mysql> GRANT ALL ON employees.* TO 'app_developer';
Query OK, 0 rows affected (0.04 sec)
```

- 创建用户。如果你不指定主机，则将采用%（任意主机）：

```
mysql> CREATE user emp_read_only IDENTIFIED BY 'emp_pass';
Query OK, 0 rows affected (0.06 sec)

mysql> CREATE user emp_writes IDENTIFIED BY 'emp_pass';
Query OK, 0 rows affected (0.04 sec)

mysql> CREATE user emp_developer IDENTIFIED BY 'emp_pass';
Query OK, 0 rows affected (0.01 sec)

mysql> CREATE user emp_read_write IDENTIFIED BY 'emp_pass';
Query OK, 0 rows affected (0.00 sec)
```

- 使用 GRANT 语句为用户分配角色。你可以为用户分配多个角色。例如，可以将读取和写入权限都分配给 emp_read_write 用户：

```
mysql> GRANT 'app_read_only' TO 'emp_read_only'@'%';
Query OK, 0 rows affected (0.04 sec)

mysql> GRANT 'app_writes' TO 'emp_writes'@'%';
Query OK, 0 rows affected (0.00 sec)

mysql> GRANT 'app_developer' TO 'emp_developer'@'%';
Query OK, 0 rows affected (0.00 sec)

mysql> GRANT 'app_read_only', 'app_writes' TO 'emp_read_write'@'%';
Query OK, 0 rows affected (0.05 sec)
```

为安全起见，请避免使用%，并限制对部署应用程序 IP 的访问。

2.12 查询数据并保存到文件和表中

我们可以使用 SELECT INTO OUTFILE 语句将输出保存到文件中。

可以指定列和行分隔符，然后可以将数据导入其他数据平台。

2.12.1　如何操作

我们可以将输出目标另存为文件或表。

另存为文件

- 要将输出结果保存到文件中，你需要拥有 FILE 权限。FILE 是一个全局特权，这意味着你不能将其限制为针对特定数据库的权限。但是，你可以限制用户查询的内容：

```
mysql> GRANT SELECT ON employees.* TO 'user_ro_file'@'%' IDENTIFIED
WITH mysql_native_password AS
'*EBD9E3BFD1489CA1EB0D2B4F29F6665F321E8C18';
Query OK, 0 rows affected, 1 warning (0.00 sec)

mysql> GRANT FILE ON *.* TO 'user_ro_file'@'%' IDENTIFIED WITH
mysql_native_password AS
'*EBD9E3BFD1489CA1EB0D2B4F29F6665F321E8C18';
Query OK, 0 rows affected, 1 warning (0.00 sec)
```

- 在 Ubuntu 系统中，默认情况下，MySQL 不允许写入文件。你应该在配置文件中设置 secure_file_priv 并重新启动 MySQL。我们将在第 4 章了解关于配置的更多信息。在 CentOS、Red Hat 系统中，secure_file_priv 被设置为/var/lib/mysql-files，这意味着所有文件都将被保存在该目录中。
- 现在，像下面这样启用写入文件的功能。打开配置文件并添加 secure_file_priv = /var/lib/mysql:

```
shell> sudo vi /etc/mysql/mysql.conf.d/mysqld.cnf
```

- 重新启动 MySQL 服务器：

```
shell> sudo systemctl restart mysql
```

以下语句会将输出结果保存为 CSV 格式：

```
mysql> SELECT first_name, last_name INTO OUTFILE 'result.csv'
```

```
        FIELDS TERMINATED BY ',' OPTIONALLY ENCLOSED BY '"'
        LINES TERMINATED BY '\n'
        FROM employees WHERE hire_date<'1986-01-01' LIMIT 10;
Query OK, 10 rows affected (0.00 sec)
```

你可以检查文件的输出结果，该文件将被创建在{secure_file_priv}/{database_name}指定的路径中，在本例中路径是/var/lib/mysql/employees/。如果目标文件已存在，语句将执行失败，所以每次执行或将文件移动到其他位置时都需要提供一个唯一的名称：

```
shell> sudo cat /var/lib/mysql/employees/result.csv
"Bezalel","Simmel"
"Sumant","Peac"
"Eberhardt","Terkki"
"Otmar","Herbst"
"Florian","Syrotiuk"
"Tse","Herber"
"Udi","Jansch"
"Reuven","Garigliano"
"Erez","Ritzmann"
"Premal","Baek"
```

另存为表

我们也可以将 SELECT 语句的结果保存到表中。即使表不存在，也可以使用 CREATE 和 SELECT 来创建表并加载数据。如果表已存在，则可以使用 INSERT 和 SELECT 加载数据。

可以将标题保存到新的 titles_only 表中：

```
mysql> CREATE TABLE titles_only AS SELECT DISTINCT title FROM titles;
Query OK, 7 rows affected (0.50 sec)
Records:7 Duplicates:0 Warnings:0
```

如果表已经存在，则可以使用 INSERT INTO SELECT 语句：

```
mysql> INSERT INTO titles_only SELECT DISTINCT title FROM titles;Query OK,
7 rows affected (0.46 sec)
```

```
Records:7 Duplicates:0 Warnings:0
```

为避免重复，可以使用 INSERT IGNORE。但是在本例中，titles_only 表中没有 PRIMARY KEY，因此 IGNORE 子句不会造成任何影响。

2.13　将数据加载到表中

前面我们将表数据转储到了文件中，反过来操作也可以，即将文件中的数据加载到表中。这种方式广泛用于加载批量数据的情况，并且是将数据加载到表中的超快速方式。你可以指定列分隔符将数据加载到相应的列中。你需要拥有表的 FILE 权限和 INSERT 权限。

2.13.1　如何操作

前面我们已将 first_name 和 last_name 保存到一个文件中。你可以使用相同的文件将数据加载到另一个表中。在加载之前需要创建表，不过如果表已经存在，则可以直接加载。表中的列应该匹配文件的字段。

创建一个表来保存数据：

```
mysql>  CREATE TABLE employee_names (
        `first_name` varchar(14) NOT NULL,
        `last_name` varchar(16) NOT NULL
        ) ENGINE=InnoDB;
Query OK, 0 rows affected (0.07 sec)
```

确保文件存在：

```
shell> sudo ls -lhtr /var/lib/mysql/employees/result.csv
-rw-rw-rw- 1 mysql mysql 180 Jun 10 14:53
/var/lib/mysql/employees/result.csv
```

使用 LOAD DATA INFILE 语句加载数据：

```
mysql> LOAD DATA INFILE 'result.csv' INTO TABLE employee_names
       FIELDS TERMINATED BY ','
       OPTIONALLY ENCLOSED BY '"'
       LINES TERMINATED BY '\n';
Query OK, 10 rows affected (0.01 sec)
```

```
Records:10  Deleted:0   Skipped:0   Warnings:0
```

该文件可以以完整路径名的形式给出，以指定其确切位置。如果以相对路径名的形式给出，则相对路径名将被解析为相对于客户机程序启动的目录。

- 如果文件开头包含一些你想忽略的行，可以用 IGNORE n Lines 指定：

```
mysql> LOAD DATA INFILE 'result.csv' INTO TABLE employee_names
       FIELDS TERMINATED BY ','
       OPTIONALLY ENCLOSED BY '"'
       LINES TERMINATED BY '\n'
       IGNORE 1 LINES;
```

- 可以用 REPLACE 或者 IGNORE 来处理重复的行：

```
mysql> LOAD DATA INFILE 'result.csv' REPLACE INTO TABLE
employee_names FIELDS TERMINATED BY ','OPTIONALLY ENCLOSED BY '"' LINES
TERMINATED BY '\n';
Query OK, 10 rows affected (0.01 sec)
Records:10 Deleted:0   Skipped:0   Warnings:<m i='2240</m>'>0

mysql> LOAD DATA INFILE 'result.csv' IGNORE INTO TABLE
employee_names FIELDS TERMINATED BY ','OPTIONALLY ENCLOSED BY '"' LINES
TERMINATED BY '\n';
Query OK, 10 rows affected (0.06 sec)
Records:10 Deleted:0   Skipped:0   Warnings:0
```

- MySQL 假定你要加载的文件在服务器上是可用的。如果你要从远程客户机连接到服务器，则可以指定 LOCAL 以装载位于客户机上的文件。本地文件将从客户端被复制到服务器上。该文件保存在服务器的常规临时路径中，在 Linux 系统中是/tmp：

```
mysql> LOAD DATA LOCAL INFILE 'result.csv' IGNORE INTO TABLE
employee_names FIELDS TERMINATED BY ','OPTIONALLY ENCLOSED BY '"'
LINES TERMINATED BY '\n';
```

2.14 表关联

到目前为止，你已经了解了从单个表插入和检索数据的操作。在本节中，我们将讨论如何关联两个或多个表来检索结果。

假设你想用 emp_no: 110022 找到员工的姓名和部门号码:

- 部门编号和名称存储在 departments 表中。
- 员工编号和其他详细信息（例如 first_name 和 last_name）存储在 employees 表中。
- 员工和部门的映射关系存储在 dept_manager 表中。

如果你不想使用 JOIN，可以这样做:

1. 从 employee 表中查找 emp_no 为 110022 的员工姓名:

```
mysql> SELECT emp.emp_no, emp.first_name, emp.last_name
FROM employees AS emp
WHERE emp.emp_no=110022;
+--------+------------+------------+
| emp_no | first_name | last_name  |
+--------+------------+------------+
| 110022 | Margareta  | Markovitch |
+--------+------------+------------+
1 row in set (0.00 sec)
```

2. 从 departments 表中查找部门编号:

```
mysql> SELECT dept_no FROM dept_manager AS dept_mgr WHERE
dept_mgr.emp_no=110022;
+---------+
| dept_no |
+---------+
| d001    |
+---------+
1 row in set (0.00 sec)
```

3. 从 departments 表中查找部门名称:

```
mysql> SELECT dept_name FROM departments dept WHERE
dept.dept_no='d001';
+-----------+
```

```
| dept_name |
+-----------+
| Marketing |
+-----------+
1 row in set (0.00 sec)
```

2.14.1 如何操作

为了避免使用三条语句查找三个不同的表的麻烦操作,你可以使用JOIN来关联它们。这里要注意的重要一点是，为了关联两个表，它们必须有一个或多个共同列。例如，你可以基于emp_no来关联employees表和 dept_manager表,因为它们都有emp_no列。虽然名称不需要匹配，但应该找出可以用来做关联的列。与之类似，dept_mgr 表和departments表都有dept_no列，可以将其作为共有列。

与列别名一样，你可以为表提供别名，并使用别名来引用该表的列。例如，你可以使用FROM employees AS emp为employees表提供别名，并使用点号引用employees表的列，例如：emp.emp_no:

```
mysql> SELECT
    emp.emp_no,
    emp.first_name,
    emp.last_name,
    dept.dept_name
FROM
    employees AS emp
JOIN dept_manager AS dept_mgr
    ON emp.emp_no=dept_mgr.emp_no AND emp.emp_no=110022
JOIN departments AS dept
    ON dept_mgr.dept_no=dept.dept_no;
+--------+------------+------------+-----------+
| emp_no | first_name | last_name  | dept_name |
+--------+------------+------------+-----------+
| 110022 | Margareta  | Markovitch | Marketing |
+--------+------------+------------+-----------+
1 row in set (0.00 sec)
```

我们来看另一个例子。假设你想了解每个部门的平均工资，你可以使用AVG函数并按照 dept_no 进行分组。要找出部门名称，可以将结果与 departments 表通过 dept_no 列进行关联：

```
mysql> SELECT
    dept_name,
    AVG(salary) AS avg_salary
FROM
    salaries
JOIN dept_emp
    ON salaries.emp_no=dept_emp.emp_no
JOIN departments
    ON dept_emp.dept_no=departments.dept_no
GROUP BY
    dept_emp.dept_no
ORDER BY
    avg_salary
DESC;
+--------------------+------------+
| dept_name          | avg_salary |
+--------------------+------------+
| Sales              | 80667.6058 |
| Marketing          | 71913.2000 |
| Finance            | 70489.3649 |
| Research           | 59665.1817 |
| Production         | 59605.4825 |
| Development        | 59478.9012 |
| Customer Service   | 58770.3665 |
| Quality Management | 57251.2719 |
| Human Resources    | 55574.8794 |
+--------------------+------------+
9 rows in set (8.29 sec)
```

通过与自己关联来识别重复项

假设你想在某个表里为某些特定列找出重复的行。例如，你想要找出哪些员工具有相同的 first_name、相同的 last_name、相同的 gender，以及相同的 hire_date。在这种情况下，你可以将 employees 表跟它自己进行关联，同时在 JOIN 子句中指定查找重复项的列。你需要为每个表设定不同的别名。

你需要为用来做关联的列添加索引。索引将在第 13 章中讨论。我们现在可以执行下面的命令来添加索引：

```
mysql> ALTER TABLE employees ADD INDEX name(first_name, last_name);
Query OK, 0 rows affected (1.95 sec)
Records:0 Duplicates:0 Warnings:0

mysql> SELECT
     emp1.*
FROM
     employees emp1
JOIN employees emp2
    ON emp1.first_name=emp2.first_name
    AND emp1.last_name=emp2.last_name
    AND emp1.gender=emp2.gender
    AND emp1.hire_date=emp2.hire_date
    AND emp1.emp_no!=emp2.emp_no
ORDER BY
    first_name, last_name;
+--------+------------+------------+-----------+--------+------------+
| emp_no | birth_date | first_name | last_name | gender | hire_date  |
+--------+------------+------------+-----------+--------+------------+
| 232772 | 1962-05-14 | Keung      | Heusch    | M      | 1986-06-01 |
| 493600 | 1964-01-26 | Keung      | Heusch    | M      | 1986-06-01 |
| 64089  | 1958-01-19 | Marit      | Kolvik    | F      | 1993-12-08 |
| 424486 | 1952-07-06 | Marit      | Kolvik    | F      | 1993-12-08 |
| 40965  | 1952-05-11 | Marsha     | Farrow    | M      | 1989-02-18 |
| 14641  | 1953-05-08 | Marsha     | Farrow    | M      | 1989-02-18 |
| 422332 | 1954-08-17 | Naftali    | Mawatari  | M      | 1985-09-14 |
| 427429 | 1962-11-06 | Naftali    | Mawatari  | M      | 1985-09-14 |
| 19454  | 1955-05-14 | Taisook    | Hutter    | F      | 1985-02-26 |
| 243627 | 1957-02-14 | Taisook    | Hutter    | F      | 1985-02-26 |
+--------+------------+------------+-----------+--------+------------+
10 rows in set (34.01  sec)
```

必须提及 emp1.emp_no!= emp2. emp_no，因为想要显示员工应拥有不同的

emp_no。否则，将显示同一个员工。

使用子查询

子查询是另一个语句中的 SELECT 语句。假设你想查找从 1986-06-26 开始担任 Senior Engineer（高级工程师）的员工的姓名。你可以从 titles 表中获取 emp_no，并从 employees 表中获取姓名。也可以使用 JOIN 查找结果。

从 titles 表中获取 emp_no：

```
mysql> SELECT emp_no FROM titles WHERE title="Senior Engineer" AND
from_date="1986-06-26";
+--------+
| emp_no |
+--------+
|  10001 |
|  84305 |
| 228917 |
| 426700 |
| 458304 |
+--------+
5 rows in set (0.14 sec)
```

获取姓名：

```
mysql> SELECT first_name, last_name FROM employees WHERE emp_no IN (< output
from preceding query>)

mysql> SELECT first_name, last_name FROM employees WHERE emp_no IN
(10001,84305,228917,426700,458304);
+------------+-----------+
| first_name | last_name |
+------------+ ----------+
| Georgi     | Facello   |
| Minghong   | Kalloufi  |
| Nechama    | Bennet    |
| Nagui      | Restivo   |
| Shuzo      | Kirkerud  |
```

```
+------------+-----------+
5 rows in set (0.00 sec)
```

MySQL 中也支持其他子句，如 EXISTS 和 EQUAL。请参阅 https:// dev.mysql.com/doc/refman/8.0/en/subqueries.html 来了解更多详情。

```
mysql> SELECT
    first_name,
    last_name
FROM
    employees
WHERE
    emp_no
IN (SELECT emp_no FROM titles WHERE title="Senior Engineer" AND
from_date="1986-06-26");
+------------+-----------+
| first_name | last_name |
+------------+-----------+
| Georgi     | Facello   |
| Minghong   | Kalloufi  |
| Nagui      | Restivo   |
| Nechama    | Bennet    |
| Shuzo      | Kirkerud  |
+------------+-----------+
5 rows in set (0.91 sec)
```

找到工资最高的员工：

```
mysql> SELECT emp_no FROM salaries WHERE salary=(SELECT MAX(salary) FROM
salaries);
+--------+
| emp_no |
+--------+
| 43624  |
+--------+
1row in set (1.54 sec)
```

SELECT MAX(salary) FROM salaries 是查找工资最高的员工的子查询，要找

到与该工资相对应的员工编号，可以在 WHERE 子句中使用该子查询。

查找表之间不匹配的行

假设你想在一个表中找到其他表中没有的行，可以通过两种方法实现这一点：使用 NOT IN 子句或使用 OUTER JOIN。

要找到匹配的行，可以使用普通 JOIN；要查找不匹配的行，可以使用 OUTER JOIN。普通 JOIN 给出 A 和 B 的一个交集；OUTER JOIN 给出 A 和 B 的匹配记录，并且用 NULL 给出与 A 不匹配的记录。如果你想要 A-B 的输出，可以使用 WHERE <JOIN COLUMN IN B> IS NULL 子句。

要了解 OUTER JOIN 的用法，首先要创建两个 employee 表并插入一些值：

```
mysql> CREATE TABLE employees_list1 AS SELECT * FROM employees WHERE
first_name LIKE 'aa%';
Query OK, 444 rows affected (0.22 sec)
Records:444 Duplicates:0 Warnings:0

mysql> CREATE TABLE employees_list2 AS SELECT * FROM employees WHERE emp_no
BETWEEN 400000 AND 500000 AND gender='F';
Query OK, 39892 rows affected (0.59 sec)
Records:39892 Duplicates:0 Warnings:0
```

我们已经知道如何找到两个列表中都存在的员工了，代码如下：

```
mysql> SELECT * FROM employees_list1 WHERE emp_no IN (SELECT emp_no FROM
employees_list2);
```

也可以使用 JOIN：

```
mysql> SELECT l1.* FROM employees_list1 l1 JOIN employees_list2 l2 ON
l1.emp_no=l2.emp_no;
```

现在要找出存在于 employees_list1 但不存在于 employees_list2 中的员工，代码如下：

```
mysql> SELECT * FROM employees_list1 WHERE emp_no NOT IN (SELECT emp_no FROM
employees_list2);
```

或者也可以使用 OUTER JOIN：

```
mysql> SELECT l1.* FROM employees_list1 l1 LEFT OUTER JOIN employees_list2
l2 ON l1.emp_no=l2.emp_no WHERE l2.emp_no IS NULL;
```

outer join 为第二个表中所有与第一个表中的行不匹配的行创建 NULL 列。如果使用 RIGHT JOIN，则为第一个表中所有与第二个表中的行不匹配的行创建 NULL 列。

你也可以使用 OUTER JOIN 来查找匹配的行。用子句 WHERE emp_no IS NOT NULL，而不是 WHERE l2.emp_no IS NULL：

```
mysql> SELECT l1.* FROM employees_list1 l1 LEFT OUTER JOIN employees_list2
l2 ON l1.emp_no=l2.emp_no WHERE l2.emp_no IS NOT NULL;
```

2.15 存储过程

假设你需要在 MySQL 中执行一系列语句，可以将所有语句封装在单个程序中，并在需要时调用这个程序，而不是每次发送所有 SQL 语句。存储过程处理的是一组 SQL 语句，且没有返回值。

除了SQL语句，还可以使用变量来存储结果并在存储过程中执行程序化的内容。例如，你可以使用 IF，CASE 子句、逻辑操作和 WHILE 循环。

- 存储的函数（function）和过程（procedure）都称为存储例程（routine）。
- 要创建存储过程，你应该具有 CREATE ROUTINE 权限。
- 存储函数具有返回值。
- 存储过程没有返回值。
- 所有代码都写在 BEGIN 和 END 块之间。
- 存储函数可以直接在 SELECT 语句中调用。
- 可以使用 CALL 语句调用存储过程。
- 由于存储例程中的语句应以分隔符（；）结尾，因此必须更改 MySQL 的分隔符，以便 MySQL 不会用正常语句解释存储例程中的 SQL 语句。创建过程结束后，可以将分隔符更改回默认值。

2.15.1 如何操作

假设你想要添加新员工，你需要更新三个表，分别是 employees 表、salaries 表和 titles 表。你可以开发一个存储过程并调用它来创建新的 employee，而不是执行三条语句。

必须传递的信息包括员工的 first_name、last_name、gender 和 birth_date 以及员工加入的部门 department。可以使用输入变量来传递这些变量，并且应该将员工编号作为输出。存储过程不返回值，但它可以更新一个变量并使用它。

下面是存储过程的一个简单示例，实现的是创建新的 employee 并更新 salary 表和 department 表：

```
/* 在创建之前，如果存在任何相同名字的存储过程，则删除已经存在的存储过程*/
DROP PROCEDURE IF EXISTS create_employee;
/* 修改分隔符为 $$ */
DELIMITER $$
/* IN 指定作为参数的变量，INOUT指定输出的变量 */
CREATE PROCEDURE create_employee (OUT new_emp_no INT, IN first_name
varchar(20), IN last_name varchar(20), IN gender enum('M','F'), IN
birth_date date, IN emp_dept_name varchar(40), IN title varchar(50))
BEGIN
    /* 为emp_dept_no和salary声明变量 */
        DECLARE emp_dept_no char(4);
        DECLARE salary int DEFAULT 60000;

    /* 查询employee表的emp_no的最大值，赋值给变量new_emp_no */
    SELECT max(emp_no) INTO new_emp_no FROM employees;
    /*增加new_emp_no */
    SET new_emp_no = new_emp_no + 1;

    /* 插入数据到employees表中 */
      /* CURDATE()函数给出当前日期*/
    INSERT INTO employees VALUES(new_emp_no, birth_date, first_name,
last_name, gender, CURDATE());

    /*找到 dept_name对应的dept_no */
```

```
    SELECT emp_dept_name;
    SELECT dept_no INTO emp_dept_no FROM departments WHERE
dept_name=emp_dept_name;
    SELECT emp_dept_no;

    /* 插入dept_emp */
    INSERT INTO dept_emp VALUES(new_emp_no, emp_dept_no, CURDATE(),
'9999-01-01');

    /* 插入titles */
    INSERT INTO titles VALUES(new_emp_no, title, CURDATE(), '9999-01-01');

    /* 以title为条件查询的薪水 */
    IF title = 'Staff'
        THEN SET salary = 100000;
    ELSEIF title = 'Senior Staff'
        THEN SET salary = 120000;
END IF;

    /* 插入salaries */
    INSERT INTO salaries VALUES(new_emp_no, salary, CURDATE(),
'9999-01-01');
END
$$
/* 把分隔符改回;*/
DELIMITER ;
```

要创建存储过程，你可以这样操作：

- 将上述代码粘贴到命令行客户端中。
- 将其保存在文件中，并使用 `mysql -u <user> -p employees <stored_procedure.sql` 将其导入 MySQL 中。
- 使用 SOURCE 从文件加载 `mysql> SOURCE stored_procedure.sql`。

要使用存储过程，请将 execute 权限授予 `emp_read_only` 用户：

```
mysql> GRANT EXECUTE ON employees.* TO 'emp_read_only'@'%';Query OK, 0 rows
affected (0.05 sec)
```

使用CALL stored_procedure（OUT变量，IN值）语句和例程的名称调用存储过程。

使用emp_read_only账户连接到MySQL：

```
shell> mysql -u emp_read_only -pemp_pass employees -A
```

把你要传递的输出值存储在@new_emp_no中，并传递所需的输入值：

```
mysql> CALL create_employee(@new_emp_no, 'John', 'Smith', 'M', '1984-06-19',
'Research', 'Staff');
Query OK, 1 row affected (0.01 sec)
```

查询存储在@new_emp_no变量中的emp_no的值：

```
mysql> SELECT @new_emp_no;
+-------------+
| @new_emp_no |
+-------------+
|      500000 |
+-------------+
1 row in set (0.00 sec)
```

检查是否在employees表、salaries表和titles 表中创建了行：

```
mysql> SELECT * FROM employees WHERE  emp_no=500000;
+--------+------------+------------+-----------+--------+------------+
| emp_no | birth_date | first_name | last_name | gender | hire_date  |
+--------+------------+------------+-----------+--------+------------+
| 500000 | 1984-06-19 | John       | Smith     | M      | 2017-06-17 |
+--------+------------+------------+-----------+--------+------------+
1 row in set (0.00 sec)

mysql> SELECT * FROM salaries WHERE emp_no=500000;
+--------+--------+------------+------------+
| emp_no | salary | from_date  | to_date    |
+--------+--------+------------+------------+
| 500000 | 100000 | 2017-06-17 | 9999-01-01 |
+--------+--------+------------+------------+
```

```
1 row in set (0.00 sec)

mysql> SELECT * FROM titles WHERE emp_no=500000;
+--------+--------+------------+------------+
| emp_no | title  | from_date  | to_date    |
+--------+-------  +------------+------------+
| 500000 | Staff  | 2017-06-17 | 9999-01-01 |
+--------+--------+------------+------------+
1 row in set (0.00 sec)
```

可以看到，即使 emp_read_only 对表没有写访问权限，也可以通过调用存储过程来写入。如果存储过程的 SQL SECURITY 创建为 INVOKER，则 emp_read_only 不能修改数据。请注意，如果你正在使用 localhost 连接，请为 localhost 用户创建权限。

要列出数据库中的所有存储过程，请执行 SHOW PROCEDURE STATUS\G。要检查现有存储例程的定义，可以执行 SHOW CREATE PROCEDURE <procedure_name> \ G。

2.15.2 延伸阅读

存储过程也可以用于增强安全性。用户需要拥有针对存储过程的 EXECUTE 权限才能执行它。

根据存储例程的定义：

- DEFINER 子句指定存储例程的创建者。如果没有指定，则获取当前用户。
- SQL SECURITY 子句指定存储例程的执行上下文。它可以是 DEFINER 或 INVOKER。

  DEFINER：即使只有 EXECUTE 权限的用户也可以调用并获取存储例程的输出，而不管该用户是否具有对基础表的操作权限。如果 DEFINER 具有权限，那就足够了。

  INVOKER：安全上下文被切换到调用存储例程的用户。在这种情况下，调用者应该可以访问基础表。

要获取更多示例和语法，请参阅文档 https://dev.mysql.com/doc/refman/8.0/en/create-procedure.html。

2.16　函数

就像存储过程一样，我们可以创建存储函数。二者的主要区别是，函数应该有一个返回值，并且可以在 SELECT 中调用函数。通常，创建存储函数是为了简化复杂的计算。

2.16.1　如何操作

下面是如何编写函数以及如何调用它的示例。假设银行工作人员想根据客户的收入水平给出信用卡额度,同时不暴露客户的实际工资,那么可以利用下面的函数查询收入水平：

```
shell> vi function.sql;
DROP FUNCTION IF EXISTS get_sal_level;
DELIMITER $$
CREATE FUNCTION get_sal_level(emp int) RETURNS VARCHAR(10)
 DETERMINISTIC
BEGIN
 DECLARE sal_level varchar(10);
 DECLARE avg_sal FLOAT;

 SELECT AVG(salary) INTO avg_sal FROM salaries WHERE emp_no=emp;

 IF avg_sal < 50000 THEN
 SET sal_level = 'BRONZE';
 ELSEIF (avg_sal >= 50000 AND avg_sal < 70000) THEN
 SET sal_level = 'SILVER';
 ELSEIF (avg_sal >= 70000 AND avg_sal < 90000) THEN
 SET sal_level = 'GOLD';
 ELSEIF (avg_sal >= 90000) THEN
 SET sal_level = 'PLATINUM';
 ELSE
 SET sal_level = 'NOT FOUND';
 END IF;
 RETURN (sal_level);
END
$$
DELIMITER ;
```

创建该函数：

```
mysql> SOURCE function.sql;
Query OK, 0 rows affected (0.00 sec)
Query OK, 0 rows affected (0.01 sec)

You have to pass the employee number and the function returns the income level.
mysql> SELECT get_sal_level(10002);
+----------------------+
| get_sal_level(10002) |
+----------------------+
| SILVER               |
+----------------------+
1 row in set (0.00 sec)

mysql> SELECT get_sal_level(10001);
+----------------------+
| get_sal_level(10001) |
+----------------------+
| GOLD                 |
+----------------------+
1 row in set (0.00 sec)

mysql> SELECT get_sal_level(1);
+------------------+
| get_sal_level(1) |
+------------------+
| NOT FOUND        |
+------------------+
1 row in set (0.00 sec)
```

要列出数据库中的所有存储函数，请执行 SHOW FUNCTION STATUS\G。要检查现有存储函数的定义，可以执行 SHOW CREATE FUNCTION <function_name>\G。

> 在函数创建中给出 DETERMINISTIC 关键字非常重要。如果一个例程对于相同的输入参数总是产生相同的结果，则认为该函数为 DETERMINISTIC，否则为 NOT DETERMINISTIC。如果在例程定义中既未给出 DETERMINISTIC，也未给出 NOT DETERMINISTIC，则默认是 NOT DETERMINISTIC。如果要声明一个函数是确定性的，则必须明确指定 DETERMINISTIC。
> 如果将一个 NON DETERMINISTIC 例程声明为 DETERMINISTIC，可能导致意想不到的结果，因为它会导致优化器选择不正确的执行计划。将 DETERMINISTIC 例程声明为 NON DETERMINISTIC 可能会导致可用的优化未被使用，降低性能。

内置函数

MySQL 提供了许多内置函数。我们前面已经使用 CURDATE() 函数获取了当前日期。

你可以在 WHERE 子句中使用 CURDATE() 函数：

```
mysql> SELECT * FROM employees WHERE hire_date = CURDATE();
```

- 例如，下面的函数给出了一周前的日期：

  ```
  mysql> SELECT DATE_ADD(CURDATE(), INTERVAL -7 DAY) AS '7 Days Ago';
  ```

- 将两个字符串相加：

  ```
  mysql> SELECT CONCAT(first_name, ' ', last_name) FROM employees
  LIMIT 1;
  +------------------------------------+
  | CONCAT(first_name, ' ', last_name) |
  +------------------------------------+
  | Aamer Anger                        |
  +------------------------------------+
  1 row in set (0.00 sec)
  ```

2.16.2　延伸阅读

请参阅 MySQL 文档获取完整方法列表，网址为 https://dev.mysql.com/

doc/refman/8.0/en/func-op-summary-ref.html。

2.17 触发器

触发器用于在触发器事件之前或之后激活某些内容。例如，可以在插入表中的每行之前或更新的每行之后激活触发器。

触发器非常有用，可以在无停机时间的情况下更改表（请参阅 10.5 节）或用于审计目的。假设你想查找某一行的前一个值，可以编写一个触发器，在更新之前将这些行保存在另一个表中。另一个表保存了以前的记录，充当审计表。

触发器动作时间可以是 BEFORE 或 AFTER，表示触发器是在每行要修改之前或之后被激活。

触发事件可以是 INSERT、DELETE 或 UPDATE。

- INSERT：无论何时通过 INSERT、REPLACE 或 LOAD DATA 语句插入新行，都会激活 INSERT 触发事件。
- UPDATE：通过 UPDATE 语句激活 UPDATE 触发事件。
- DELETE：通过 DELETE 或 REPLACE 语句激活 DELETE 触发事件。

从 MySQL 5.7 开始，一个表可以同时具有多个触发器。例如，一个表可以有两个 BEFORE INSERT 触发器。必须使用 FOLLOWS 或 PRECEDES 指定先行的触发器。

2.17.1 如何操作

例如，假设你希望在将薪水插入 salaries 表之前对其进行四舍五入。NEW 指的是正在插入的新值：

```
shell> vi before_insert_trigger.sql
DROP TRIGGER IF EXISTS salary_round;
DELIMITER $$
CREATE TRIGGER salary_round BEFORE INSERT ON salaries
FOR EACH ROW
BEGIN
        SET NEW.salary=ROUND(NEW.salary);
```

```
END
$$
DELIMITER ;
```

通过导入文件创建触发器：

```
mysql> SOURCE before_insert_trigger.sql;
Query OK, 0 rows affected (0.06 sec)
Query OK, 0 rows affected (0.00 sec)
```

通过在薪水中插入浮点数来测试触发器：

```
mysql> INSERT INTO salaries VALUES(10002, 100000.79, CURDATE(),
'9999-01-01');
Query OK, 1 row affected (0.04 sec)
```

可以看到薪水被四舍五入：

```
mysql> SELECT * FROM salaries WHERE emp_no=10002 AND from_date=CURDATE();
+--------+--------+------------+------------+
| emp_no | salary | from_date  | to_date    |
+--------+--------+ -----------+------------+
|10002   | 100001 | 2017-06-18 | 9999-01-01 |
+--------+--------+ -----------+------------+
1 row in set (0.00 sec)
```

与之类似，你还可以创建一个 BEFORE UPDATE 触发器来完成薪水的四舍五入操作。另一个示例：假设你要记录 salaries 表中新增的薪水记录，首先创建一个审记表：

```
mysql> CREATE TABLE salary_audit (emp_no int, user varchar(50),
date_modified date);
```

请注意，以下触发器由 PRECEDES salary_round 指定在 salary_round 触发器之前执行：

```
shell> vi before_insert_trigger.sql
DELIMITER $$
CREATE TRIGGER salary_audit
BEFORE INSERT
   ON salaries FOR EACH ROW PRECEDES salary_round
```

```
BEGIN
   INSERT INTO salary_audit VALUES(NEW.emp_no, USER(), CURDATE());
END;$$
DELIMITER ;
```

将其插入 salaries 表：

```
mysql> INSERT INTO salaries VALUES(10003, 100000.79, CURDATE(),
'9999-01-01');
Query OK, 1 row affected (0.06 sec)
```

通过查询 salary_audit 表来找出谁插入了薪水：

```
mysql> SELECT * FROM salary_audit WHERE emp_no=10003;
+--------+----------------+---------------+
| emp_no | user           | date_modified |
+--------+----------------+---------------+
| 10003  | root@localhost | 2017-06-18    |
+--------+----------------+---------------+
1 row in set (0.00 sec)
```

如果 salary_audit 表被删除或不可用，则 salaries 表上的所有插入都将被阻止。如果你不想执行审计，则应先删除触发器，然后再删除表。
由于上述复杂性，触发器产生的开销会影响写入速度。
要检查所有触发器，请执行 SHOW TRIGGERS\G。
要检查现有触发器的定义，请执行 SHOW CREATE TRIGGER <trigger_name>。

2.17.2 延伸阅读

更多详细信息请参阅 MySQL 文档，网址为 https://dev.mysql.com/doc/refman/8.0/en/trigger-syntax.html。

2.18 视图

视图是一个基于SQL语句的结果集的虚拟表。它就像一个真正的表一样也具有行和列，

但是有一些限制，这些将在后面讨论。视图隐藏了 SQL 的复杂性，更重要的是，它提供了额外的安全性。

2.18.1 如何操作

假设你只想提供对 salaries 表的 emp_no 列和 salary 列，且 from_date 在 2002-01-01 之后的数据的访问权限。为此，你可以通过提供所需结果的 SQL 语句创建视图。

```
mysql> CREATE ALGORITHM=UNDEFINED
DEFINER=`root`@`localhost`
SQL SECURITY DEFINER VIEW salary_view
AS
SELECT emp_no, salary FROM salaries WHERE from_date > '2002-01-01';
```

现在，salary_view 视图已创建，你可以像查询其他表一样查询它：

```
mysql> SELECT emp_no, AVG(salary) as avg FROM salary_view GROUP BY emp_no
ORDER BY avg DESC LIMIT 5;
```

可以看到该视图可以访问特定行（即 from_date>'2002-01-01'），而不是所有行。我们可以使用视图来限制用户对特定行的访问。

要列出所有视图，请执行：

```
mysql> SHOW FULL TABLES WHERE TABLE_TYPE LIKE 'VIEW';
```

要检查视图的定义，请执行：

```
mysql> SHOW CREATE VIEW salary_view\G
```

你可能已经注意到了 current_dept_emp 和 dept_emp_latest_date 视图，这些视图是 employees 数据库的一部分。你可以探索其定义并查看其功能。

我们可以更新没有子查询、JOINS、GROUP BY 子句、union 等的简单视图。如果基础表有默认值，那么 salary_view 就是一个可以被更新或插入的简单视图：

```
mysql> UPDATE salary_view SET salary=100000 WHERE emp_no=10001;
Query OK, 1 row affected (0.01 sec)
Rows matched:2 Changed:1 Warnings:0
```

```
mysql> INSERT INTO salary_view VALUES(10001,100001);
ERROR 1423 (HY000):Field of view 'employees.salary_view' underlying table
doesn't have a default value
```

如果该表有一个默认值，即使它不符合视图中的过滤器条件，你也可以向其中插入一行。为了避免这种情况，为了只允许插入符合视图条件的行，必须在定义里面提供 WITH CHECK OPTION。

VIEW 算法：

- MERGE：MySQL 将输入查询和视图定义合并到一个查询中，然后执行组合查询。仅允许在简单视图上使用 MERGE 算法。
- TEMPTABLE：MySQL 将结果存储在临时表中，然后对这个临时表执行输入查询。
- UNDEFINED（默认）：MySQL 自动选择 MERGE 或 TEMPTABLE 算法。MySQL 把 MERGE 算法作为首选的 TEMPTABLE 算法，因为 MERGE 算法效率更高。

2.19 事件

就像 Linux 服务器上的 cron 一样，MySQL 的 EVENTS 是用来处理计划任务的。MySQL 使用称为事件调度线程的特殊线程来执行所有预定事件。默认情况下，事件调度线程是未启用（版本低于 8.0.3）的状态，如要启用它，执行以下命令：

```
mysql> SET GLOBAL event_scheduler = ON;
```

2.19.1 如何操作

假设你不再需要保留一个月之前的薪水审计记录，则可以设定一个每日运行的事件，用它从 salary_audit 表中删除一个月之前的记录。

```
mysql> DROP EVENT IF EXISTS purge_salary_audit;
DELIMITER $$
CREATE EVENT IF NOT EXISTS purge_salary_audit
ON SCHEDULE
  EVERY 1 WEEK
  STARTS CURRENT_DATE
```

```
    DO BEGIN
        DELETE FROM salary_audit WHERE date_modified < DATE_ADD(CURDATE(),
INTERVAL -7 day);
    END $$
DELIMITER ;
```

一旦事件被创建，它将自动完成清除一个月之前的薪水审计记录的工作。

- 要检查事件，请执行：

```
mysql> SHOW EVENTS\G
*************************** 1.row ***************************
                  Db:employees
                Name:purge_salary_audit
             Definer:root@localhost
           Time zone:SYSTEM
                Type:RECURRING
          Execute at:NULL
      Interval value:1
      Interval field:MINUTE
              Starts:2017-06-18 00:00:00
                Ends:NULL
              Status:ENABLED
          Originator:0
character_set_client:utf8
collation_connection:utf8_general_ci
  Database Collation:utf8mb4_0900_ai_ci
1 row in set (0.00 sec)
```

- 要检查事件的定义，请执行：

```
mysql> SHOW CREATE EVENT purge_salary_audit\G
```

- 要禁用/启用该事件，请执行以下操作：

```
mysql> ALTER EVENT purge_salary_audit DISABLE;
mysql> ALTER EVENT purge_salary_audit ENABLE;
```

访问控制

所有存储的程序（过程、函数、触发器和事件）和视图都有一个 DEFINER。如果未指定 DEFINER，则创建该对象的用户将被选为 DEFINER。

存储例程(包括过程和函数)和视图具有值为 DEFINER 或 INVOKER 的 SQL SECURITY 特性，来指定对象是在 definer 还是在 invoker 上下文中执行。触发器和事件没有 SQL SECURITY 特性，并且始终在 definer 上下文中执行。服务器根据需要自动调用这些对象，因此不存在调用用户。

2.19.2 延伸阅读

有很多方法可以用来安排事件，请参阅 https://dev.mysql.com/doc/refman/8.0/en/event-scheduler.html 以获取更多详细信息。

2.20 获取有关数据库和表的信息

你可能已经注意到了在数据库列表中有一个 information_schema 数据库。information_ schema 数据库是由所有数据库对象的元数据组成的视图集合，你可以连接到 information_schema 并浏览所有表，本章介绍了其中最广泛使用的表。你可以查询 information_schema 表或使用 SHOW 命令，它们实质上是相同的。

INFORMATION_SCHEMA 查询作为数据字典表的视图来实现。INFORMATION_SCHEMA 表中有下面两种类型的元数据。

- **静态表元数据**：TABLE_SCHEMA、TABLE_NAME、TABLE_TYPE 和 ENGINE。这些统计信息将直接从数据字典中读取。
- **动态表元数据**：AUTO_INCREMENT、AVG_ROW_LENGTH 和 DATA_FREE。动态元数据会频繁更改（例如，AUTO_INCREMENT 值将在每次 INSERT 后增长）。在很多情况下，动态元数据在一些需要精确计算的情况下也会产生一些开销，并且准确性可能对常规查询不会有好处。考虑到 DATA_FREE 统计量的情况（该统计量显示表中的空闲字节数），缓存值通常足够了。

在 MySQL 8.0 中，动态表元数据将默认被缓存。这可以通过 information_schema_

stats 设置（默认缓存）进行配置，并且可以更改为 SET @@ GLOBAL.information_schema_stats ='LATEST'，以便始终直接从存储引擎中检索动态信息（代价是略长的查询执行时间）。

作为替代方案，用户也可以在表上执行 ANALYZE TABLE 来更新缓存的动态统计信息。

大多数表具有引用数据库名称的 TABLE_SCHEMA 列和引用表名称的 TABLE_NAME 列。

更多详细信息请参阅 https://mysqlserverteam.com/mysql-8-0-improvements-to-information_schema/。

2.20.1　如何操作

检查所有表的列表：

```
mysql> USE INFORMATION_SCHEMA;
mysql> SHOW TABLES;
```

TABLES

TABLES 表包含有关表的所有信息，例如哪个数据库属于 TABLE_SCHEMA，以及行数（TABLE_ROWS）、ENGINE、DATA_LENGTH、INDEX_LENGTH 和 DATA_FREE：

```
mysql> DESC INFORMATION_SCHEMA.TABLES;
+-----------------+-----------------------------------------------------
-----------+------+-----+------------------+---------------------------
+
| Field           | Type
| Null | Key | Default                | Extra                         |
+-----------------+-----------------------------------------------------
-----------+------+-----+------------------+---------------------------
+
| TABLE_CATALOG   | varchar(64)
| NO   |      | NULL                  |                               |
| TABLE_SCHEMA    | varchar(64)
| NO   |      | NULL                  |                               |
| TABLE_NAME      | varchar(64)
| NO   |      | NULL                  |                               |
```

```
| TABLE_TYPE      | enum('BASE TABLE','VIEW','SYSTEM VIEW')
| NO    |       | NULL                    |                                |
| ENGINE          | varchar(64)
| YES   |       | NULL                    |                                |
| VERSION         | int(2)
| YES   |       | NULL                    |                                |
| ROW_FORMAT      |
enum('Fixed','Dynamic','Compressed','Redundant','Compact','Paged') | YES  |
| NULL                |                                |
| TABLE_ROWS      | bigint(20) unsigned
| YES   |       | NULL                    |                                |
| AVG_ROW_LENGTH  | bigint(20) unsigned
| YES   |       | NULL                    |                                |
| DATA_LENGTH     | bigint(20) unsigned
| YES   |       | NULL                    |                                |
| MAX_DATA_LENGTH | bigint(20) unsigned
| YES   |       | NULL                    |                                |
| INDEX_LENGTH    | bigint(20) unsigned
| YES   |       | NULL                    |                                |
| DATA_FREE       | bigint(20) unsigned
| YES   |       | NULL                    |                                |
| AUTO_INCREMENT  | bigint(20) unsigned
| YES   |       | NULL                    |                                |
| CREATE_TIME     | timestamp
| NO    |       | CURRENT_TIMESTAMP  | on update CURRENT_TIMESTAMP |
| UPDATE_TIME     | timestamp
| YES   |       | NULL                    |                                |
| CHECK_TIME      | timestamp
| YES   |       | NULL                    |                                |
| TABLE_COLLATION | varchar(64)
| YES   |       | NULL                    |                                |
| CHECKSUM        | bigint(20) unsigned
| YES   |       | NULL                    |                                |
| CREATE_OPTIONS  | varchar(256)
| YES   |       | NULL                    |                                |
| TABLE_COMMENT   | varchar(256)
| YES   |       | NULL                    |                                |
```

```
+----------------+------------------------------------------------------
------------+------+-----+------------------+--------------------------
+
21 rows in set (0.00 sec)
```

例如，假设你想知道 employees 数据库中的 DATA_LENGTH、INDEX_LENGTH 和 DATE_FREE，代码如下：

```
mysql> SELECT SUM(DATA_LENGTH)/1024/1024 AS DATA_SIZE_MB,
SUM(INDEX_LENGTH)/1024/1024 AS INDEX_SIZE_MB, SUM(DATA_FREE)/1024/1024 AS
DATA_FREE_MB FROM INFORMATION_SCHEMA.TABLES WHERE TABLE_SCHEMA='employees';
+--------------+--------------+ --------------+
| DATA_SIZE_MB | INDEX_SIZE_MB | DATA_FREE_MB  |
+--------------+--------------+ --------------+
|17.39062500   |  14.62500000 | 11.00000000   |
+--------------+--------------+ --------------+
1row in set (0.01 sec)
```

COLUMNS

该表列出了每个表的所有列及其定义：

```
mysql> SELECT * FROM COLUMNS WHERE TABLE_NAME='employees'\G
```

FILES

你已经看到 MySQL 将 InnoDB 数据存储在字典中的目录（与数据库名称相同）内的.ibd 文件中。要获取有关这些文件的更多信息，可以查询 FILES 表：

```
mysql> SELECT * FROM FILES WHERE FILE_NAME LIKE
'./employees/employees.ibd'\G
~~~
EXTENT_SIZE:1048576
AUTOEXTEND_SIZE:4194304
DATA_FREE:13631488
~~~
```

你应该关注 DATA_FREE，它表示未分配的数据段，以及由于碎片而在数据段内部空闲的数据段。重建表时，可以释放 DATA_FREE 中显示的字节。

INNODB_TABLESPACES

INNODB_TABLESPACES 表也提供了该文件的大小：

```
mysql> SELECT * FROM INNODB_TABLESPACES WHERE NAME='employees/employees'\G
*************************** 1.row ***************************
 SPACE:118
 NAME:employees/employees
 FLAG:16417
 ROW_FORMAT:Dynamic
 PAGE_SIZE:16384
 ZIP_PAGE_SIZE:0
 SPACE_TYPE:Single
 FS_BLOCK_SIZE:4096
 FILE_SIZE:32505856
ALLOCATED_SIZE:32509952
1 row in set (0.00 sec)
```

你可以在文件系统中验证相同的内容：

```
shell> sudo ls -ltr /var/lib/mysql/employees/employees.ibd
-rw-r----- 1 mysql mysql 32505856 Jun 20 16:50
/var/lib/mysql/employees/employees.ibd
```

INNODB_TABLESTATS

INNODB_TABLESTATS 表提供了索引大小和近似行数：

```
mysql> SELECT * FROM INNODB_TABLESTATS WHERE NAME='employees/employees'\G
*************************** 1.row ***************************
 TABLE_ID:128
 NAME:employees/employees
 STATS_INITIALIZED:Initialized
 NUM_ROWS:299468
 CLUST_INDEX_SIZE:1057
 OTHER_INDEX_SIZE:545
 MODIFIED_COUNTER:0
 AUTOINC:0
 REF_COUNT: 1
```

```
1 row in set (0.00 sec)
```

PROCESSLIST

最常用的视图之一是 PROCESSLIST，它列出了服务器上运行的所有查询：

```
mysql> SELECT * FROM PROCESSLIST\G
*************************** 1.row ***************************
     ID: 85
   USER: event_scheduler
   HOST: localhost
     DB: NULL
COMMAND: Daemon
   TIME: 44
  STATE: Waiting for next activation
   INFO: NULL
*************************** 2.row ***************************
     ID:26231
   USER:root
   HOST:localhost
     DB:information_schema
COMMAND:Query
   TIME:0
  STATE:executing
   INFO:SELECT * FROM PROCESSLIST
2 rows in set (0.00 sec)
```

你也可以执行 SHOW PROCESSLIST 来获得相同的输出结果。

其他表：ROUTINES 包含函数和存储例程的定义，TRIGGERS 包含触发器的定义，VIEWS 包含视图的定义。

2.20.2 延伸阅读

要了解 INFORMATION_SCHEMA 中的改进，请参阅 http://mysqlserverteam.com/mysql-8-0-improvements-to-information_schema/。

第3章 使用MySQL（进阶）

在本章中，我们将介绍以下内容：

- 使用JSON
- 公用表表达式（CTE）
- 生成列（generated column）
- 窗口函数

3.1 引言

在本章中，我们将讲解MySQL最新引入的功能。

3.2 使用JSON

正如在第2章中所看到的，要在MySQL中存储数据，就必须定义数据库和表结构（schema），这是一个主要的限制。为了应对这一点，从MySQL 5.7开始，MySQL支持了**JavaScript**对象表示（**JavaScript Object Notation**，**JSON**）数据类型。之前，这类数据不是单独的数据类型，会被存储为字符串。新的JSON数据类型提供了自动验证的JSON文档以及优化的存储格式。

JSON文档以二进制格式存储，它提供以下功能：

- 对文档元素的快速读取访问。

- 当服务器再次读取 JSON 文档时，不需要重新解析文本获取该值。
- 通过键或数组索引直接查找子对象或嵌套值，而不需要读取文档中的所有值。

3.2.1　如何操作

如果希望存储关于员工的更多详细信息，可以使用 JSON 保存它们：

```
CREATE TABLE emp_details(
  emp_no int primary key,
  details json
);
```

插入 JSON

```
INSERT INTO emp_details(emp_no, details)
VALUES ('1',
'{ "location": "IN", "phone": "+11800000000", "email": "abc@example.com",
"address": { "line1": "abc", "line2": "xyz street", "city": "Bangalore",
"pin": "560103"} }'
);
```

检索 JSON

可以使用 -> 和 ->> 运算符检索 JSON 列的字段：

```
mysql> SELECT emp_no, details->'$.address.pin' pin FROM emp_details;
+--------+----------+
| emp_no | pin      |
+--------+----------+
| 1      | "560103" |
+--------+----------+
1 row in set (0.00 sec)
```

如果不用引号检索数据，可以使用 ->> 运算符：

```
mysql> SELECT emp_no, details->>'$.address.pin' pin FROM emp_details;
+--------+----------+
| emp_no |pin       |
+--------+----------+
```

```
| 1        | 560103    |
+--------+----------+
1 row in set (0.00 sec)
```

JSON 函数

MySQL 提供了许多处理 JSON 数据的函数，让我们看看最常用的几种函数。

优雅浏览

想要以优雅的格式显示 JSON 值，请使用 `JSON_PRETTY()` 函数：

```
mysql> SELECT emp_no, JSON_PRETTY(details) FROM emp_details \G
*************************** 1. row ***************************
 emp_no: 1
JSON_PRETTY(details): {
 "email": "abc@example.com",
 "phone": "+11800000000",
 "address": {
 "pin": "560103",
 "city": "Bangalore",
 "line1": "abc",
 "line2": "xyz street"
 },
 "location": "IN"
}
1 row in set (0.00 sec)
```

不用 `JSON_PRETTY()` 函数：

```
mysql> SELECT emp_no, details FROM emp_details \G
*************************** 1. row ***************************
 emp_no: 1
details: {"email": "abc@example.com", "phone": "+11800000000", "address":
{"pin": "560100", "city": "Bangalore", "line1": "abc", "line2": "xyz
street"}, "location": "IN"}
1 row in set (0.00 sec)
```

查找

可以在 WHERE 子句中使用 col->>path 运算符来引用 JSON 的某一列：

```
mysql> SELECT emp_no FROM emp_details WHERE
details->>'$.address.pin'="560103"; +--------+
| emp_no |
+--------+
| 1      |
+--------+
1 row in set (0.00 sec)
```

也可以使用 JSON_CONTAINS 函数查询数据。如果找到了数据，则返回 1，否则返回 0：

```
mysql> SELECT JSON_CONTAINS(details->>'$.address.pin', "560103") FROM
emp_details;
+-----------------------------------------------------+
| JSON_CONTAINS(details->>'$.address.pin', "560103") |
+-----------------------------------------------------+
| 1                                                   |
+-----------------------------------------------------+
1 row in set (0.00 sec)
```

如何查询一个 key？假设要检查 address.line1 是否存在：

```
mysql> SELECT JSON_CONTAINS_PATH(details, 'one', "$.address.line1") FROM
emp_details;
+-----------------------------------------------------------------------
+
|JSON_CONTAINS_PATH(details, 'one', "$.address.line1")
|
+-----------------------------------------------------------------------
+
|1
|
+-----------------------------------------------------------------------
+
1 row in set (0.01 sec)
```

这里，one 表示至少应该存在一个键。假设要检查 address.line1 或者 address.line2 是否存在，代码如下：

```
mysql> SELECT JSON_CONTAINS_PATH(details, 'one', "$.address.line1",
"$.address.line5") FROM emp_details;
+--------------------------------------------------------------------
+
JSON_CONTAINS_PATH(details, 'one', "$.address.line1", "$.address.line2")
+--------------------------------------------------------------------
+
1
+--------------------------------------------------------------------
+
```

如果要检查 address.line1 和 address.line5 是否同时存在，可以使用 all，而不是 one：

```
mysql> SELECT JSON_CONTAINS_PATH(details, 'all', "$.address.line1",
"$.address.line5") FROM emp_details;
+--------------------------------------------------------------------
+
JSON_CONTAINS_PATH(details, 'all', "$.address.line1", "$.address.line5")
+--------------------------------------------------------------------
+
0
+--------------------------------------------------------------------
+
1 row in set (0.00 sec)
```

修改

可以使用三种不同的函数来修改数据：JSON_SET()、JSON_INSERT() 和 JSON_REPLACE()。在 MySQL 8 之前的版本中，我们还需要对整个列进行完整的更新，这并不是最佳的方法。

- JSON_SET()：替换现有值并添加不存在的值。
  假设要替换员工的 pin 码，并添加昵称的详细信息，代码如下：

```
mysql> UPDATE
    emp_details
SET
     details = JSON_SET(details, "$.address.pin", "560100",
"$.nickname", "kai")
WHERE
    emp_no = 1;
Query OK, 1 row affected (0.03 sec)
Rows matched: 1 Changed: 1 Warnings: 0
```

- JSON_INSERT ()：插入值，但不替换现有值。

  假设你希望添加新列而不更新现有值，则可以使用 JSON_INSERT ()：

```
mysql> UPDATE emp_details SET details=JSON_INSERT(details,
"$.address.pin", "560132", "$.address.line4", "A Wing") WHERE emp_no =
1;
Query OK, 1 row affected (0.00 sec)
Rows matched: 1 Changed: 1 Warnings: 0
```

  在这种情况下，pin 不会被更新，只会添加一个新的字段 address.line4。

- JSON_REPLACE ()：仅替换现有值。

  假设只需要替换现有字段，不需要添加新字段，代码如下：

```
mysql> UPDATE emp_details SET details=JSON_REPLACE(details,
"$.address.pin", "560132", "$.address.line5", "Landmark") WHERE emp_no
= 1;
Query OK, 1 row affected (0.04 sec)
Rows matched: 1 Changed: 1 Warnings: 0
```

  在这种情况下，line5 不会被添加，只有 pin 会被更新。

删除

JSON_REMOVE 能从 JSON 文档中删除数据。

假设你不再需要地址中的 line5，删除它的代码如下：

```
mysql> UPDATE emp_details SET details=JSON_REMOVE(details,
"$.address.line5") WHERE emp_no = 1;
```

```
Query OK, 1 row affected (0.04 sec)
Rows matched: 1 Changed: 1 Warnings: 0
```

其他函数

其他函数如下。

- `JSON_KEYS()`：获取 JSON 文档中的所有键。

```
mysql> SELECT JSON_KEYS(details) FROM emp_details WHERE emp_no = 1;
*************************** 1. row ***************************
JSON_KEYS(details): ["email", "phone", "address", "nickname",
"location"]
```

- `JSON_LENGTH ()`：给出 JSON 文档中的元素数。

```
mysql> SELECT JSON_LENGTH(details) FROM emp_details WHERE emp_no = 1;
*************************** 1. row ***************************
JSON_LENGTH(details): 5
```

3.2.2 延伸阅读

可以在`https://dev.mysql.com/doc/refman/8.0/en/json-function-reference.html`上查看完整的函数列表。

3.3 公用表表达式（CTE）

MySQL 8 支持公用表表达式，包括非递归和递归两种。

公用表表达式允许使用命名的临时结果集，这是通过允许在 `SELECT` 语句和某些其他语句前面使用 `WITH` 子句来实现的。

为什么需要 CTE

不能在同一查询中两次引用派生表，因为那样的话，查询会根据派生表的引用次数计算两次或多次，这会引发严重的性能问题。使用 CTE 后，子查询只会计算一次。

3.3.1　如何操作

以下各节将对非递归和递归 CTE 进行详述。

非递归 CTE

公用表表达式（CTE）与派生表类似，但它的声明会放在查询块之前，而不是 FROM 子句中。

派生表

```
SELECT... FROM (subquery) AS derived, t1 ...
```

CTE

```
SELECT... WITH derived AS (subquery) SELECT ... FROM derived, t1 ...
```

CTE 可能在 SELECT/UPDATE/DELETE 之前，包括 WITH derived AS (subquery) 的子查询，例如：

```
DELETE FROM t1 WHERE t1.a IN (SELECT b FROM derived);
```

假设你想了解每年的工资较前一年的增长百分比，如果没有 CTE，你需要编写两个子查询，这两个子查询本质上是相同的，但 MySQL 并不能识别出它们是相同的，这就会导致子查询被执行两次。

```
mysql> SELECT
    q1.year,
    q2.year AS next_year,
    q1.sum,
    q2.sum AS next_sum,
    100*(q2.sum-q1.sum)/q1.sum AS pct
FROM
    (SELECT year(from_date) as year, sum(salary) as sum FROM salaries GROUP
BY year)    AS q1,
   (SELECT year(from_date) as year, sum(salary) as sum FROM   salaries
GROUP BY year) AS q2
WHERE q1.year = q2.year-1;
+------+----------+------------+------------+----------+
```

```
| year | next_year | sum         | next_sum    | pct      |
+------+-----------+-------------+-------------+----------+
| 1985 |      1986 | 972864875   | 2052895941  | 111.0155 |
| 1986 |      1987 | 2052895941  | 3156881054  | 53.7770  |
| 1987 |      1988 | 3156881054  | 4295598688  | 36.0710  |
| 1988 |      1989 | 4295598688  | 5454260439  | 26.9732  |
| 1989 |      1990 | 5454260439  | 6626146391  | 21.4857  |
| 1990 |      1991 | 6626146391  | 7798804412  | 17.6974  |
| 1991 |      1992 | 7798804412  | 9027872610  | 15.7597  |
| 1992 |      1993 | 9027872610  | 10215059054 | 13.1502  |
| 1993 |      1994 | 10215059054 | 11429450113 | 11.8882  |
| 1994 |      1995 | 11429450113 | 12638817464 | 10.5812  |
| 1995 |      1996 | 12638817464 | 13888587737 | 9.8883   |
| 1996 |      1997 | 13888587737 | 15056011781 | 8.4056   |
| 1997 |      1998 | 15056011781 | 16220495471 | 7.7343   |
| 1998 |      1999 | 16220495471 | 17360258862 | 7.0267   |
| 1999 |      2000 | 17360258862 | 17535667603 | 1.0104   |
| 2000 |      2001 | 17535667603 | 17507737308 | -0.1593  |
| 2001 |      2002 | 17507737308 | 10243358658 | -41.4924 |
+------+-----------+-------------+-------------+----------+
17 rows in set (3.22 sec)
```

如果使用非递归 CTE，派生查询只执行一次并重用：

```
mysql>
WITH CTE AS
    (SELECT year(from_date) AS year, SUM(salary) AS sum FROM salaries GROUP
BY year)
SELECT
    q1.year, q2.year as next_year, q1.sum, q2.sum as next_sum, 100*(q2.sum-
q1.sum)/q1.sum as pct FROM
    CTE AS q1,
    CTE AS q2
WHERE
    q1.year = q2.year-1;
+------+-----------+-------------+-------------+----------+
| year | next_year | sum         | next_sum    | pct      |
+------+-----------+-------------+-------------+----------+
```

```
| 1985 |       1986 | 972864875   | 2052895941  | 111.0155 |
| 1986 |       1987 | 2052895941  | 3156881054  | 53.7770  |
| 1987 |       1988 | 3156881054  | 4295598688  | 36.0710  |
| 1988 |       1989 | 4295598688  | 5454260439  | 26.9732  |
| 1989 |       1990 | 5454260439  | 6626146391  | 21.4857  |
| 1990 |       1991 | 6626146391  | 7798804412  | 17.6974  |
| 1991 |       1992 | 7798804412  | 9027872610  | 15.7597  |
| 1992 |       1993 | 9027872610  | 10215059054 | 13.1502  |
| 1993 |       1994 | 10215059054 | 11429450113 | 11.8882  |
| 1994 |       1995 | 11429450113 | 12638817464 | 10.5812  |
| 1995 |       1996 | 12638817464 | 13888587737 | 9.8883   |
| 1996 |       1997 | 13888587737 | 15056011781 | 8.4056   |
| 1997 |       1998 | 15056011781 | 16220495471 | 7.7343   |
| 1998 |       1999 | 16220495471 | 17360258862 | 7.0267   |
| 1999 |       2000 | 17360258862 | 17535667603 | 1.0104   |
| 2000 |       2001 | 17535667603 | 17507737308 | -0.1593  |
| 2001 |       2002 | 17507737308 | 10243358658 | -41.4924 |
+------+------------+-------------+-------------+----------+
17 rows in set (1.63 sec)
```

你可能已经注意到，在使用 CTE 后，结果与之前相同，查询时间缩短了 50%，可读性变好，而且可以被多次引用。

派生查询不能引用其他派生查询：

```
SELECT ...
 FROM (SELECT ... FROM ...) AS d1, (SELECT ... FROM d1 ...) AS d2 ...
ERROR: 1146 (42S02): Table 'db.d1' doesn't exist
```

CTE 可以引用其他 CTE：

```
WITH d1 AS (SELECT ... FROM ...), d2 AS (SELECT ... FROM d1 ...)
SELECT
 FROM d1, d2 ...
```

递归 CTE

递归 CTE 是一种特殊的 CTE，其子查询会引用自己的名字。WITH 子句必须以 WITH

RECURSIVE 开头。递归 CTE 子查询包括两部分：seed 查询和 recursive 查询，由 UNION [ALL] 或 UNION DISTINCT 分隔。

seed SELECT 被执行一次以创建初始数据子集；recursive SELECT 被重复执行以返回数据的子集，直到获得完整的结果集。当迭代不会生成任何新行时，递归会停止。这对挖掘层次结构（父/子或部分/子部分）非常有用：

```
WITH RECURSIVE cte AS
(SELECT ... FROM table_name /* seed SELECT */
UNION ALL
SELECT ... FROM cte, table_name) /* "recursive" SELECT */
SELECT ... FROM cte;
```

假设要打印从 1 到 5 的所有数字：

```
mysql> WITH RECURSIVE cte (n) AS
( SELECT 1 /* seed query */
  UNION ALL
  SELECT n + 1 FROM cte WHERE n < 5 /* recursive query */
)
SELECT * FROM cte;
+---+
| n |
+---+
| 1 |
| 2 |
| 3 |
| 4 |
| 5 |
+---+
5 rows in set (0.00 sec)
```

在每次迭代中，SELECT 都会生成一个带有新值的行，比前一行的值 n 多 1。第一次迭代在初始行集合（1）上运行并生成值为 1 + 1 = 2 的行；第二次迭代对第一次迭代的行集合（2）进行操作并生成值为 2 + 1 = 3 的行。以此类推，一直持续到 n 不再小于 5 时，递归结束。

假设要执行分层数据遍历，以便为每个员工生成一个组织结构图（即从 CEO 到每个员工的路径），也可以使用递归CTE！

创建带有manager_id的测试表：

```
mysql> CREATE TABLE employees_mgr (
 id INT PRIMARY KEY NOT NULL,
 name VARCHAR(100) NOT NULL,
 manager_id INT NULL,
 INDEX (manager_id),
FOREIGN KEY (manager_id) REFERENCES employees_mgr (id)
);
```

插入示例数据：

```
mysql> INSERT INTO employees_mgr VALUES
(333, "Yasmina", NULL), # Yasmina is the CEO (manager_id is NULL)
(198, "John", 333), # John has ID 198 and reports to 333 (Yasmina)
(692, "Tarek", 333),
(29, "Pedro", 198),
(4610, "Sarah", 29),
(72, "Pierre", 29),
(123, "Adil", 692);
```

执行递归CTE：

```
mysql> WITH RECURSIVE employee_paths (id, name, path) AS
(
 SELECT id, name, CAST(id AS CHAR(200))
 FROM employees_mgr
 WHERE manager_id IS NULL
 UNION ALL
 SELECT e.id, e.name, CONCAT(ep.path, ',', e.id)
 FROM employee_paths AS ep JOIN employees_mgr AS e
 ON ep.id = e.manager_id
)
SELECT * FROM employee_paths ORDER BY path;
```

产生以下结果：

```
+------+---------+-----------------+
```

```
| id     | name     | path                 |
+------+---------+-----------------+
| 333    | Yasmina | 333                 |
| 198    | John     | 333,198             |
| 29     | Pedro    | 333,198,29          |
| 4610 | Sarah    | 333,198,29,4610 |
| 72     | Pierre   | 333,198,29,72      |
| 692    | Tarek    | 333,692             |
| 123    | Adil     | 333,692,123         |
+------+---------+-----------------+
7 rows  in set (0.00 sec)
```

WITH RECURSIVE employee_paths (id, name, path) AS 是 CTE 的名称，列是(id, name, path)。

SELECT id, name, CAST(id AS CHAR(200)) FROM employees_mgr WHERE manager_id IS NULL 是查询 CEO 的 seed 查询（没有在 CEO 之上的管理者）。

SELECT e.id, e.name, CONCAT(ep.path, ',', e.id) FROM employee_paths AS ep JOIN employees_mgr AS e ON ep.id = e.manager_id 是递归查询。

递归查询生成的每一行，会查找直接向前一行生成的员工做汇报的所有员工。对于每个员工，该行的信息包括员工 ID、姓名和员工管理链，该链是在最后添加了员工 ID 的管理链。

3.4 生成列（generated column）

生成列（generated column）的值是根据列定义中包含的表达式计算得出的。生成列包含下面两种类型。

- virtual 生成列：当从表中读取记录时，将计算该列。
- stored 生成列：当向表中写入新记录时，将计算该列并将其作为常规列存储在表中。

virtual 生成列比 stored 生成列更有用，因为一个虚拟的列不占用任何存储空间。你可以使用触发器模拟 stored 生成列的行为。

3.4.1　如何操作

假设你的应用程序从 employees 表中检索数据时，使用 full_name 表示 concat（'first_name', '', 'last_name'），而不是使用表达式来表示，你可以使用虚拟列实时计算 full_name。可以添加另一列，其后跟随着表达式：

```
mysql> CREATE TABLE `employees` (
  `emp_no` int(11) NOT NULL,
  `birth_date` date NOT NULL,
  `first_name` varchar(14) NOT NULL,
  `last_name` varchar(16) NOT NULL,
  `gender` enum('M','F') NOT NULL,
  `hire_date` date NOT NULL,
  `full_name` VARCHAR(30) AS (CONCAT(first_name,' ',last_name)),
 PRIMARY KEY (`emp_no`),
  KEY `name` (`first_name`,`last_name`)
) ENGINE=InnoDB DEFAULT CHARSET=utf8mb4;
```

请注意，应该根据虚拟列修改插入语句。你可以这样使用 full insert：

```
mysql> INSERT INTO employees (emp_no, birth_date, first_name, last_name,
gender, hire_date) VALUES (123456, '1987-10-02', 'ABC' , 'XYZ', 'F',
'2008-07-28');
```

如果要在 INSERT 语句中包含 full_name，就只能将其指定为 DEFAULT。所有其他值都会引发 ERROR 3105 (HY000):错误。不允许在 employees 表中为生成的列 full_name 指定值：

```
mysql> INSERT INTO employees (emp_no, birth_date, first_name, last_name,
gender, hire_date, full_name) VALUES (123456, '1987-10-02', 'ABC' , 'XYZ',
'F', '2008-07-28', DEFAULT);
```

可以直接从 employees 表中选择 full_name：

```
mysql> SELECT * FROM employees WHERE emp_no=123456;
+--------+------------+------------+-----------+------+----------+----
------+
```

```
| emp_no | birth_date | first_name | last_name | gender| hire_date|
full_name |
+--------+------------+------------+-----------+------+----------+----
------+
| 123456 | 1987-10-02 | ABC         | XYZ        | F     | 2017-11-23 | ABC
XYZ        |
+--------+------------+------------+-----------+------+----------+----
------+
1 row in set (0.00 sec)
```

如果你已经创建了表并希望添加新的生成列，请执行 ALTER TABLE 语句，这将在第 10 章中详细介绍。

例如：

```
mysql> ALTER TABLE employees ADD hire_date_year YEAR AS (YEAR(hire_date))
VIRTUAL;
```

请参阅 `https://dev.mysql.com/doc/refman/8.0/en/create-table-generated-columns.html` 了解更多关于生成列的内容。13.4 节和 13.13 节将讲述虚拟列的其他用途。

3.5 窗口函数

对于查询中的每一行，可以使用窗口函数，利用与该行相关的行执行计算。这是通过使用 `OVER` 和 `WINDOW` 子句来完成的。

以下是可以执行计算的列表。

- `CUME_DIST()`：累积分布值。
- `DENSE_RANK()`：分区内当前行的等级（无间隔）。
- `FIRST_VALUE()`：窗口帧中第一行的参数值。
- `LAG()`：落后于分区内当前行的那一行的参数值。
- `LAST_VALUE()`：窗口帧中最末行的参数值。
- `LEAD()`：领先于分区内当前行的那一行的参数值。
- `NTH_VALUE()`：窗口帧中的第 n 行的参数值。

- NTILE()：分区内当前行的桶的编号。
- PERCENT_RANK()：百分比排名值。
- RANK()：分区中当前行的等级（有间隔）。
- ROW_NUMBER()：分区内当前行的编号。

3.5.1　如何操作

窗口函数的使用方式有多种。在下面的小节中，我们就来看看这些方式。为了使这些示例有效，请先添加 hire_date_year 虚拟列：

```
mysql> ALTER TABLE employees ADD hire_date_year YEAR AS (YEAR(hire_date))
VIRTUAL;
```

行号

可以用每行的行号来排列结果：

```
mysql> SELECT CONCAT(first_name, " ", last_name) AS full_name, salary,
ROW_NUMBER() OVER(ORDER BY salary  DESC) AS 'Rank'  FROM employees JOIN
salaries ON salaries.emp_no=employees.emp_no LIMIT 10;
+-------------------+--------+------+
| full_name         | salary | Rank |
+-------------------+--------+------+
| Tokuyasu Pesch    | 158220 | 1    |
| Tokuyasu Pesch    | 157821 | 2    |
| Honesty Mukaidono | 156286 | 3    |
| Xiahua Whitcomb   | 155709 | 4    |
| Sanjai Luders     | 155513 | 5    |
| Tsutomu Alameldin | 155377 | 6    |
| Tsutomu Alameldin | 155190 | 7    |
| Tsutomu Alameldin | 154888 | 8    |
| Tsutomu Alameldin | 154885 | 9    |
| Willard Baca      | 154459 | 10   |
+-------------------+--------+------+
10 rows in set (6.24 sec)
```

分割结果

可以在 OVER 子句中分割结果。假设你想找出每年的工资排列情况，方法如下：

```
mysql> SELECT hire_date_year, salary, ROW_NUMBER() OVER(PARTITION BY
hire_date_year ORDER BY salary DESC) AS 'Rank' FROM employees JOIN salaries
ON salaries.emp_no=employees.emp_no ORDER BY salary DESC LIMIT 10;
+----------------+--------+------+
| hire_date_year | salary | Rank |
+----------------+--------+------+
|           1985 | 158220 |    1 |
|           1985 | 157821 |    2 |
|           1986 | 156286 |    1 |
|           1985 | 155709 |    3 |
|           1987 | 155513 |    1 |
|           1985 | 155377 |    4 |
|           1985 | 155190 |    5 |
|           1985 | 154888 |    6 |
|           1985 | 154885 |    7 |
|           1985 | 154459 |    8 |
+----------------+--------+------+
10 rows in set (8.04 sec)
```

可以注意到，排在前面的员工中有两位分别是 1986 年和 1987 年入职的员工，其他都是 1985 年入职的员工。

命名窗口

可以根据需要对一个窗口命名，并多次使用它，无须每次都重新定义：

```
mysql> SELECT hire_date_year, salary, RANK() OVER w AS 'Rank' FROM
employees join salaries ON salaries.emp_no=employees.emp_no WINDOW w AS
(PARTITION BY hire_date_year ORDER BY salary DESC) ORDER BY salary DESC
LIMIT 10;
+----------------+--------+------+
| hire_date_year | salary | Rank |
+----------------+--------+------+
|           1985 | 158220 |    1 |
```

```
|            1985 | 157821 |    2 |
|            1986 | 156286 |    1 |
|            1985 | 155709 |    3 |
|            1987 | 155513 |    1 |
|            1985 | 155377 |    4 |
|            1985 | 155190 |    5 |
|            1985 | 154888 |    6 |
|            1985 | 154885 |    7 |
|            1985 | 154459 |    8 |
+----------------+--------+------+
10 rows in set (8.52 sec)
```

第一个、最后一个和第 *n* 个值

你可以选择窗口结果中的第一个、最后一个和第 *n* 个值。如果该行不存在，则返回 NULL。

假设你想从窗口中找到第一个、最后一个和第 3 个值，代码如下：

```
mysql> SELECT hire_date_year, salary, RANK() OVER w AS 'Rank',
FIRST_VALUE(salary) OVER w AS 'first',
NTH_VALUE(salary, 3) OVER w AS 'third',
LAST_VALUE(salary) OVER w AS 'last'
FROM employees join salaries ON salaries.emp_no=employees.emp_no
WINDOW w AS (PARTITION BY hire_date_year ORDER BY salary DESC)
ORDER BY salary DESC LIMIT 10;
+----------------+--------+------+--------+--------+--------+
| hire_date_year | salary | Rank | first  | third  | last   |
+----------------+--------+------+--------+--------+--------+
|           1985 | 158220 |    1 | 158220 |   NULL | 158220 |
|           1985 | 157821 |    2 | 158220 |   NULL | 157821 |
|           1986 | 156286 |    1 | 156286 |   NULL | 156286 |
|           1985 | 155709 |    3 | 158220 | 155709 | 155709 |
|           1987 | 155513 |    1 | 155513 |   NULL | 155513 |
|           1985 | 155377 |    4 | 158220 | 155709 | 155377 |
|           1985 | 155190 |    5 | 158220 | 155709 | 155190 |
|           1985 | 154888 |    6 | 158220 | 155709 | 154888 |
|           1985 | 154885 |    7 | 158220 | 155709 | 154885 |
|           1985 | 154459 |    8 | 158220 | 155709 | 154459 |
```

```
+----------------+--------+------+--------+--------+--------+
10 rows in set (12.88 sec)
```

要了解更多关于窗口函数的其他用例，请参阅 https://mysqlserverteam.com/mysql-8-0-2-introducing-window-functions 和 https://dev.mysql.com/doc/refman/8.0/en/window-function-descriptions.html#function_row-number。

第 4 章 配置 MySQL

在本章中，我们将介绍以下内容：

- 使用配置文件
- 使用全局变量和会话变量
- 在启动脚本中使用参数
- 配置参数
- 更改数据目录

4.1 引言

MySQL 有两种类型的参数。

- **静态参数**：重启 MySQL 服务器后才能使之生效。
- **动态参数**：可以在不重新启动 MySQL 服务器的情况下即时更改它。

变量可以通过以下方式设置。

- **配置文件**：MySQL 有一个配置文件，我们可以在其中指定数据的位置、MySQL 可以使用的内存以及其他各种参数。
- **启动脚本**：可以直接将参数传递给 `mysqld` 进程。启动脚本仅在调用服务器时才有效。

- **使用 SET 命令**（仅限动态变量）：这将持续到服务器重新启动时。你还需要在配置文件中设置变量，以便在重新启动时保持更改持久化。另一种使更改持久化的方法是在 `PERSIST` 关键字或`@@persist` 之前加上变量名称。

4.2 使用配置文件

默认配置文件是`/etc/my.cnf`（在 Red Hat 和 CentOS 上）和`/etc/mysql/ my.cnf`（在 Debian 系统上）。在你最喜欢的编辑器中打开该文件并根据需要修改参数，其中的主要参数将在本节讨论。

4.2.1 如何操作

配置文件包含由 `section_name` 指定的部分。所有与 section 相关的参数都可以放在 `section_name` 下面，例如：

```
[mysqld] <---section name
<parameter_name> = <value> <---parameter values
[client]
<parameter_name> = <value>
[mysqldump]
<parameter_name> = <value>
[mysqld_safe]
<parameter_name> = <value>
[server]
<parameter_name> = <value>
```

- `[mysql]`：该部分由 `mysql` 命令行客户端读取。
- `[client]`：该部分由所有连接的客户端读取（包括 `mysql cli`）。
- `[mysqld]`：该部分由 `mysql` 服务器读取。
- `[mysqldump]`：该部分由名为 `mysqldump` 的备份工具读取。
- `[mysqld_safe]`:该部分由`mysqld_safe`进程读取(MySQL 服务器启动脚本)。

除此之外，`mysqld_safe` 进程会从选项文件中的`[mysqld]`和`[server]`部分读取所有选项。

例如，`mysqld_safe` 进程从 `mysqld` 部分读取 `pid-file` 选项。

```
shell> sudo vi /etc/my.cnf
[mysqld]
pid-file = /var/lib/mysql/mysqld.pid
```

在使用 systemd 的系统中，mysqld_safe 将不会被安装。要配置启动脚本，需要在 /etc/systemd/system/mysqld.service.d/override.conf 中设置值。

例如：

```
[Service]
LimitNOFILE=max_open_files
PIDFile=/path/to/pid/file
LimitCore=core_file_limit
Environment="LD_PRELOAD=/path/to/malloc/library"
Environment="TZ=time_zone_setting"
```

4.3　使用全局变量和会话变量

正如在前面的章节中看到的，你可以通过连接到 MySQL 并执行 SET 命令来设置参数。

根据变量的作用域可以将变量分为两种：

- **全局变量**：适用于所有新连接。
- **会话变量**：仅适用于当前连接（会话）。

4.3.1　如何操作

如果你想记录所有执行时间超过 1 秒的查询，则可以执行：

```
mysql> SET GLOBAL long_query_time = 1;
```

要使更改在重新启动时保持持久化，请使用：

```
mysql> SET PERSIST long_query_time = 1;
Query OK, 0 rows affected (0.01 sec)
```

或者使用：

```
mysql> SET @@persist.long_query_time = 1; Query OK, 0 rows affected (0.00
sec)
```

持久化的全局系统变量设置存储在数据目录中的 mysqld-auto.cnf 中。

假设你只想为此会话记录查询，而不是为所有连接记录查询，可以使用以下命令：

```
mysql> SET SESSION long_query_time = 1;
```

4.4 在启动脚本中使用参数

假设你希望使用启动脚本（而不是通过 systemd）来启动 MySQL，尤其是在发生测试或临时更改时，那么可以将变量传递给脚本，而不是在配置文件中更改它。

4.4.1 如何操作

```
shell> /usr/local/mysql/bin/mysqld --basedir=/usr/local/mysql --
datadir=/usr/local/mysql/data --plugin-dir=/usr/local/mysql/lib/plugin --
user=mysql --log-error=/usr/local/mysql/data/centos7.err --pid-
file=/usr/local/mysql/data/centos7.pid --init-file=/tmp/mysql-init &
```

可以看到--init-file 参数被传递给了服务器。服务器在启动之前执行该文件中的 SQL 语句。

4.5 配置参数

本节将介绍安装完成后需要配置的基本内容。除了这些需要配置的内容，剩下的全部可以保留默认值或稍后根据负载进行调整。

4.5.1 如何操作

让我们看看细节。

数据目录

由 MySQL 服务器管理的数据存储在名为数据目录的目录下。数据目录的每个子目录都是一个数据库目录，并对应服务器管理的数据库。默认情况下，数据目录有三个子目录。

- mysql：MySQL 系统数据库。
- performance_schema：提供用于在运行时检查服务器的内部执行情况的信息。
- sys：提供一组对象，帮助更轻松地解释 performance schema 信息。

除此之外，数据目录包含日志文件、InnoDB 表空间和 InnoDB 日志文件、SSL 和 RSA 密钥文件、mysqld 的 pid，以及存储持久化全局系统变量设置的 mysqld-auto.cnf。

要设置数据目录，需要更改 datadir 值，或将 datadir 的值添加到配置文件中。datadir 的值默认是/var/lib/mysql：

```
shell> sudo vi /etc/my.cnf
[mysqld]
datadir = /data/mysql
```

你可以将其设置为你要存储数据的任何位置，但应该将数据目录的所有权更改为 mysql。

确保带有数据目录的磁盘卷有足够的空间来容纳所有数据。

innodb_buffer_pool_size

这是 InnoDB 最重要的调优参数，它决定 InnoDB 存储引擎可以使用多少内存空间来缓存内存中的数据和索引。将其值设置得太低可能会降低 MySQL 服务器的性能，而将其值设置得太高会增加 MySQL 进程的内存消耗。MySQL 8 中的 innodb_buffer_pool_size 是动态的，这意味着可以改变 innodb_buffer_pool_size，而不用重新启动服务器。

下面是如何调试 innodb_buffer_pool_size 的简单指南。

1. 找出数据集的大小。不要将 innodb_buffer_pool_size 的值设置得高于数据集的值。假设有一个 12 GB RAM 的机器，数据集是 3 GB，那么你可以设置 innodb_buffer_pool_size 的值为 3 GB。如果数据集的大小增长了，可以在需要时增加它，而无须重新启动 MySQL。

2. 通常，数据集的大小比可用 RAM 大得多。对于整个 RAM，你可以为操作系统分配一些，为其他进程分配一些，为 MySQL 内的 per-thread 缓冲区分配一些，为 InnoDB 之外的 MySQL 服务器分配一些，其余的可以分配给 InnoDB 缓冲池。这是一个通用性很强的表，假设它是一个专用的 MySQL 服务器，所有表都是 InnoDB 的，并且 per-thread 缓冲区都保留默认值，那么你就可以开始使用它了。如果系统内存不足，可以动态减小缓冲池的大小。下表列出了 RAM 与对应的缓冲池大小。

RAM	缓冲池大小（范围）
4 GB	1 GB~2 GB
8 GB	4 GB~6 GB
12 GB	6 GB~10 GB
16 GB	10 GB~12 GB
32 GB	24 GB~28 GB
64 GB	45 GB~56 GB
128 GB	108 GB~116 GB
256 GB	220 GB~245 GB

innodb_buffer_pool_instances

可以将 InnoDB 缓冲池划分为不同的区域，以便在不同线程读取和写入缓存页面时减少争用，从而提高并发性。例如，如果缓冲池大小为 64 GB，innodb_buffer_pool_instances 的值为 32，则缓冲区将被分为 32 个区域，每个区域的大小为 2 GB。

如果缓冲池大小超过 16 GB，则可以设置实例，以便每个区域至少获得 1 GB 的空间。

innodb_log_file_size

innodb_log_file_size 是重做日志空间的大小，用于数据库崩溃时重放已提交的事务。innodb_log_file_size 的默认值为 48 MB，这可能不足以满足生产工作负载的需求。开始的时候，可以将其值设置为 1 GB 或 2 GB。此更改需要重新启动服务器才能生效。停止 MySQL 服务器的运行并确保它没有被错误地关闭。在 my.cnf 中进行更改并启动服务器。在早期版本中，你需要停止服务器的运行，删除日志文件，然后启动服务器。在 MySQL 8 中，这一系列操作是自动进行的。修改重做日志文件将在第 11.2 节中进行介绍。

4.6　更改数据目录

数据量会随着时间的推移而增长，当它超出文件系统的容量时，需要添加磁盘或将数据目录移动到更大的卷中。

4.6.1　如何操作

1. 检查当前的数据目录。默认情况下，数据目录是/var/lib/mysql：

```
mysql> show variables like '%datadir%';
+---------------+-----------------+
| Variable_name | Value           |
+---------------+-----------------+
| datadir       | /var/lib/mysql/ |
+---------------+-----------------+
1  row in set (0.04 sec)
```

2. 停止 mysql，并确保它成功停止了：

```
shell> sudo systemctl stop mysql
```

3. 检查状态：

```
shell> sudo systemctl status mysql
```

它应该显示 stopped MySQL Community Server。

4. 在新位置创建目录并将所有权更改到 mysql 下：

```
shell> sudo mkdir -pv /data
shell> sudo chown -R mysql:mysql /data/
```

5. 将文件移动到新的数据目录中：

```
shell> sudo rsync -av /var/lib/mysql /data
```

6. 在 Ubuntu 系统中，如果启用了 AppArmor，则需要配置 Access Control（访问控制）：

```
shell> vi /etc/apparmor.d/tunables/alias
alias /var/lib/mysql/ -> /data/mysql/,
shell> sudo systemctl restart apparmor
```

7. 启动 MySQL 服务器并确认数据目录已更改：

```
shell> sudo systemctl start mysql
mysql> show variables like '%datadir%';
+---------------+--------------+
| Variable_name | Value        |
+---------------+--------------+
| datadir       | /data/mysql/ |
+---------------+--------------+
1 row in set (0.00 sec)
```

8. 验证数据是否完整并删除旧的数据目录：

```
shell> sudo rm -rf /var/lib/mysql
```

如果 MySQL 无法启动并显示错误——MySQL data dir not found at/var/lib/mysql,Please Creat one，那么执行 sudo mkdir/var/lib/mysql/mysql-p。

如果没有找到 MySQL 系统数据库，运行 mysql_install_db 工具，它会创建所需的目录。

第5章 事务

在本章中，我们将介绍以下内容：

- 执行事务
- 使用保存点
- 隔离级别
- 锁

5.1 引言

在下面的内容中，我们将讨论 MySQL 中的事务和各种隔离级别。事务就是一组应该一起成功或一起失败的 SQL 语句。事务还应该具备原子性、一致性、隔离性和持久性（Atomicity、Consistency、Isolation 和 Durability，简称 ACID）的属性。举一个非常基本的例子，从账户 A 到账户 B 转账，假设账户 A 有 600 美元，账户 B 有 400 美元，希望从账户 A 转 100 美元到账户 B 上。

银行将从账户 A 中扣除 100 美元，并使用以下 SQL 代码（用于说明）向账户 B 添加 100 美元：

```
mysql> SELECT balance INTO @a.bal FROM account WHERE account_number='A';
```

以编程方式检查@a.bal 是否大于或等于 100：

```
mysql> UPDATE account SET balance=@a.bal-100 WHERE account_number='A';
mysql> SELECT balance INTO @b.bal FROM account WHERE account_number='B';
```

以编程方式检查`@b.bal`是否为非空值：

```
mysql> UPDATE account SET balance=@b.bal+100 WHERE account_number='B';
```

这4条SQL语句应该是单个事务的一部分，并且满足以下ACID属性。

- **原子性**：所有的SQL语句要么全部成功，要么全部失败，不会存在部分更新。如果数据库在运行两个SQL语句之后没有服从这个属性，那么账户`A`就会凭空损失100美元。
- **一致性**：事务只能以允许的方式改变受其影响的数据。在这个例子中，如果`account_number`与账户`B`不存在，则整个事务应该被回滚。
- **隔离性**：同时发生的事务（并发事务）不应该导致数据库处于不一致的状态中。系统中每个事务都应该像唯一事务一样执行。任何事务都不应影响其他事务的存在。假设`A`向`B`转账的同时，`A`完全转移所有600美元，两个事务应该独立进行，在进行转账前确认好余额。
- **持久性**：无论数据库或系统是否发生故障，数据都会永久保存在磁盘上，并且不会丢失。

`InnoDB`是MySQL中的默认存储引擎，`InnoDB`支持事务处理，而MyISAM不支持事务处理。

5.2 执行事务

我们通过创建dummy表和示例数据来了解本节的主题：

```
mysql> CREATE DATABASE bank;
mysql> USE bank;
mysql> CREATE TABLE account(account_number varchar(10) PRIMARY KEY, balance
int);
mysql> INSERT INTO account VALUES('A',600),('B',400);
```

5.2.1 如何操作

要启动一个事务（一组SQL），请执行`START TRANSACTION`或`BEGIN`语句：

```
mysql> START TRANSACTION;
```

或:

```
mysql> BEGIN;
```

然后执行你希望在事务中包含的所有语句，例如从账户 A 转 100 美元到账户 B:

```
mysql> SELECT balance INTO @a.bal FROM account WHERE account_number='A';
```

以编程方式检查@a.bal 是否大于或者等于 100:

```
mysql> UPDATE account SET balance=@a.bal-100 WHERE account_number='A'; mysql>
SELECT balance INTO @b.bal FROM account WHERE account_number='B';
```

以编程方式检查@b.bal 是否为非空值:

```
mysql> UPDATE account SET balance=@b.bal+100 WHERE account_number='B';
```

确保所有 SQL 语句成功执行后，执行 COMMIT 语句，该语句将完成事务并提交数据:

```
mysql> COMMIT;
```

如果遇到错误并希望中止事务，可以发送 ROLLBACK 语句而非 COMMIT 语句。

例如，如果账户 A 想要给一个并不存在的账户转账，则应取消交易并将金额退还给账户 A:

```
mysql> BEGIN;

mysql> SELECT balance INTO @a.bal FROM account WHERE account_number='A';

mysql> UPDATE account SET balance=@a.bal-100 WHERE account_number='A';

mysql> SELECT balance INTO @b.bal FROM account WHERE account_number='C';
Query OK, 0 rows affected, 1 warning (0.07 sec)

mysql> SHOW WARNINGS;
+---------+------+-----------------------------------------------------+
| Level   | Code | Message                                             |
+---------+------+-----------------------------------------------------+
| Warning | 1329 | No data - zero rows fetched, selected, or processed|
+---------+------+-----------------------------------------------------+
1 row in set (0.02 sec)
```

```
mysql> SELECT @b.bal;
+--------+
| @b.bal |
+--------+
| NULL   |
+--------+
1 row in set (0.00 sec)

mysql> ROLLBACK;
Query OK, 0 rows affected (0.01 sec)
```

autocommit

默认情况下，autocommit 的状态是 ON，这意味着所有单独的语句一旦被执行就会被提交，除非该语句在 BEGIN...COMMIT 块中。如果 autocommit 的状态为 OFF，则需要明确发出 COMMIT 语句来提交事务。要禁用 autocommit，请执行：

```
mysql> SET autocommit=0;
```

DDL 语句，如数据库的 CREATE 或 DROP 语句，以及表或存储例程的 CREATE、DROP 或 ALTER 语句，都是无法回滚的。

某些语句，如 DDL 语句、LOAD DATA INFILE、ANALYZE TABLE 以及与 replication 相关的语句等，会导致隐式 COMMIT。有关这些语句的更多详细信息，请参阅 https://dev.mysql.com/doc/refman/8.0/en/implicit-commit.html。

5.3 使用保存点

使用保存点可以回滚到事务中的某些点，而且无须中止事务。你可以使用 SAVEPOINT 标识符为事务设置名称，并使用 ROLLBACK TO 标识语句将事务回滚到指定的保存点而不中止事务。

5.3.1 如何操作

假设账户 A 想向到多个账户转账，即使向其中一个账户转账的操作失败，向其他账户转账的操作也不应该回滚：

```
mysql> BEGIN;
Query OK, 0 rows affected (0.00 sec)

mysql> SELECT balance INTO @a.bal FROM account WHERE account_number='A';
Query OK, 1 row affected (0.01 sec)

mysql> UPDATE account SET balance=@a.bal-100 WHERE account_number='A';
Query OK, 1 row affected (0.01 sec)
Rows matched: 1 Changed: 1 Warnings: 0

mysql> UPDATE account SET balance=balance+100 WHERE account_number='B';
Query OK, 1 row affected (0.00 sec)
Rows matched: 1 Changed: 1 Warnings: 0

mysql> SAVEPOINT transfer_to_b;
Query OK, 0 rows affected (0.00 sec)

mysql> SELECT balance INTO @a.bal FROM account WHERE account_number='A';
Query OK, 1 row affected (0.00 sec)

mysql> UPDATE account SET balance=balance+100 WHERE account_number='C';
Query OK, 0 rows affected (0.00 sec)
Rows matched: 0 Changed: 0 Warnings: 0

###由于没有行被更新，意味着没有账户'C'，所以可以将事务回滚至成功传输到B的保存点（SAVEPOINT）。然后'A'将得到从C扣除的100美元。如果不想使用保存点，则应在两个交易中执行这些操作。

mysql> ROLLBACK TO transfer_to_b;
Query OK, 0 rows affected (0.00 sec)

mysql> COMMIT;
```

```
Query OK, 0 rows affected (0.00 sec)

mysql> SELECT balance FROM account WHERE account_number='A';
+---------+
| balance |
+---------+
| 400     |
+---------+
1 row in set (0.00 sec)

mysql> SELECT balance FROM account WHERE account_number='B';
+---------+
| balance |
+---------+
| 600     |
+---------+
1 row in set (0.00 sec)
```

5.4 隔离级别

当两个或多个事务同时发生时，隔离级别定义了一个事务与其他事务在资源或者数据修改方面的隔离程度。有 4 种类型的隔离级别，要更改隔离级别，需要设置 `tx_isolation` 变量，该变量是动态的并具有会话级别的作用范围。

5.4.1 如何操作

要更改隔离级别，请执行 `SET @@ transaction_isolation='READ-COMMITTED';` 语句。

读取未提交（read uncommitted）

当前事务可以读取由另一个未提交的事务写入的数据，这也称为**脏读**（**dirty read**）。

例如，想要在账户 A 中增加一些金额，并将其转到账户 B。假设两个交易同时发生，流程将会是下面这样的。

账户 A 最初有 400 美元，我们想为账户 A 添加 500 美元，然后从账户 A 向账户 B 转账 500 美元，详细流程见下表。

# 事务 1（增加存款）	# 事务 2（转账）
BEGIN;	BEGIN;
UPDATE account SET balance=balance+500 WHERE account_number='A';	--
--	SELECT balance INTO @a.bal FROM account WHERE account_number='A'; **# 账户 A 中的金额变为 900 美元**
ROLLBACK; **# 假设由于某种原因事务回滚**	--
	#由于在先前的 SELECT 中账户 A 中有 900 美元，所以将账户 A 中的 900 美元转到账户 B UPDATE account SET balance=balance-900 WHERE account_number='A';
--	**# 账户 B 收到款** UPDATE account SET balance=balance+900 WHERE account_number='B';
--	# 事务 2 成功完成 COMMIT;

可以注意到，事务 2 已经读取了未提交或从事务 1 回滚的数据，导致账户 A 中的值在此事务之后成为负的，这显然不是我们所期望看到的。

读提交（read committed）

当前事务只能读取另一个事务提交的数据，这也称为**不可重复读取（non-repeatable read）**。

再来看刚才的例子，假设账户 A 有 400 美元，账户 B 有 600 美元。下表展示了读提交的过程。

# 事务 1（增加存款）	# 事务 2（转账）
BEGIN;	BEGIN;
UPDATE account SET balance=balance+500 WHERE account_number='A';	--
--	SELECT balance INTO @a.bal FROM account WHERE account_number='A'; **# 账户 A 中有 400 美元，因为事务 1 还没有提交数据**
COMMIT;	--
--	SELECT balance INTO @a.bal FROM account WHERE account_number='A'; **# 账户 A 中的金额变为 900 美元，因为事务 1 已经提交了数据**

可以注意到，在同一个事务中，相同的 SELECT 语句获取了不同的结果。

可重复读取（repeatable read）

一个事务通过第一条语句只能看到相同的数据，即使另一个事务已提交数据。在同一事务中，读取通过第一次读取建立快照是一致的。一个例外是，一个事务可以读取在同一事务中更改的数据。

当事务开始并执行第一次读取数据时，将创建读取视图并保持打开状态，直到事务结束。为了在事务结束之前提供相同的结果集，InnoDB 使用行版本控制和 UNDO 信息。假设事务 1 选择了几行,另一个事务删除了这些行并提交了数据。如果事务 1 处于打开状态，它应该能够看到自己在开始时选择的行。已被删除的行保留在 UNDO 日志空间中以履行事务 1。一旦事务 1 操作完成，那些行便被标记为从 UNDO 日志中删除。这称为**多版本并发控制（MVCC）**。

再来看刚才的例子，假设账户 A 中有 400 美元，账户 B 中有 600 美元，下表展示了整个过程。

#事务1（增加存款）	#事务2（转账）
BEGIN;	BEGIN;
--	SELECT balance INTO @a.bal FROM account WHERE account_number='A'; #账户A里有400美元
UPDATE account SET balance=balance+500 WHERE account_number='A';	--
--	SELECT balance INTO @a.bal FROM account WHERE account_number='A'; **#尽管事务1已经提交，账户A中仍然是400美元**
COMMIT;	--
--	COMMIT;
--	SELECT balance INTO @a.bal FROM account WHERE account_number='A'; **#账户A中的金额变为900美元，因为这是一个全新的事务**

这仅适用于SELECT语句，不一定适用于DML语句。如果插入或修改某些行并提交该事务，那么从另一个并发REPEATABLE READ事务发出的DELETE或UPDATE语句，可能会影响那些刚刚提交的行，即使会话无法查询这些语句。如果事务更新或删除由不同事务提交的行，则这些更改对当前事务变为可见。具体例子参见下表。

#事务1	#事务2
BEGIN;	BEGIN;
SELECT * FROM account; **#2行被返回**	--
--	INSERT INTO account VALUES('C',1000); **#新account被创建**
--	COMMIT;
SELECT * FROM account WHERE account_number='C';	--

# 由于 **MVCC**，没有行被返回	
DELETE FROM account WHERE account_number='C'; **# 令人惊讶的是，account C 被删除了**	--
--	SELECT * FROM account; **# 因为事务 1 尚未提交，所以返回 3 行**
COMMIT;	--
--	SELECT * FROM account; **# 因为事务 1 已被提交，所以返回 2 行**

下表是另一个例子：

# 事务 1	# 事务 2
BEGIN;	BEGIN;
SELECT * FROM account; **# 2 行被返回**	--
--	INSERT INTO account VALUES('D',1000);
--	COMMIT;
SELECT * FROM account; **# 由于 MVCC，返回了 3 行**	--
UPDATE account SET balance=1000 WHERE account_number='D'; **# 令人惊讶的是，account D 得到更新**	--
SELECT * FROM account; **# 令人惊讶的是，返回了 4 行**	--

序列化（serializable）

通过把选定的所有行锁起来，序列化可以提供最高级别的隔离。此级别与 REPEATABLE READ 类似，但如果禁用 autocommit，则 InnoDB 会将所有普通 SELECT 语句隐式转换为 SELECT...LOCK IN SHARE MODE；如果启用 autocommit，则 SELECT 就是它自己的事务，如下表所示。

#事务 1	#事务 2
BEGIN;	BEGIN;
SELECT * FROM account WHERE account_number='A';	--
--	UPDATE account SET balance=1000 WHERE account_number='A'; **#将一直等待，直到事务 1 在 A 行上的锁被释放**
COMMIT;	--
--	**# UPDATE 现在会成功**

另一个例子参见下表：

#事务 1	#事务 2
BEGIN;	BEGIN;
SELECT * FROM account WHERE account_number='A'; **#选择 A 的值**	--
--	INSERT INTO account VALUES('D',2000); **#插入 D**
SELECT * FROM account WHERE account_number='D'; #等待，直到事务 2 操作完成	--
--	COMMIT;
#现在前面的 select 语句返回 D 的值	--

所以，序列化会等待被锁的行，并且总是读取最新提交的数据。

5.5　锁

有以下两种类型的锁。

- **内部锁**：MySQL 在自身服务器内部执行内部锁，以管理多个会话对表内容的争用。
- **外部锁**：MySQL 为客户会话提供选项来显式地获取表锁，以阻止其他会话访问表。

内部锁又可以分为下面两种类型。

- **行级锁**：行级锁是细粒度的。只有被访问的行会被锁定。这允许通过多个会话同时进行写访问，使其适用于多用户、高度并发和 OLTP 的应用程序。只有 InnoDB 支持行级锁。
- **表级锁**：MySQL 对 MyISAM、MEMORY 和 MERGE 表使用表级锁，一次只允许一个会话更新这些表。这种锁定级别使得这些存储引擎更适用于只读的或以读取操作为主的或单用户的应用程序。

请参阅 https://dev.mysql.com/doc/refman/8.0/en/internal-locking.html 和 https://dev.mysql.com/doc/refman/8.0/en/innodb-locking.html 来了解更多关于 InnoDB 锁的信息。

外部锁：可以使用 LOCK TABLE 和 UNLOCK TABLES 语句来控制锁定。

READ 和 WRITE 的表锁定解释如下。

- READ：当一个表被锁定为 READ 时，多个会话可以从表中读取数据而不需要获取锁。此外，多个会话可以在同一个表上获得锁，这就是为什么 READ 锁也被称为**共享锁**。当 READ 锁被保持时，没有会话可以将数据写入表格中（包括持有该锁的会话）。如果有任何写入尝试，该操作将处于等待状态，直到 READ 锁被释放。
- WRITE：当一个表被锁定为 WRITE 时，除持有该锁的会话之外，其他任何会话都不能读取或向表中写入数据。除非现有锁被释放，否则其他任何会话都不能获得任何锁。这就是为什么 WRITE 锁被称为**排他锁**。如果有任何读取/写入尝试，该操作将处于等待状态，直到 WRITE 锁被释放。

当执行 UNLOCK TABLES 语句时或当会话终止时，所有锁都会被释放。

5.5.1 如何操作

锁定表的语法如下：

```
mysql> LOCK TABLES table_name [READ | WRITE]
```

要解锁表，请使用：

```
mysql> UNLOCK TABLES;
```

要锁定所有数据库中的所有表，请执行以下语句。在获取数据库的一致快照时需要使用该语句，它会冻结对数据库的所有写入操作：

```
mysql> FLUSH TABLES WITH READ LOCK;
```

锁队列

除共享锁（一个表可以有多个共享锁）之外，没有两个锁可以一起加在一个表上。如果一个表已经有一个共享锁，此时有一个排他锁要进来，那么它将被保留在队列中，直到共享锁被释放。当排他锁在队列中时，所有后续的共享锁也会被阻塞并保留在队列中。

当 InnoDB 从表中读取/写入数据时会获取元数据锁。如果第二个事务请求 WRITE LOCK，该事务将被保留在队列中，直到第一个事务完成。如果第三个事务想要读取数据，就必须等到第二个事务完成。

事务 1：

```
mysql> BEGIN;
Query OK, 0 rows affected (0.00 sec)

mysql> SELECT * FROM employees LIMIT  10;
+--------+------------+------------+-----------+--------+------------+
| emp_no | birth_date | first_name | last_name | gender | hire_date  |
+--------+------------+------------+-----------+--------+------------+
| 10001  | 1953-09-02 | Georgi     | Facello   | M      | 1986-06-26 |
| 10002  | 1964-06-02 | Bezalel    | Simmel    | F      | 1985-11-21 |
| 10003  | 1959-12-03 | Parto      | Bamford   | M      | 1986-08-28 |
| 10004  | 1954-05-01 | Chirstian  | Koblick   | M      | 1986-12-01 |
| 10005  | 1955-01-21 | Kyoichi    | Maliniak  | M      | 1989-09-12 |
| 10006  | 1953-04-20 | Anneke     | Preusig   | F      | 1989-06-02 |
| 10007  | 1957-05-23 | Tzvetan    | Zielinski | F      | 1989-02-10 |
| 10008  | 1958-02-19 | Saniya     | Kalloufi  | M      | 1994-09-15 |
| 10009  | 1952-04-19 | Sumant     | Peac      | F      | 1985-02-18 |
| 10010  | 1963-06-01 | Duangkaew  | Piveteau  | F      | 1989-08-24 |
+--------+------------+------------+-----------+--------+------------+
10 rows in set (0.00 sec)
```

请注意 COMMIT 未执行，该事务仍保持开放状态。

事务 2：

```
mysql> LOCK TABLE employees WRITE;
```

该语句必须等到事务 1 完成才会被执行。

事务 3：

```
mysql> SELECT * FROM employees LIMIT 10;
```

事务 3 不会给出任何结果，因为队列中有排他锁（它正在等待事务 2 完成操作），而且它阻塞了该表上的所有操作。

可以通过另一个会话的 SHOW PROCESSLIST 来查看现在的状态：

```
mysql> SHOW PROCESSLIST;
+----+------+-----------+-----------+---------+------+----------------
------------+---------------------------------+
| Id | User | Host      | db        | Command | Time | State
| Info                              |
+----+------+-----------+-----------+---------+------+----------------
------------+---------------------------------+
| 20 | root | localhost | employees | Sleep   | 48   |
| NULL                              |
| 21 | root | localhost | employees | Query   | 34   | Waiting for table
metadata lock | LOCK TABLE employees WRITE        |
| 22 | root | localhost | employees | Query   | 14   | Waiting for table
metadata lock | SELECT * FROM employees LIMIT 10 |
| 23 | root | localhost | employees | Query   | 0    | starting
| SHOW PROCESSLIST                  |
+----+------+-----------+-----------+---------+------+----------------
------------+---------------------------------+
4 rows in set (0.00 sec)
```

可以注意到事务 2 和事务 3 都在等待事务 1 的完成。

要了解更多关于元数据锁的信息，请参考 https://dev.mysql.com/doc/refman/8.0/en/metadata-locking.html。使用 FLUSH TABLES WITH READ LOCK 也可以观察到相同情况。

事务 1:

```
mysql> BEGIN;
Query OK, 0 rows affected (0.00 sec)

mysql> SELECT * FROM employees LIMIT 10;
+--------+------------+------------+-----------+--------+------------+
| emp_no | birth_date | first_name | last_name | gender | hire_date  |
+--------+------------+------------+-----------+--------+------------+
| 10001  | 1953-09-02 | Georgi     | Facello   | M      | 1986-06-26 |
| 10002  | 1964-06-02 | Bezalel    | Simmel    | F      | 1985-11-21 |
| 10003  | 1959-12-03 | Parto      | Bamford   | M      | 1986-08-28 |
| 10004  | 1954-05-01 | Chirstian  | Koblick   | M      | 1986-12-01 |
| 10005  | 1955-01-21 | Kyoichi    | Maliniak  | M      | 1989-09-12 |
| 10006  | 1953-04-20 | Anneke     | Preusig   | F      | 1989-06-02 |
| 10007  | 1957-05-23 | Tzvetan    | Zielinski | F      | 1989-02-10 |
| 10008  | 1958-02-19 | Saniya     | Kalloufi  | M      | 1994-09-15 |
| 10009  | 1952-04-19 | Sumant     | Peac      | F      | 1985-02-18 |
| 10010  | 1963-06-01 | Duangkaew  | Piveteau  | F      | 1989-08-24 |
+--------+------------+------------+-----------+--------+------------+
10 rows in  set (0.00 sec)
```

请注意 COMMIT 未执行，该事务仍保持开放状态。

事务 2:

```
mysql> FLUSH TABLES WITH READ LOCK;
```

事务 3:

```
mysql> SELECT * FROM employees LIMIT 10;
```

事务 3 也不会给出任何结果，因为 FLUSH TABLES 在锁定之前需要等待表中的所有操作完成，而且它阻止了该表上的所有操作。

可以通过在另一个会话中检查 SHOW PROCESSLIST 来查看现在的状态。

```
mysql> SHOW PROCESSLIST;
+----+------+-----------+-----------+---------+------+-----------------
----+----------------------------------------------------+
```

```
| Id | User | Host      | db        | Command | Time | State
| Info                                                  |
+----+------+-----------+-----------+---------+------+------------------
----+-----------------------------------------------------+
| 20 | root | localhost | employees | Query   | 7    | Creating sort index
| SELECT * FROM employees ORDER BY first_name  DESC |
| 21 | root | localhost | employees | Query   | 5    | Waiting for table
flush | FLUSH TABLES WITH READ LOCK                         |
| 22 | root | localhost | employees | Query   | 3    | Waiting for table
flush | SELECT * FROM employees LIMIT 10              |
| 23 | root | localhost | employees | Query   | 0    | starting
| SHOW      PROCESSLIST                                     |
+----+------+-----------+-----------+---------+------+------------------
----+-----------------------------------------------------+
4 rows in set (0.00 sec)
```

为了保证备份的一致性，所有备份方法都使用 FLUSH TABLES WITH READ LOCK，然而如果表上存在长时间运行的事务，这种做法可能会带来危险。

第6章 二进制日志

在本章中，我们将介绍以下内容：

- 使用二进制日志
- 二进制日志的格式
- 从二进制日志中提取语句
- 忽略要写入二进制日志的数据库
- 迁移二进制日志

6.1 引言

二进制日志包含数据库的所有更改记录，包括数据和结构两方面。二进制日志不记录 `SELECT` 或 `SHOW` 等不修改数据的操作。运行带有二进制日志的服务器会带来轻微的性能影响。二进制日志能保证数据库出故障时数据是安全的。只有完整的事件或事务会被记录或回读。

为什么应该使用二进制日志？

- **复制**：使用二进制日志，可以把对服务器所做的更改以流式方式传输到另一台服务器上。从（slave）服务器充当镜像副本，也可用于分配负载。接受写入的服务器称为主（master）服务器。
- **时间点恢复**：假设你在星期日的 00:00 进行了备份，而数据库在星期日的 08:00 出现故障。使用备份可以恢复到周日 00:00 的状态；而使用二进制日志可以恢复到周日 08:00 的状态。

6.2 使用二进制日志

要启用二进制日志，必须设置 log_bin 和 server_id 并重新启动服务器。可以在 log_bin 内提及 path 和 base 名称。例如，log_bin 设置为/data/mysql/binlogs/server1，二进制日志存储在/data/mysql/binlogs 文件夹中名为 server1.000001、server1.000002 等的日志文件中。每当服务器启动或刷新日志时，或者当前日志的大小达到 max_binlog_size 时，服务器都会在系列中创建一个新文件。每个二进制日志的位置都在 server1.index 文件中被维护。

6.2.1 如何操作

让我们看看如何使用二进制日志。相信你会喜欢这个学习过程。

启用二进制日志

1. 启用二进制日志并设置 server_id。在自己常用的编辑器中打开 MySQL 配置文件并添加以下代码。选择 server_id，使其在基础架构中对每个 MySQL 服务器都是唯一的。
   也可以简单地把 log_bin 变量放在 my.cnf 中，不赋予任何值。在这种情况下，二进制日志是在数据目录中创建的。可以使用主机名作为目录名称。

```
shell> sudo vi /etc/my.cnf
[mysqld]
log_bin = /data/mysql/binlogs/server1
server_id = 100
```

2. 重新启动 MySQL 服务器：

```
shell> sudo systemctl restart mysql
```

3. 验证是否创建了二进制日志：

```
mysql> SHOW VARIABLES LIKE 'log_bin%';
+---------------------------------+------------------------------
---+
| Variable_name                   | Value
|
```

```
+--------------------------------+---------------------------------
---+
| log_bin                        | ON
|
| log_bin_basename               | /data/mysql/binlogs/server1
|
| log_bin_index                  |
/data/mysql/binlogs/server1.index |
log_bin_trust_function_creators   | OFF
|
log_bin_use_v1_row_events      | OFF
|
+--------------------------------+---------------------------------
---+
5  rows in set (0.00 sec)

mysql> SHOW MASTER LOGS;
+----------------+-----------+
| Log_name       | File_size |
+----------------+-----------+
| server1.000001 |154        |
+----------------+-----------+
1  row in set (0.00 sec)

shell> sudo ls -lhtr /data/mysql/binlogs
total 8.0K
-rw-r----- 1 mysql mysql 34 Aug 15 05:01 server1.index
-rw-r----- 1 mysql mysql 154 Aug 15 05:01 server1.000001
```

4. 执行 SHOW BINARY LOGS;或 SHOW MASTER LOGS;，以显示服务器的所有二进制日志。

5. 执行命令 SHOW MASTER STATUS;以获取当前的二进制日志位置：

```
mysql> SHOW MASTER STATUS;
+----------------+----------+--------------+------------------+----
--------------+++++++++++++++++++++++++++++++++++++-+
| File           | Position | Binlog_Do_DB | Binlog_Ignore_DB |
```

```
Executed_Gtid_Set |
+----------------+----------+-------------+-----------------+----
---------------+
| server1.000002 |     3273 |             |                 |
|
+----------------+----------+-------------+-----------------+----
---------------+
1 row in set (0.00 sec)
```

一旦 server1.000001 达到 max_binlog_size（默认为 1 GB），一个新文件 server1.000002 就会被创建，并被添加到 server1.index 中。可以使用 SET @@global.max_binlog_size = 536870912 动态设置 max_binlog_size。

禁用会话的二进制日志

有些情况下我们不希望将执行语句复制到其他服务器上。为此，可以使用以下命令来禁用该会话的二进制日志：

```
mysql> SET SQL_LOG_BIN = 0;
```

在这条语句后的所有 SQL 语句都不会被记录到二进制日志中，不过这仅仅是针对该会话的。

要重新启用二进制日志，可以执行以下操作：

```
mysql> SET SQL_LOG_BIN = 1;
```

移至下一个日志

可以使用 FLUSH LOGS 命令关闭当前的二进制日志并打开一个新的二进制日志：

```
mysql> SHOW BINARY LOGS;
+----------------+-----------+
| Log_name       | File_size |
+----------------+-----------+
| server1.000001 |       154|
+----------------+-----------+
1 row in set (0.00 sec)
mysql> FLUSH LOGS;
```

```
Query OK, 0 rows affected (0.02 sec)

mysql> SHOW BINARY LOGS;
+----------------+-----------+
| Log_name       | File_size |
+----------------+-----------+
| server1.000001 | 198       |
| server1.000002 | 154       |
+----------------+-----------+
2 rows in set (0.00 sec)
```

清理二进制日志

随着写入次数的增多，二进制日志会消耗大量空间。如果放任不管，这些写入操作将很快占满磁盘空间，因此清理它们至关重要。

1. 使用 binlog_expire_logs_seconds 和 expire_logs_days 设置日志的到期时间。
   如果想以天为单位设置到期时间，请设置 expire_logs_days。例如，如果要删除两天之前的所有二进制日志，请 SET @@global.expire_logs_days = 2。如果将该值设置为 0，则禁用设置会自动到期。
   如果想以更细的粒度来设置到期时间，可以使用 binlog_expire_logs_seconds 变量，它能够以秒为单位来设置二进制日志过期时间。
   这个变量的效果和 expire_logs_days 的效果是叠加的。例如，如果 expire_logs_days 是 1 并且 binlog_expire_logs_seconds 是 43200，那么二进制日志就会每 1.5 天清除一次。这与将 binlog_expire_logs_seconds 设置为 129600、将 expire_logs_days 设置为 0 的效果是相同的。在 MySQL 8.0 中，binlog_expire_logs_seconds 和 expire_logs_days 必须设置为 0，以禁止自动清除二进制日志。

2. 要手动清除日志，请执行 PURGE BINARY LOGS TO'<file_name>'。例如，有 server1.000001、server1.000002、server1.000003 和 server1.000004 文件，如果执行 PURGE BINARY LOGS TO'server1.000004'，则从 server1.000001 到 server1.000003 的所有文件都会被删除，但文件 server1.000004 不会被删除：

```
mysql> SHOW BINARY LOGS;
+----------------+-----------+
| Log_name       | File_size |
+----------------+-----------+
| server1.000001 |        198|
| server1.000002 |        198|
| server1.000003 |        198|
| server1.000004 |        154|
+----------------+-----------+

4 rows in set (0.00 sec)
mysql> PURGE BINARY LOGS TO 'server1.000004';
Query OK, 0 rows affected (0.00 sec)

mysql> SHOW BINARY LOGS;
+----------------+-----------+
| Log_name       | File_size |
+----------------+-----------+
| server1.000004 |        154|
+----------------+-----------+
1row in set (0.00sec)
```

除了指定某个日志文件，还可以执行命令 `PURGE BINARY LOGS BEFORE'2017-08-03 15:45:00'`。除了使用 `BINARY`，还可以使用 `MASTER`。

`mysql> PURGE MASTER LOGS TO 'server1.000004'`可以实现和之前语句一样的效果。

3. 要删除所有二进制日志并再次从头开始，请执行 `RESET MASTER`：

```
mysql> SHOW BINARY LOGS;
+----------------+-----------+
| Log_name       | File_size |
+----------------+-----------|
| server1.000004 |        154|
+----------------+-----------+
1 row in set (0.00 sec)
```

```
mysql> RESET MASTER;
Query OK, 0 rows affected (0.01 sec)

mysql> SHOW BINARY LOGS;
+----------------+-----------+
| Log_name       | File_size |
+----------------+-----------+
|server1.000001  |        154|
+----------------+-----------+
1 row in set (0.00 sec)
```

使用 replication 清除二进制日志是非常不安全的方法。清除二进制日志的安全方法是使用 mysqlbinlogpurge 脚本，我们将在第12章中介绍。

6.3 二进制日志的格式

二进制日志可以写成下面三种格式。

1. STATEMENT：记录实际的 SQL 语句。
2. ROW：记录每行所做的更改。
   例如，更新语句更新 10 行，所有 10 行的更新信息都会被写入日志。而在基于语句的复制中，只有更新语句会被写入日志，默认格式是 ROW。
3. MIXED：当需要时，MySQL 会从 STATEMENT 切换到 ROW。

有些语句在不同服务器上执行时可能会得到不同的结果。例如，UUID()函数的输出就因服务器而异。这些语句被称为非确定性的语句，基于这些语句的复制是不安全的。在这些情况下，当设置 MIXED 格式时，MySQL 服务器会切换为基于行的格式。

请参阅 https://dev.mysql.com/doc/refman/8.0/en/binary-log-mixed.html 来了解更多有关不安全声明的内容，以及在什么情况下 MySQL 服务器需要切换格式。

6.3.1 如何操作

可以使用兼具全局和会话范围作用域的动态变量 binlog_format 来设置格式。在全局范围进行设置可使所有客户端使用指定的格式：

```
mysql> SET GLOBAL binlog_format = 'STATEMENT';
```

或：

```
mysql> SET GLOBAL binlog_format = 'ROW';
```

请参阅 https://dev.mysql.com/doc/refman/8.0/en/replication-sbr-rbr.html 以了解各种格式的优缺点。

1. MySQL 8.0 使用第 2 版的二进制日志行事件，MySQL 5.6.6 之前的 MySQL Server 版本无法读取这些事件。将 log-bin-use-v1-row-events 设置为 1，便可以使用第 1 版，这样 MySQL 5.6.6 之前的版本就可以读取这些事件了。log-bin-use-v1-row-events 的默认值是 0。

   ```
   mysql> SET @@GLOBAL.log_bin_use_v1_row_events=0;
   ```

2. 当创建一个存储函数时，你必须声明它是确定性的或者它不修改数据，否则二进制日志可能不安全。默认情况下，为了接受 CREATE FUNCTION 语句，必须至少明确指定 DETERMINISTIC、NO SQL 或 READS SQL DATA 中的一个。否则会发生错误：

   ```
   ERROR 1418 (HY000): This function has none of DETERMINISTIC, NO
   SQL, or READS SQL DATA in its declaration and binary logging is
   enabled (you *might* want to use the less safe
   log_bin_trust_function_creators variable)
   ```

可以在例程内编写非确定性语句，并仍然声明为 DETERMINISTIC（不是一种很好的做法），如果要复制未声明为 DETERMINISTIC 的例程，可以设置 log_bin_trust_function_creators 变量：

```
mysql> SET GLOBAL log_bin_trust_function_creators = 1;
```

6.3.2 延伸阅读

请参阅 https://dev.mysql.com/doc/refman/8.0/en/stored-programs-logging.html 以了解更多存储程序如何被复制的相关信息。

6.4 从二进制日志中提取语句

可以使用 mysqlbinlog 实用程序（MySQL 已包含）从二进制日志中提取内容，并将其应用到其他服务器。

6.4.1 准备工作

使用各种二进制格式执行几条语句。如果把 binlog_format 设置为 GLOBAL 级别（全局范围），必须断开并重新连接，以使更改生效。如果想保持连接，请把 binlog_format 设置为 SESSION 级别（会话范围）。

更改为基于语句的复制（**SBR**）：

```
mysql> SET @@GLOBAL.BINLOG_FORMAT='STATEMENT';
Query OK, 0 rows affected (0.00 sec)
```

更新几行：

```
mysql> BEGIN;
Query OK, 0 rows affected (0.00 sec)

mysql> UPDATE salaries SET salary=salary*2 WHERE emp_no<10002;
Query OK, 18 rows affected (0.00 sec)
Rows matched: 18    Changed: 18 Warnings: 0

mysql> COMMIT;
Query OK, 0 rows affected (0.00 sec)
```

更改为基于行的复制（**RBR**）：

```
mysql> SET @@GLOBAL.BINLOG_FORMAT='ROW';
Query OK, 0 rows affected (0.00 sec)
```

更新几行：

```
mysql> BEGIN;
Query OK, 0 rows affected (0.00 sec)

mysql> UPDATE salaries SET salary=salary/2 WHERE emp_no<10002;
Query OK, 18 rows affected (0.00 sec)
Rows matched: 18    Changed: 18 Warnings: 0

mysql> COMMIT;
Query OK, 0 rows affected (0.00 sec)
```

改为 MIXED 格式：

```
mysql> SET @@GLOBAL.BINLOG_FORMAT='MIXED';
Query OK, 0 rows affected (0.00 sec)
```

更新几行：

```
mysql> BEGIN;
Query OK, 0 rows affected (0.00 sec)

mysql> UPDATE salaries SET salary=salary*2 WHERE emp_no<10002;
Query OK, 18 rows affected (0.00 sec)
Rows matched: 18 Changed: 18 Warnings: 0

mysql> INSERT INTO departments VALUES('d010',UUID());
Query OK, 1 row affected (0.00 sec)

mysql> COMMIT;
Query OK, 0 rows affected (0.00 sec)
```

6.4.2 如何操作

要显示日志 server1.000001 的内容，请执行以下操作：

```
shell> sudo mysqlbinlog /data/mysql/binlogs/server1.000001
```

你会得到类似下面这样的输出结果：

```
# at 226
#170815 12:49:24 server id 200 end_log_pos 312 CRC32 0x9197bf88 Query
thread_id=5 exec_time=0 error_code=0
BINLOG '
~
~
```

在第一行中，# at 后面的数字表示二进制日志文件中事件的起始位置（文件偏移量）。第二行包含了语句在服务器上被启用的时间戳。时间戳后面跟着 server ID、end_log_pos、thread_id、exec_time 和 error_code。

- server id：产生该事件的服务器的 server_id 值（在这个例子中为 200）。
- end_log_pos：下一个事件的开始位置。
- thread_id：指示哪个线程执行了该事件。
- exec_time：在主服务器上，它代表执行事件的时间；在从服务器上，它代表从服务器的最终执行时间与主服务器的开始执行时间之间的差值，这个差值可以作为备份相对于主服务器滞后多少的指标。
- error_code：代表执行事件的结果。零意味着没有错误发生。

回顾

1. 我们在基于语句的复制中执行了 UPDATE 语句，而且在二进制日志中记录了相同的语句。除了保存在服务器上，会话变量也被保存在二进制日志中，以在从库上复制相同的行为：

```
# at 226
#170815 13:28:38 server id 200 end_log_pos 324 CRC32 0x9d27fc78
Query thread_id=8 exec_time=0 error_code=0
SET TIMESTAMP=1502803718/*!*/;
SET @@session.pseudo_thread_id=8/*!*/;
SET @@session.foreign_key_checks=1, @@session.sql_auto_is_null=0,
@@session.unique_checks=1, @@session.autocommit=1/*!*/;
SET @@session.sql_mode=1436549152/*!*/;
SET @@session.auto_increment_increment=1,
@@session.auto_increment_offset=1/*!*/;
/*!\C utf8 *//*!*/;
```

```
SET
@@session.character_set_client=33,@@session.collation_connection=33
,@@session.collation_server=255/*!*/;
SET @@session.lc_time_names=0/*!*/;
SET @@session.collation_database=DEFAULT/*!*/;

BEGIN
/*!*/;
# at 324
#170815 13:28:38 server id 200 end_log_pos 471 CRC32 0x35c2ba45
Query thread_id=8 exec_time=0 error_code=0
use `employees`/*!*/;
SET TIMESTAMP=1502803718/*!*/;
UPDATE salaries SET salary=salary*2 WHERE emp_no<10002
/*!*/;
# at 471
#170815 13:28:40 server id 200end_log_pos 502 CRC32 0xb84cfeda
Xid = 53
COMMIT/*!*/;
```

2. 当使用基于行的复制（RBR）时，会以二进制格式对整行（而不是语句）进行保存，而且二进制格式不能读取。此外，你可以观察长度，单个更新语句会生成很多数据。请查看 6.4.2 节来了解如何查看二进制格式。

```
BEGIN
/*!*/;
# at 660
#170815 13:29:02 server id 200 end_log_pos 722 CRC32 0xe0a2ec74
Table_map:`employees`.`salaries` mapped to number 165
# at 722
#170815 13:29:02 server id 200 end_log_pos 1298 CRC32 0xf0ef8b05
Update_rows: table id 165 flags: STMT_END_F

BINLOG '
HveSWRPIAAAAPgAAANICAAAAAKUAAAAAAAAEACWVtcGxveWVlcwAIc2FsYXJpZXMABAM
DCgoAAAEBAHTsouA=HveSWR/IAAAAQAIAABIFAAAAAKUAAAAAAAAEAAgAE///wEScAAF
SrAwDahA/ahg/wEScAAKrVAQDahA/ahg/wEScAAFjKAwDahg/ZiA/wEScAACzlAQDah
```

```
g/ZiA/wEScAAGgIBADZiA/Zig/wEScAADQEAgDZiA/Zig/wEScAAJAQBADZig/ZjA/w
EScAAEgIAgDZig/ZjA/wEScAAEQWBADZjA/Zjg/wEScAACILAgDZjA/Zjg/wEScAABh
WBADZjg/YkA/wEScAAAwrAgDZjg/YkA/wEScAAHSJBADYkA/Ykg/wEScAALpEAgDYkA
/Ykg/wEScAAFiYBADYkg/YlA/wEScAACxMAgDYkg/YlA/wEScAAGijBADYlA/Ylg/wE
ScAALRRAgDYlA/Ylg/wEScAAFCxBADYlg/XmA/wEScAAKhYAgDYlg/XmA/wEScAADTi
BADXmA/Xmg/wEScAABpxAgDXmA/Xmg/wEScAAATyBADXmg/XnA/wEScAAAJ5AgDXmg/
XnA/wEScAACTzBADXnA/Xng/wEScAAJJ5AgDXnA/Xng/wEScAANQuBQDXng/WoA/wES
cAAGqXAgDXng/WoA/wEScAAOAxBQDWoA/Wog/wEScAAPCYAgDWoA/Wog/wEScAAKQxB
QDWog/WpA/wEScAANKYAgDWog/WpA/wEScAAIAaBgDWpA8hHk7wEScAAEANAwDWpA8h
Hk7wEScAAIAaBgDSwg8hHk7wEScAAEANAwDSwg8hHk4Fi+/w
'/*!*/;
# at 1298
#170815 13:29:02 server id 200 end_log_pos 1329 CRC32 0xa6dac5dc
Xid = 56
COMMIT/*!*/;
```

3. 当使用 MIXED 格式时，UPDATE 语句被记录为 SQL 语句，而 INSERT 语句以基于行的格式被记录，因为 INSERT 有非确定性的 UUID() 函数：

```
BEGIN
/*!*/;
# at 1499
#170815 13:29:27 server id 200 end_log_pos 1646 CRC32 0xc73d68fb Query
thread_id=8 exec_time=0 error_code=0
SET TIMESTAMP=1502803767/*!*/;
UPDATE salaries SET salary=salary*2 WHERE emp_no<10002
/*!*/;
# at 1646
#170815 13:29:50 server id 200 end_log_pos 1715 CRC32 0x03ae0f7e
Table_map: `employees`.`departments` mapped to number 166
# at 1715
#170815 13:29:50 server id 200 end_log_pos 1793 CRC32 0xa43c5dac
Write_rows: table id 166 flags: STMT_END_F
BINLOG
'TveSWRPIAAAARQAAALMGAAAAAKYAAAAAAAMACWVtcGxveWVlcwALZGVwYXJ0bWVudH
MAAv4PBP4QoAAAAgP8/wB+D64DTveSWR7IAAAATgAAAAEHAAAAAKYAAAAAAAEAAgAC/
/wEZDAxMSRkMDNhMjQwZS04MWJkLTExZTctODQxMC00MjAxMGE5NDAwMDKsXTyk
```

```
'/*!*/;
# at 1793
#170815 13:29:50 server id 200end_log_pos 1824 CRC32 0x4f63aa2e
Xid = 59
COMMIT/*!*/;
```

提取的日志可以被传送给 MySQL 以回放事件。在重放二进制日志时最好使用 force 选项，这样即使 force 选项卡在某个点上，执行也不会停止。稍后，你可以查找错误并手动修复数据。

```
shell> sudo mysqlbinlog /data/mysql/binlogs/server1.000001 | mysql -f -h
<remote_host> -u <username> -p
```

或者也可以先保存到文件中，稍后执行：

```
shell> sudo mysqlbinlog /data/mysql/binlogs/server1.000001 >
server1.binlog_extract
shell> cat server1.binlog_extract | mysql -h <remote_host> -u <username> -p
```

根据时间和位置进行抽取

我们可以通过指定位置从二进制日志中提取部分数据。假设你想做时间点恢复。假如在 2017-08-19 12:18:00 执行了 DROP DATABASE 命令，并且最新的可用备份是在 2017-08-19 12:00:00 做的，该备份已经恢复。现在，需要恢复从 12:00:01 至 2017-08-19 12:17:00 的数据。请记住，如果提取完整的日志，它还将包含 DROP DATABASE 命令，该命令将再次擦除数据。

可以通过--start-datetime 和--stop-datatime 选项来指定提取数据的时间窗口。

```
shell> sudo mysqlbinlog /data/mysql/binlogs/server1.000001 --start-
datetime="2017-08-19 00:00:01" --stop-datetime="2017-08-19 12:17:00" >
binlog_extract
```

使用时间窗口的缺点是，你会失去灾难发生那一刻的事务。为避免这种情况，必须在二进制日志中使用事件的文件偏移量。

一个连续的备份会保存它已完成备份的所有 binlog 文件的偏移量。备份恢复后，必须从备份的偏移量中提取 binlog。我们将在第 7 章中详细了解备份。

假设备份的偏移量为 471，执行 DROP DATABASE 命令的偏移量为 1793。可以使用 --start-position 和--stop-position 选项来提取偏移量之间的日志：

```
shell> sudo mysqlbinlog /data/mysql/binlogs/server1.000001 --start-
position=471 --stop-position=1793 > binlog_extract
```

请确保 DROP DATABASE 命令在提取的 binlog 中不再出现。

基于数据库进行提取

使用--database 选项可以过滤特定数据库的事件。如果多次提交，则只有最后一个选项会被考虑。这对于基于行的复制非常有效。但对于基于语句的复制和 MIXED，只有当选择默认数据库时才会提供输出。

以下命令从 employees 数据库中提取事件：

```
shell> sudo mysqlbinlog /data/mysql/binlogs/server1.000001 --
database=employees > binlog_extract
```

正如 MySQL 8 文档中所解释的，假设二进制日志是通过使用基于语句的日志记录执行这些语句而创建的：

```
mysql>
INSERT INTO test.t1 (i) VALUES(100);
INSERT INTO db2.t2 (j) VALUES(200);

USE test;
INSERT INTO test.t1 (i) VALUES(101);
INSERT INTO t1 (i) VALUES(102);
INSERT INTO db2.t2 (j) VALUES(201);

USE db2;
INSERT INTO test.t1 (i) VALUES(103);
INSERT INTO db2.t2 (j) VALUES(202);
INSERT INTO t2 (j) VALUES(203);
```

mysqlbinlog --database=test 不输出前两个 INSERT 语句，因为没有默认数据库。

mysqlbinlog --database=test 输出 USE test 后面的三条 INSERT 语句，但不是 USE db2 后面的三条 INSERT 语句。

因为没有默认数据库，mysqlbinlog --database=db2 不输出前两条 INSERT 语句。

mysqlbinlog --database=db2 不会输出 USE test 后的三条 INSERT 语句，但会输出在 USE db2 之后的三条 INSERT 语句。

提取行事件显示

默认情况下，基于行的复制日志显示为二进制格式。要查看行信息，必须将--verbose 或-v 选项传递给 mysqlbinlog。行事件的二进制格式以注释的伪 SQL 语句的形式显示，其中的行以###开始。可以看到，单个更新语句被改写为了每行的 UPDATE 语句。

```
shell> mysqlbinlog /data/mysql/binlogs/server1.000001 --start-position=660
--stop-position=1298 --verbose
~
~
# at 660
#170815 13:29:02 server id 200 end_log_pos 722 CRC32 0xe0a2ec74
Table_map: `employees`.`salaries` mapped to number 165
# at 722
#170815 13:29:02 server id 200 end_log_pos 1298 CRC32 0xf0ef8b05
Update_rows: table id 165 flags: STMT_END_F

BINLOG '
HveSWRPIAAAAPgAAANICAAAAAKUAAAAAAAEACWVtcGxveeWVlcwAIc2FsYXJpZXMABAMDCgoA
AAEB
AHTsouA=
~
~
'/*!*/;
### UPDATE `employees`.`salaries`
### WHERE
###   @1=10001
###   @2=240468
```

```
###   @3='1986:06:26'
###   @4='1987:06:26'
### SET
###   @1=10001
###   @2=120234
###   @3='1986:06:26'
###   @4='1987:06:26'
~
~
### UPDATE `employees`.`salaries`
### WHERE
###   @1=10001
###   @2=400000
###   @3='2017:06:18'
###   @4='9999:01:01'
### SET
###   @1=10001
###   @2=200000
###   @3='2017:06:18'
###   @4='9999:01:01'
SET @@SESSION.GTID_NEXT= 'AUTOMATIC' /* added by mysqlbinlog */ /*!*/;
DELIMITER ;
# End of log file
/*!50003 SET COMPLETION_TYPE=@OLD_COMPLETION_TYPE*/; /*!50530 SET
@@SESSION.PSEUDO_SLAVE_MODE=0*/;
```

如果你只想查看没有二进制行信息的伪 SQL 语句，请指定--base64-output ="decode-rows"以及--verbose：

```
shell> sudo mysqlbinlog /data/mysql/binlogs/server1.000001 --start-
position=660 --stop-position=1298 --verbose --base64-output="decode-rows"
/*!50530 SET @@SESSION.PSEUDO_SLAVE_MODE=1*/;
/*!50003 SET @OLD_COMPLETION_TYPE=@@COMPLETION_TYPE,COMPLETION_TYPE=0*/;
DELIMITER /*!*/;
# at 660
#170815 13:29:02 server id 200 end_log_pos 722 CRC32 0xe0a2ec74
Table_map: `employees`.`salaries` mapped to number 165
# at 722
```

```
#170815 13:29:02 server id 200 end_log_pos 1298 CRC32 0xf0ef8b05
Update_rows: table id 165 flags: STMT_END_F
### UPDATE `employees`.`salaries`
### WHERE
###   @1=10001
###   @2=240468
###   @3='1986:06:26'
###   @4='1987:06:26'
### SET
###   @1=10001
###   @2=120234
###   @3='1986:06:26'
###   @4='1987:06:26'
~
```

重写数据库名称

假设你想将生产服务器上的 employees 数据库的二进制日志恢复为开发服务器上的 employees_dev，可以使用--rewrite-db ='from_name-> to_name'选项。这会将所有 from_name 重写为 to_name。

要转换多个数据库，请多次指定该选项：

```
shell> sudo mysqlbinlog /data/mysql/binlogs/server1.000001 --start-position=1499 --stop-position=1646 --rewrite-db='employees->employees_dev'
~
# at 1499
#170815 13:29:27 server id 200 end_log_pos 1646 CRC32 0xc73d68fb  Query
thread_id=8 exec_time=0     error_code=0
use `employees_dev`/*!*/;
~
~
UPDATE salaries SET salary=salary*2 WHERE emp_no<10002
/*!*/;
SET @@SESSION.GTID_NEXT= 'AUTOMATIC' /* added by mysqlbinlog */ /*!*/;
DELIMITER ;
# End of log file
~
```

可以看到上面使用了 use `employees_dev` / *! * /;语句。因此，在恢复时，所有更改将应用于 employees_dev 数据库。

> 正如 MySQL 文档中所解释的那样：
> 当与--database 选项一起使用时，先是--rewrite-db 选项会被应用，然后--database 选项会被用来重写数据库名称。如果选项的提供顺序不同，在这种情况下不会产生任何影响。这意味着，例如，如果 mysqlbinlog 以 --rewrite-db ='mydb-> yourdb' - database = yourdb 开头，那么对数据库 mydb 和 yourdb 中的任何表的所有更新都将包含在输出中。
> 另一方面，如果它是以--rewrite-db ='mydb-> yourdb' --database = mydb 开头的，那么 mysqlbinlog 根本不会输出任何语句，因为在应用--database 选项之前，所有对 mydb 的更新都已被重写到 yourdb 中，所以没有更新可以匹配 --database = mydb 条件。

在恢复时禁用二进制日志

在恢复二进制日志的过程中，如果你不希望 mysqlbinlog 进程创建二进制日志，则可以使用--disable-log-bin 选项：

```
shell> sudo mysqlbinlog /data/mysql/binlogs/server1.000001 --start-
position=660 --stop-position=1298 --disable-log-bin > binlog_restore
```

可以看到 SQL_LOG_BIN = 0 被写入 binlog 恢复文件中，这将防止创建 binlog。

```
/*!32316 SET @OLD_SQL_LOG_BIN=@@SQL_LOG_BIN, SQL_LOG_BIN=0*/;
```

显示二进制日志文件中的事件

除了使用 mysqlbinlog，还可以使用 SHOW BINLOG EVENTS 命令来显示事件。

以下命令将显示 server1.000008 二进制日志中的事件。如果未指定 LIMIT，则显示所有事件：

```
mysql> SHOW BINLOG EVENTS IN 'server1.000008' LIMIT 10;
+--------------- +---- +---------------- +----------+----------+-------
----------------------------------+
| Log_name        | Pos  | Event_type     | Server_id| End_log_pos | Info
|
+--------------- +---- +---------------- +----------+----------+-------
----------------------------------+
| server1.000008 |    4| Format_desc      |      200 |         123 | Server
ver: 8.0.3-rc-log,     Binlog ver: 4   |
| server1.000008 | 123 | Previous_gtids  |      200 |         154 |
|
| server1.000008 | 154 | Anonymous_Gtid  |       200|         226 | SET
@@SESSION.GTID_NEXT= 'ANONYMOUS'         |
| server1.000008 | 226 | Query           |      200 |         336 | drop
database company /* xid=4134 */          |
| server1.000008 | 336 | Anonymous_Gtid  |      200 |         408 | SET
@@SESSION.GTID_NEXT= 'ANONYMOUS'         |
| server1.000008 | 408 | Query           |      200 |         485 | BEGIN
|
| server1.000008 | 485 | Table_map       |      200 |         549 |
table_id: 975 (employees.emp_details)    |
| server1.000008 | 549 | Write_rows      |       200|         804 |
table_id: 975 flags: STMT_END_F                |
| server1.000008 | 804 | Xid             |       200|         835 | COMMIT
/* xid=9751 */                   |
| server1.000008 | 835 | Anonymous_Gtid |        200|         907 | SET
@@SESSION.GTID_NEXT= 'ANONYMOUS'         |
+--------------- +---- +---------------- +----------+----------+-------
----------------------------------+
10 rows in set (0.00 sec)
```

也可以指定位置和偏移量：

```
mysql> SHOW BINLOG EVENTS IN 'server1.000008' FROM 123 LIMIT 2,1;
+----------------+-----+-----------+----------+------------+--------
--------------------------+
| Log_name        | Pos | Event_type | Server_id | End_log_pos | Info
|
```

```
+----------------+-----+------------+----------+------------+---------
-------------------------+
| server1.000008 | 226 | Query      |      200 |        336 | drop
database company   /* xid=4134 */ |
+----------------+-----+------------+----------+------------+---------
-------------------------+
1 row in set (0.00 sec)
```

6.5　忽略要写入二进制日志的数据库

可以通过在 my.cnf 中指定 --binlog-do-db = db_name 选项，来选择将哪些数据库写入二进制日志。要指定多个数据库，就必须使用此选项的多个实例。由于数据库的名字可以包含逗号，因此如果提供逗号分隔列表，则该列表将被视为单个数据库的名字。需要重新启动 MySQL 服务器才能使更改生效。

6.5.1　如何操作

打开 my.cnf 并添加以下行：

```
shell> sudo vi /etc/my.cnf
[mysqld]
binlog_do_db=db1
binlog_do_db=db2
```

binlog-do-db 上的行为从基于语句的日志记录更改为基于行的日志记录，就像 mysqlbinlog 实用程序中的--database 选项一样。

在基于语句的日志记录中，只有那些默认数据库（即用 USE 选择的）的语句才会被写入二进制日志。使用 binlog-do-db 选项时应该非常小心，因为它的工作方式与你在使用基于语句的日志记录时的方式不同。请仔细阅读 MySQL 文档中提到的以下示例。

例 1

如果服务器以--binlog-do-db = sales 启动并且执行以下语句，则该 UPDATE 语句不会被记录：

```
mysql> USE prices;
```

```
mysql> UPDATE sales.january SET amount=amount+1000;
```

这种只检查默认数据库的行为主要是因为，单从语句来看，很难知道它是否应该被复制。如果没有必要，只检查默认数据库比检查所有数据库耗时更短。

例 2

如果服务器以--binlog-do-db = sales 启动，则即使在设置--binlog-do-db 时未包括 prices，也会记录以下 UPDATE 语句：

```
mysql> USE sales;
mysql> UPDATE prices.discounts SET percentage = percentage + 10;
```

因为当 UPDATE 语句执行时，sales 是默认数据库，所以会记录该 UPDATE 语句。

在基于行的日志记录中，它仅限定于数据库 db_name。只记录对 db_name 数据库中的表的更改，默认数据库对此没有影响。

与基于行的日志记录方式相比，基于语句的日志记录方式在执行--binlog-do-db 处理方面的另一个重要区别是，后者涉及引用多个数据库的语句。假设服务器以--binlog-do-db = db1 启动，并执行以下语句：

```
mysql> USE db1;
mysql> UPDATE db1.table1 SET col1 = 10, db2.table2 SET col2 = 20;
```

如果正在使用基于语句的日志记录，则这两个表的更新都会被写入二进制日志。但是，如果使用基于行的方式，则只会记录对 table1 的更改，不会记录对 table2 的更改，因为 table2 位于不同的数据库中。

同样，可以使用--binlog-ignore-db = db_name 选项来指定不写入二进制日志的数据库。

更多有关信息，请参阅 MySQL 8 文档，网址为 https://dev.mysql.com/doc/refman/8.0/en/replication-rules.html。

6.6 迁移二进制日志

由于二进制日志占用越来越多的空间，有时你可能希望更改二进制日志的位置，可以按照以下步骤操作。单独更改 log_bin 是不够的，必须迁移所有二进制日志并在索引文件中更新位置。mysqlbinlogmove 工具可以自动执行这些任务，简化你的工作。

6.6.1 如何操作

你需要先安装 MySQL 工具集，以使用 mysqlbinlogmove 脚本。相关安装步骤，请参阅第 1 章。

1. 停止 MySQL 服务器的运行：

```
shell> sudo systemctl stop mysql
```

2. 启动 mysqlbinlogmove 工具。如果要将二进制日志从/data/mysql/binlogs 更改为/binlogs，则应使用以下命令。如果 base name 不是默认名称，则必须通过--bin-log-basename 选项设定 base name：

```
shell> sudo mysqlbinlogmove --bin-log-basename=server1 --binlog-dir=/data/mysql/binlogs /binlogs
#
Moving bin-log files...
- server1.000001
- server1.000002
- server1.000003
- server1.000004
- server1.000005
#
#...done.
#
```

3. 编辑 my.cnf 文件并更新 log_bin 的新位置：

```
shell> sudo vi /etc/my.cnf
[mysqld]
log_bin=/binlogs
```

4. 启动 MySQL 服务器：

```
shell> sudo systemctl start mysql
```

新位置在 AppArmor 或 SELinux 中更新。

如果有大量的二进制日志，服务器的停机时间会很长。为了避免这种情况，可以使用 `--server` 选项来重新定位所有二进制日志——但是当前正在使用的二进制日志（具有较高序列号）除外。然后停止服务器的运行，使用上述方法，并重新定位最后一个二进制日志，这会快很多，因为只有一个文件存在。然后你可以更改 `my.cnf` 并启动服务器。

例如：

```
shell> sudo mysqlbinlogmove --server=root:pass@host1:3306 /new/location
```

第 7 章 备份

在本章中，我们将介绍以下内容：

- 使用 `mysqldump` 进行备份
- 使用 `mysqlpump` 进行备份
- 使用 `mydumper` 进行备份
- 使用普通文件进行备份
- 使用 XtraBackup 进行备份
- 锁定实例进行备份
- 使用二进制日志进行备份

7.1 引言

建立数据库后，下一个重要的事情就是设置备份。在本章中，我们将学习如何设置各种类型的备份。做备份的方法主要有两种，一种是逻辑备份，它将所有数据库、表结构、数据和存储例程导出到一组可以再次执行的 SQL 语句中，以重新创建数据库的状态；另一种是物理备份，它包含了系统上的所有文件，这里的系统是指数据库用于存储所有数据库实体的系统。

- **逻辑备份工具**：`mysqldump`、`mysqlpump` 和 `mydumper`（不随 MySQL 提供）。
- **物理备份工具**：XtraBackup（不随 MySQL 提供）和普通文件备份。

对于时间点恢复，备份应该能够提供开始做备份之前的二进制日志的位置。这被称为连续的备份。

强烈建议从一个从（slave）服务器备份到 mount 于其上的文件中。

7.2 使用 mysqldump 进行备份

mysqldump 是一个广泛使用的逻辑备份工具。它提供了多种选项来包含或排除数据库、选择要备份的特定数据、仅备份不包含数据的 schema，或者只备份存储的例程而不包括其他任何东西，等等。

7.2.1 如何操作

mysqldump 与 mysql 二进制文件是一起提供的，因此不需要单独安装 mysqldump。本节将介绍使用 mysqldump 进行备份的大部分生产环境场景。

语法如下：

```
shell> mysqldump [options]
```

在该选项中，你可以指定连接到数据库的用户名、密码和主机名，如下所示：

```
--user <user_name> --password <password>
```

或

```
-u <user_name> -p<password>
```

在本章中，并没有在每个例子中都使用--user 和--password，这是为了让读者把注意力集中在其他重要的选项上。

完整备份所有数据库

完整备份所有数据库可以通过以下方式完成：

```
shell> mysqldump --all-databases > dump.sql
```

--all-databases 选项支持所有数据库和所有表的备份。>运算符将输出重定向到 dump.sql 文件。在 MySQL 8 之前，存储过程和事件存储在 mysql.proc 和 mysql.event 表中。从 MySQL 8 开始，相应对象的定义存储在数据字典表中，但这些表不会被备份。要将存储过程和事件包含在使用 --all-databases 创建的备份中，请使用--routines 和--events 选项。

包含存储过程和事件的代码如下：

```
shell> mysqldump --all-databases --routines --events > dump.sql
```

可以打开 dump.sql 文件以查看它的结构。前几行是转储时的会话变量。接下来是 CREATE DATABASE 语句，后面跟着 USE DATABASE 命令。再接下来是 DROP TABLE IF EXISTS 语句，后面跟着 CREATE TABLE，然后是插入数据的 INSERT 语句。由于数据被存储为 SQL 语句，因此这称为**逻辑备份**。

你会注意到，当还原 dump 时，DROP TABLE 语句将在创建表之前清除所有表。

时间点恢复

要获得时间点恢复，应该指定--single-transaction 和--master-data。

--single-transaction 选项在执行备份之前，通过将事务隔离模式更改为 REPEATABLE READ 模式，并执行 START TRANSACTION 来提供一致的备份。--single-transaction 选项仅适用于诸如 InnoDB 之类的事务表，因为它在 START TRANSACTION 执行时能保存数据库的一致状态而不阻塞任何应用程序。

--master-data 选项将服务器的二进制日志的位置输出到 dump 文件。如果--master-data = 2，它将打印为注释。它也使用 FLUSH TABLES WITH READ LOCK 语句来获取二进制日志的快照。正如第 5 章中所解释的那样，当存在任何复杂事务时，这样做可能非常危险：

```
shell> mysqldump --all-databases --routines --events --single-transaction
--master-data > dump.sql
```

保存主库二进制日志位置

备份始终在从服务器上进行。要获取备份时主服务器的二进制日志位置，可以使用--dump-slave 选项。如果你正在从主服务器上进行二进制日志备份，请使用此选项。否则，请使用--master-data 选项：

```
shell> mysqldump --all-databases --routines --events --single-transaction
- -dump-slave > dump.sql
```

输出将如下所示：

```
--
Position to start replication or point-in-time recovery from (the master of
this slave)
--
CHANGE MASTER TO MASTER_LOG_FILE='centos7-bin.000001', MASTER_LOG_POS=463;
```

指定数据库和表

要仅备份指定的数据库，请执行以下操作：

```
shell> mysqldump --databases employees > employees_backup.sql
```

要仅备份指定的表，请执行以下操作：

```
shell> mysqldump --databases employees --tables employees >
employees_backup.sql
```

忽略表

要忽略某些表，可以使用`--ignore-table = database.table`选项。如果指定多个要忽略的表，请多次使用该指令：

```
shell> mysqldump --databases employees --ignore-table=employees.salary >
employees_backup.sql
```

指定行

`mysqldump`可以帮助你过滤备份的数据。假设你想对2000年之后加入的员工的信息进行备份：

```
shell> mysqldump --databases employees --tables employees --databases
employees --tables employees --where="hire_date>'2000-01-01'" >
employees_after_2000.sql
```

可以使用`LIMIT`子句来限制结果集：

```
shell> mysqldump --databases employees --tables employees --databases
employees --tables employees --where="hire_date >= '2000-01-01' LIMIT 10"
> employees_after_2000_limit_10.sql
```

从远程服务器备份

有时，你可能没有 SSH 访问数据库服务器的权限（就像云实例的情况，例如 Amazon 的 RDS）。在这些情况下，可以使用 `mysqldump` 从远程服务器备份到本地服务器。为此，你需要使用`--hostname` 选项提及主机名。确保用户具有适当的连接和执行备份的权限：

```
shell> mysqldump --all-databases --routines --events --triggers --hostname
<remote_hostname> > dump.sql
```

用于重建另一个具有不同 schema 的服务器的备份

我们有时候会希望在另一台服务器上拥有不同的 schema。在这种情况下，你必须备份和恢复 schema，根据需要更改 schema，然后备份并恢复数据。使用数据更改 schema 可能需要很长时间，具体取决于数据量。请注意，只有当修改后的 schema 与插入语句兼容时，此方法才有效。修改后的表可以有多余的列，但它应该包含原始表中的所有列。

仅备份不包含数据的 schema

可以使用`--no-data` 仅备份不包含数据的 schema：

```
shell> mysqldump --all-databases --routines --events --triggers --no-data >
schema.sql
```

仅备份不包含 schema 的数据

可以使用以下选项仅备份不包含 schema 的数据。

`--complete-insert` 将在 `INSERT` 语句中打印列名，如果修改的表中有更多列，这样做是有好处的：

```
shell> mysqldump --all-databases --no-create-db --no-create-info --
complete-insert > data.sql
```

用于与其他服务器合并数据的备份

可以通过备份来替换旧数据，或在发生冲突时保留旧数据。

用新数据替换

假设你想要将数据从生产数据库恢复到已有一些数据的开发服务器。如果要将生产数据合并到开发过程中，可以使用--replace选项，该选项将使用REPLACE INTO语句而不是INSERT语句。还应该包含--skip-add-drop-table选项，该选项不会将DROP TABLE语句写入dump文件。如果拥有相同数量的表和结构，则还可以包含--no-create-info选项，该选项将跳过dump文件中的CREATE TABLE语句：

```
shell> mysqldump --databases employees --skip-add-drop-table --no-create-info --replace > to_development.sql
```

如果在生产服务器中有一些额外的表，则之前的dump在恢复时将会失败，因为该表在开发服务器上不存在。在这种情况下，你不应该添加--no-create-info选项，而应在恢复时使用force选项。否则，在CREATE TABLE中恢复将失败，表示该表已存在。不幸的是，mysqldump没有提供CREATE TABLE IF NOT EXISTS选项。

忽略数据

在写入dump文件时可以使用INSERT IGNORE语句代替REPLACE。这将保留服务器上的现有数据并插入新数据。

7.3 使用mysqlpump进行备份

mysqlpump是一个非常类似于mysqldump的程序，但它带有一些额外的功能。

7.3.1 如何操作

使用mysqlpump进行备份的方法有很多，我们来仔细看看每一个方法。

并行处理

可以通过指定线程数量（根据CPU数量）加速备份过程。例如，使用8个线程进行完整备份：

```
shell> mysqlpump --default-parallelism=8 > full_backup.sql
```

甚至可以指定每个数据库的线程数。在我们的案例中，employees 数据库比 company 数据库大很多。所以可以为 employees 数据库指定 4 个线程，为 company 数据库指定 2 个线程：

```
shell> mysqlpump -u root --password --parallel-schemas=4:employees --
default-parallelism=2 > full_backup.sql
Dump progress: 0/6 tables, 250/331145 rows
Dump progress: 0/34 tables, 494484/3954504 rows
Dump progress: 0/42 tables, 1035414/3954504 rows
Dump progress: 0/45 tables, 1586055/3958016 rows
Dump progress: 0/45 tables, 2208364/3958016 rows
Dump progress: 0/45 tables, 2846864/3958016 rows
Dump progress: 0/45 tables, 3594614/3958016 rows
Dump completed in 6957
```

再来看一个分配线程的示例，其中 3 个线程用于 db1 和 db2 数据库，2 个线程用于 db3 和 db4 数据库，还有 4 个线程用于其他数据库：

```
shell> mysqlpump --parallel-schemas=3:db1,db2 --parallel-schemas=2:db3,db4
--default-parallelism=4 > full_backup.sql
```

你会注意到有一个进度条可以帮助估计时间。

使用正则表达式排除/包含数据库对象

对以 prod 结尾的所有数据库进行备份：

```
shell> mysqlpump --include-databases=%prod --result-file=db_prod.sql
```

假设某些数据库中有一些测试表，我们希望将它们从备份中排除。可以使用 --exclude-tables 选项来指定，该选项将排除所有数据库中名称为 test 的表：

```
shell> mysqlpump --exclude-tables=test --result-
file=backup_excluding_test.sql
```

每个包含和排除选项的值都是适当对象类型以逗号分隔的名称列表。允许在对象名称中使用通配符：

- %匹配零个或多个字符的任何序列。
- _匹配任何单个字符。

除了数据库和表，还可以包含或排除触发器、例程、事件和用户，例如--include-routines，-include-events 和-exclude-triggers。

要详细了解包含和排除选项，请参阅 https://dev.mysql.com/doc/refman/8.0/en/mysqlpump.html#mysqlpump-filtering。

备份用户

在 mysqldump 中，你不会在 CREATE USER 或 GRANT 语句中获得用户的备份；相反，你必须备份 mysql.user 表。使用 mysqlpump，可以将用户账户备份为账户管理语句（CREATE USER 和 GRANT），而不是将用户账户插入 mysql 系统数据库中：

```
shell> mysqlpump --exclude-databases=% --users > users_backup.sql
```

还可以通过指定--exclude-users 选项来排除某些用户：

```
shell> mysqlpump --exclude-databases=% --exclude-users=root --users >
users_backup.sql
```

压缩备份

可以通过压缩备份来减少磁盘空间和网络带宽的占用。可以使用--compress-output = lz4 或--compress-output = zlib。

请注意，你需要有相应的解压缩工具：

```
shell> mysqlpump -u root -pxxxx --compress-output=lz4 > dump.lz4
```

执行下面的语句进行解压缩：

```
shell> lz4_decompress dump.lz4 dump.sql
```

使用 zlib 执行此下面的语句：

```
shell> mysqlpump -u root -pxxxx --compress-output=zlib > dump.zlib
```

执行下面的语句进行解压缩：

```
shell> zlib_decompress dump.zlib dump.sql
```

加速重新加载

你会注意到，在输出中，辅助索引从 CREATE TABLE 语句中省略了。这将加速恢复过程。我们将使用 ALTER TABLE 语句在 INSERT 结尾处添加这些索引。

索引将在第 13 章中介绍。

以前我们可以 dump mysql 系统数据库中的所有表。从 MySQL 8 开始，mysqldump 和 mysqlpump 仅能 dump 该数据库中的非数据字典表。

7.4　使用 mydumper 进行备份

mydumper 是一个类似 mysqlpump 的逻辑备份工具。

与 mysqldump 相比，mydumper 在以下方面具有优势。

- 并行（因此速度更快）和性能（避免使用复杂的字符集转换例程，因而代码总体上很高效）。
- 一致性。mydumper 维护所有线程的快照，提供准确的主库和从库日志位置等。mysqlpump 不保证一致性。
- 更易于管理输出（将表和元数据文件分离，并且方便查看/解析数据）。mysqlpump 将所有内容写入一个文件，这限制了加载部分数据库对象的选项。
- 使用正则表达式包含和排除数据库对象。
- 有用于终止阻塞备份和所有后续查询的长事务的选项。

mydumper 是一款开源的备份工具，需要单独安装。在本节中，将介绍其在 Debian 和 Red Hat 系统中的安装步骤及使用情况。

7.4.1　如何操作

让我们从该方法的安装开始，然后一起学习与备份相关的很多内容。

安装

在 Ubuntu / Debain 系统上：

```
shell> sudo apt-get install libglib2.0-dev libmysqlclient-dev zlib1g-dev
libpcre3-dev cmake git
```

在 Red Hat / CentOS / Fedora 系统上：

```
shell> yum install glib2-devel mysql-devel zlib-devel pcre-devel cmake gcc-
c++ git
shell> cd /opt
shell> git clone https://github.com/maxbube/mydumper.git shell> cd mydumper
shell> cmake .

shell> make
Scanning dependencies of target mydumper
[ 25%] Building C object CMakeFiles/mydumper.dir/mydumper.c.o
[ 50%] Building C object CMakeFiles/mydumper.dir/server_detect.c.o
[ 75%] Building C object CMakeFiles/mydumper.dir/g_unix_signal.c.o

shell> make install
[ 75%] Built target mydumper
[100%] Built target myloader
Linking C executable CMakeFiles/CMakeRelink.dir/mydumper
Linking C executable CMakeFiles/CMakeRelink.dir/myloader
Install the project...
-- Install configuration: ""
-- Installing: /usr/local/bin/mydumper
-- Installing: /usr/local/bin/myloader
```

另外，使用YUM或APT，你可以在找到发布的mydumper，网址为https://github.com/maxbube/mydumper/releases，代码如下：

```
#YUM
shell> sudo yum install -y
"https://github.com/maxbube/mydumper/releases/download/v0.9.3/mydumper-0.9.
3-41.el7.x86_64.rpm"
```

```
#APT
shell> wget
"https://github.com/maxbube/mydumper/releases/download/v0.9.3/mydumper_0
.9.3-41.jessie_amd64.deb"

shell> sudo dpkg -i mydumper_0.9.3-41.jessie_amd64.deb
shell> sudo apt-get install -f
```

完全备份

以下命令会将所有数据库备份到/backups 文件夹中：

```
shell> mydumper -u root --password=<password> --outputdir /backups
```

多个文件将在/backups 文件夹中被创建。每个数据库的 CREATE DATABASE 语句均为<database_name> -schema-create.sql，每个表都有自己的 schema 和数据文件。schema 文件存储为<database_name>.<table> -schema.sql，数据文件存储为<database_name>.<table>.sql。

视图存储为<database_name>.<table>-schema-view.sql。存储的例程、触发器和事件存储为<database_name>-schema-post.sql（如果目录未创建，则使用 sudo mkdir-pv/ backups）：

```
shell> ls -lhtr /backups/company*
-rw-r--r-- 1 root root 69 Aug 13 10:11 /backups/company-schema-create.sql
-rw-r--r-- 1 root root 180 Aug 13 10:11 /backups/company.payments.sql
-rw-r--r-- 1 root root 239 Aug 13 10:11 /backups/company.new_customers.sql
-rw-r--r-- 1 root root 238 Aug 13 10:11 /backups/company.payments-schema.sql
-rw-r--r-- 1 root root 303 Aug 13 10:11 /backups/company.new_customers-
schema.sql
-rw-r--r-- 1 root root 324 Aug 13 10:11 /backups/company.customers-
schema.sql
```

如果有任何查询超过 60 秒，mydumper 将失败并出现以下错误提示：

```
** (mydumper:18754): CRITICAL **: There are queries in PROCESSLIST running
longer than 60s, aborting dump,
```

```
use --long-query-guard to change the guard value, kill queries (--kill-
long-queries) or use different server for dump
```

为了避免这种情况，可以使用--kill-long-queries 选项，或将--long-query-guard 设置为更大的值。

--kill-long-queries 选项会结束所有长于 60 秒（或--long-query-guard 设置的值）的查询。请注意--kill-long-queries 还会因为一个 bug 而杀死复制线程（https://bugs.launchpad.net/mydumper/+bug/1713201）：

```
shell> sudo mydumper --kill-long-queries --outputdir /backups
** (mydumper:18915): WARNING **: Using trx_consistency_only, binlog
coordinates will not be accurate if you are writing to non transactional
tables.
** (mydumper:18915): WARNING **: Killed a query that was running for 368s
```

一致的备份

备份目录中的元数据文件包含用于一致备份的二进制日志坐标。

在主服务器上，备份目录中的元数据文件会捕获二进制日志位置：

```
shell> sudo cat /backups/metadata
Started dump at: 2017-08-20 12:44:09
SHOW MASTER STATUS:
   Log: server1.000008
   Pos: 154
       GTID:
```

在从服务器上，备份目录中的元数据文件会捕获主服务器和从服务器的二进制日志位置：

```
shell> cat /backups/metadata
Started dump at: 2017-08-26 06:26:19
SHOW MASTER STATUS:
 Log: server1.000012
 Pos: 154
 GTID:
SHOW SLAVE STATUS:
```

```
    Host: 35.186.158.188
    Log: master-bin.000013
    Pos: 4633
    GTID:
Finished dump at: 2017-08-26 06:26:24
```

备份单独表

以下命令会将 employees 数据库的 employees 表备份到/backups 目录中：

```
shell> mydumper -u root --password=<password> -B employees -T employees --
triggers --events --routines --outputdir /backups/employee_table

shell> ls -lhtr /backups/employee_table/
total 17M
-rw-r--r-- 1 root root 71 Aug 13 10:35 employees-schema-create.sql
-rw-r--r-- 1 root root 397 Aug 13 10:35 employees.employees-schema.sql
-rw-r--r-- 1 root root 3.4K Aug 13 10:35 employees-schema-post.sql
-rw-r--r-- 1 root root 75 Aug 13 10:35 metadata
-rw-r--r-- 1 root root 17M Aug 13 10:35 employees.employees.sql
```

这些文件的命名规则如下：

- employees-schema-create.sql 包含 CREATE DATABASE 语句。
- employees.employees-schema.sql 包含 CREATE TABLE 语句。
- employees-schema-post.sql 包含 ROUTINES、TRIGGERS 和 EVENTS。
- employees.employees.sql 包含 INSERT 语句形式表示的实际数据。

使用正则表达式来备份特定的数据库

可以使用 regex 选项包含/排除特定数据库。以下命令将从备份中排除 mysql 和 test 数据库：

```
shell> mydumper -u root --password=<password> --regex '^(?!(mysql|test))' -
-outputdir /backups/specific_dbs
```

采用 mydumper 备份大表

为了加速大表的转储和恢复，可以将它分成小块。块的大小可以通过它包含的行数来指定，每个块将被写入一个单独的文件中。

```
shell> mydumper -u root --password=<password> -B employees -T employees --
triggers --events --routines --rows=10000 -t 8 --trx-consistency-only --
outputdir /backups/employee_table_chunks
```

- `-t`：指定线程的数量。
- `--trx-consistency-only`：如果只使用事务表，例如 InnoDB，那么使用此选项将使锁定最小化。
- `--rows`：将表分成这些行的块。

对于每一个块，都会创建一个名为 `database_name>.<table_name>. <number>.sql` 的文件，其中数字用零填充，补足 5 位：

```
shell> ls -lhr /backups/employee_table_chunks
total 17M
-rw-r--r-- 1 root root 71 Aug 13 10:45 employees-schema-create.sql
-rw-r--r-- 1 root root 75 Aug 13 10:45 metadata
-rw-r--r-- 1 root root 397 Aug 13 10:45 employees.employees-schema.sql
-rw-r--r-- 1 root root 3.4K Aug 13 10:45 employees-schema-post.sql
-rw-r--r-- 1 root root 633K Aug 13 10:45 employees.employees.00008.sql
-rw-r--r-- 1 root root 634K Aug 13 10:45 employees.employees.00002.sql
-rw-r--r-- 1 root root 1.3M Aug 13 10:45 employees.employees.00006.sql
-rw-r--r-- 1 root root 1.9M Aug 13 10:45 employees.employees.00004.sql
-rw-r--r-- 1 root root 2.5M Aug 13 10:45 employees.employees.00000.sql
-rw-r--r-- 1 root root 2.5M Aug 13 10:45 employees.employees.00001.sql
-rw-r--r-- 1 root root 2.6M Aug 13 10:45 employees.employees.00005.sql
-rw-r--r-- 1 root root 2.6M Aug 13 10:45 employees.employees.00009.sql
-rw-r--r-- 1 root root 2.6M Aug 13 10:45 employees.employees.00010.sql
```

无阻塞备份

为了提供一致的备份，mydumper 通过执行 `FLUSH TABLES WITH READ LOCK` 来获取 `GLOBAL LOCK`。

我们已经在第 5 章中介绍过，如果有任何长时间运行的事务，使用 FLUSH TABLES WITH READ LOCK 是多么危险。为了避免这种情况，可以传递--kill-long-queries 选项来终止阻塞查询，而不是中止 mydumper。

- --trx-consistency-only：相当于 mysqldump 的--single-transaction，但具有 binlog 位置。显然，这个位置只适用于事务表。使用这个选项的好处之一是全局读锁只保持在线程的协调过程中，所以一旦事务开始，该选项就会被释放。
- --use-savepoints：减少元数据锁定问题（需要 SUPER 权限）。

压缩备份

可以指定--compress 选项来进行压缩备份：

```
shell> mydumper -u root --password=<password> -B employees -T employees -t
8 --trx-consistency-only --compress --outputdir /backups/employees_compress

shell> ls -lhtr /backups/employees_compress
total 5.3M
-rw-r--r-- 1 root root 91 Aug 13 11:01 employees-schema-create.sql.gz
-rw-r--r-- 1 root root 263 Aug 13 11:01 employees.employees-schema.sql.gz
-rw-r--r-- 1 root root 75 Aug 13 11:01 metadata
-rw-r--r-- 1 root root 5.3M Aug 13 11:01 employees.employees.sql.gz
```

仅备份数据

可以使用--no-schemas 选项来跳过 schema 并且仅备份数据：

```
shell> mydumper -u root --password=<password> -B employees -T employees -t
8 --no-schemas --compress --trx-consistency-only --outputdir
/backups/employees_data
```

7.5 使用普通文件进行备份

这是一种物理备份方法，可以通过直接复制数据目录中的文件来进行备份。由于在复制文件时写入了新数据，因此备份将不一致并且无法使用。为了避免这种情况，必须先关闭 MySQL，复制文件，然后启动 MySQL。此方法不适用于每日备份，但非常适合在维护时段进行升级或降级时使用，或者在进行主机交换时使用。

7.5.1 如何操作

1. 关闭 MySQL 服务器：

```
shell> sudo service mysqld stop
```

2. 将文件复制到数据目录中（你的目录可能不同）：

```
shell> sudo rsync -av /data/mysql /backups
or do rsync over ssh to remote server
shell> rsync -e ssh -az /data/mysql/
backup_user@remote_server:/backups
```

3. 启动 MySQL 服务器：

```
shell> sudo service mysqld start
```

7.6 使用 XtraBackup 进行备份

XtraBackup 是由 Percona 提供的开源备份软件。它能在不关闭服务器的情况下复制普通文件。但为了避免不一致，它会使用 REDO 日志文件。XtraBackup 被许多公司广泛用作标准备份工具。与逻辑备份工具相比，其优势是备份速度非常快，恢复速度也非常快。

以下是 Percona XtraBackup 的工作原理（摘自 Percona XtraBackup 文档）。

1. XtraBackup 复制 `InnoDB` 数据文件，这会导致内部不一致的数据，但是它会对文件执行崩溃恢复，以使其再次成为一个一致的可用数据库。

2. 这样做是可行的，因为 `InnoDB` 维护一个 REDO 日志，也称为事务日志。REDO 日志包含了 `InnoDB` 数据每次更改的记录。当 `InnoDB` 启动时，REDO 日志会检查数据文件和事务日志，并执行两个步骤。它将已提交的事务日志条目应用于数据文件，并对任何修改了数据但未提交的事务执行 undo 操作。

3. Percona XtraBackup 会在启动时记住**日志序列号**（**LSN**），然后复制数据文件。这需要一些时间来完成，如果文件正在改变，那么它们会在不同的时间点反映数据库的状态。同时，Percona XtraBackup 运行一个后台进程，用于监视事务日志文件，并从中复制更改。Percona XtraBackup 需要持续这样做，因为事务日志是以循环方式写入的，并且可以在一段时间后重新使用。Percona XtraBackup 开始执行后，需要复制每次数据文件更改对应的事务日志记录。

7.6.1 如何操作

在撰写本书时，Percona XtraBackup 暂不支持 MySQL 8。最终，Percona 将发布支持 MySQL 8 的新版 XtraBackup，那时只需覆盖安装即可。

安装

下面是安装步骤。

在 CentOS / Red Hat / Fedora 系统上

1. 安装 `mysql-community-libs-compat`：

```
shell> sudo yum install -y mysql-community-libs-compat
```

2. 安装 Percona 库：

```
shell> sudo yum install
http://www.percona.com/downloads/percona-release/redhat/0.1-4/perco
na-release-0.1-4.noarch.rpm
```

你应该会看到如下输出：

```
Retrieving
http://www.percona.com/downloads/percona-release/redhat/0.1-4/perco
na-release-0.1-4.noarch.rpm
Preparing...
########################################### [100%]
   1:percona-release
########################################### [100%]
```

3. 测试库：

```
shell> yum list | grep xtrabackup
holland-xtrabackup.noarch 1.0.14-3.el7 epel
percona-xtrabackup.x86_64 2.3.9-1.el7 percona-release-x86_64
percona-xtrabackup-22.x86_64 2.2.13-1.el7 percona-release-x86_64
percona-xtrabackup-22-debuginfo.x86_64 2.2.13-1.el7 percona-
release-x86_64
percona-xtrabackup-24.x86_64 2.4.8-1.el7 percona-release-x86_64
percona-xtrabackup-24-debuginfo.x86_64 2.4.8-1.el7 percona-release-
x86_64
percona-xtrabackup-debuginfo.x86_64 2.3.9-1.el7 percona-release-
x86_64
percona-xtrabackup-test.x86_64 2.3.9-1.el7 percona-release-x86_64
percona-xtrabackup-test-22.x86_64 2.2.13-1.el7 percona-release-
x86_64
percona-xtrabackup-test-24.x86_64 2.4.8-1.el7 percona-release-
x86_64
```

4. 安装 XtraBackup：

```
shell> sudo yum install percona-xtrabackup-24
```

在 Debian / Ubuntu 系统上

1. 从 Percona 获取库软件包：

```
shell> wget https://repo.percona.com/apt/percona-
release_0.1-4.$(lsb_release -sc)_all.deb
```

2. 用 dpkg 安装下载的软件包。要做到这一点，需要以 root 身份或使用 sudo 运行以下命令：

```
shell> sudo dpkg -i percona-release_0.1-4.$(lsb_release - sc)_all.deb
```

一旦安装了这个软件包，就应该添加 Percona 软件仓库。可以在 /etc/apt/sources.list.d/percona-release.list 文件中检查软件仓库设置。

3. 请记得更新本地缓存：

```
shell> sudo apt-get update
```

4. 之后可以安装软件包：

```
shell> sudo apt-get install percona-xtrabackup-24
```

7.7 锁定实例进行备份

从 MySQL 8 开始，我们可以锁定实例进行备份了，这将允许在线备份期间的 DML，并阻止可能导致快照不一致的所有操作。

7.7.1 如何操作

在开始备份之前，请锁定需要备份的实例：

```
mysql> LOCK INSTANCE FOR BACKUP;
```

执行备份，完成后解锁实例：

```
mysql> UNLOCK INSTANCE;
```

7.8 使用二进制日志进行备份

我们知道二进制日志是时间点恢复所需要的。在本节中，你将了解如何备份二进制日志。该进程将二进制日志从数据库服务器流式传输到远程备份服务器。既可以从从服务器也可以从主服务器进行二进制日志备份。如果你正在从主服务器进行二进制日志备份，并在从服务器进行实际备份，则应使用`--dump-slave`获取相应的主日志位置。如果你使用的是`mydumper`或 XtraBackup，则主和从二进制日志位置会被同时提供。

7.8.1 如何操作

1. 在服务器上创建一个复制用户，并设置一个强密码：

```
mysql> GRANT REPLICATION SLAVE ON *.* TO 'binlog_user'@'%'
IDENTIFIED BY 'binlog_pass';
```

```
Query OK, 0 rows affected, 1 warning (0.03 sec)
```

2. 检查服务器上的二进制日志：

```
mysql> SHOW BINARY LOGS;
+----------------+-----------+
| Log_name       | File_size |
+----------------+-----------+
| server1.000008 |       2451|
| server1.000009 |        199|
| server1.000010 |       1120|
| server1.000011 |        471|
| server1.000012 |        154|
+----------------+-----------+
5 rows in set (0.00 sec)
```

你可以在服务器上找到第一个可用的二进制日志，可以从这里开始备份。在这个例子中，它是 `server1.000008`。

3. 登录到备份服务器并执行以下命令，会将二进制日志从 MySQL 服务器复制到备份服务器。你可以使用 `nohup` 或 `disown`：

```
shell> mysqlbinlog -u <user> -p<pass> -h <server> --read-from-
remote-server --stop-never
--to-last-log --raw server1.000008 &
shell> disown -a
```

4. 验证是否正在备份二进制日志：

```
shell> ls -lhtr server1.0000*
-rw-r-----. 1  mysql  mysql 2.4K   Aug 25 12:22 server1.000008
-rw-r-----. 1  mysql  mysql  199   Aug 25 12:22 server1.000009
-rw-r-----. 1  mysql  mysql 1.1K   Aug 25 12:22 server1.000010
-rw-r-----. 1  mysql  mysql   471  Aug 25 12:22 server1.000011
-rw-r-----. 1  mysql  mysql   154  Aug 25 12:22 server1.000012
```

第8章 恢复数据

在本章中，我们将介绍以下内容：

- 从 mysqldump 和 mysqlpump 中恢复
- 使用 myloader 从 mydumper 中恢复
- 从普通文件备份中恢复
- 执行时间点恢复

8.1 引言

在本章中，我们将学习各种备份恢复方法。这里假定备份和二进制日志（binary log）在服务器上是可用的。

8.2 从 mysqldump 和 mysqlpump 中恢复

逻辑备份工具 mysqldump 和 mysqlpump 将数据写入单个文件。

8.2.1 如何操作

登录备份所在的服务器：

```
shell> cat /backups/full_backup.sql | mysql -u <user> -p
or
shell> mysql -u <user> -p < /backups/full_backup.sql
```

要在远程服务器上恢复，可以使用-h <主机名>选项：

```
shell> cat /backups/full_backup.sql | mysql -u <user> -p -h
<remote_hostname>
```

当恢复一个备份时，该备份的语句将被记录到二进制日志中，这可能会拖慢恢复过程。如果不希望恢复过程被写入二进制日志，则可以使用 SET SQL_LOG_BIN = 0;选项在session（会话）级别关闭这个功能：

```
shell> (echo "SET SQL_LOG_BIN=0;";cat /backups/full_backup.sql) | mysql -u
<user> -p -h <remote_hostname>
```

或者使用：

```
mysql> SET SQL_LOG_BIN=0; SOURCE full_backup.sql
```

8.2.2 更多建议

1. 由于备份恢复需要很长时间，因此建议在 screen 会话内启动恢复过程，即使断开与服务器的连接，备份的恢复也会继续。

2. 有时候，在恢复期间可能会出现故障。如果将--force 选项传递给 MySQL，恢复过程将继续：

```
shell> (echo "SET SQL_LOG_BIN=0;";cat /backups/full_backup.sql) |
mysql -u <user> -p -h <remote_hostname> -f
```

8.3 使用 myloader 从 mydumper 中恢复

myloader 是多线程恢复 mydumper 备份集的工具。myloader 与 mydumper 是一起的，不需要单独安装。在本节中，我们将学习恢复备份的各种方法。

8.3.1 如何操作

myloader 的常用选项有：要连接的 MySQL 服务器的主机名（默认值为 localhost）、用户名、密码和端口。

恢复完整的数据库

```
shell> myloader --directory=/backups --user=<user> --password=<password> --
queries-per-transaction=5000 --threads=8 --compress-protocol --overwrite-
tables
```

各选项的含义如下所述。

- `--overwrite-tables`：这个选项会删除已经存在的表。
- `--compress-protocol`：该选项在 MySQL 连接上使用压缩。
- `--threads`：该选项指定要使用的线程数量，默认值是 4。
- `--queries-per-transaction`：指定每个事务的查询数量，默认值是 1000。
- `--directory`：指定要导入的转储目录。

恢复单个数据库

可以指定`--source-db <db_name>`，仅恢复单个数据库。

假设你想恢复 `company` 数据库：

```
shell> myloader --directory=/backups --queries-per-transaction=5000 --
threads=6 --compress-protocol --user=<user> --password=<password> --source-
db company --overwrite-tables
```

恢复单个表

`mydumper` 将每个表的备份写入单独的`.sql` 文件。你可以选择这个`.sql` 文件并恢复：

```
shell> mysql -u <user> -p<password> -h <hostname> company -A -f <
company.payments.sql
```

如果这个表被拆分为 chunk（块），则可以将与这个表相关的所有 chunk 和信息复制到一个目录并指定其位置。

复制所需的文件：

```
shell> sudo cp /backups/employee_table_chunks/employees.employees.* \
/backups/employee_table_chunks/employees.employees-schema.sql \
/backups/employee_table_chunks/employees-schema-create.sql \
/backups/employee_table_chunks/metadata \
```

```
/backups/single_table/
```

使用 myloader 加载，它会自动检测 chunk 并加载它们：

```
shell> myloader --directory=/backups/single_table/ --queries-per-
transaction=50000 --threads=6 --compress-protocol --overwrite-tables
```

8.4 从普通文件备份中恢复

从普通文件备份中恢复，需要先关闭 MySQL 服务器，替换所有文件，更改权限，然后再启动 MySQL。

8.4.1 如何操作

1. 停止 MySQL 服务器的运行：

```
shell> sudo systemctl stop mysql
```

2. 将文件移至数据目录：

```
shell> sudo mv /backup/mysql /var/lib
```

3. 将所有权更改为 mysql：

```
shell> sudo chown -R mysql:mysql /var/lib/mysql
```

4. 启动 MySQL：

```
shell> sudo systemctl start mysql
```

为了最大限度地减少停机时间，如果磁盘上有足够的空间，可以将备份复制到/var/lib/mysql2，然后停止 MySQL 的运行，重命名目录并启动服务器：

```
shell> sudo mv /backup/mysql /var/lib/mysql2
shell> sudo systemctl stop mysql
shell> sudo mv /var/lib/mysql2 /var/lib/mysql
shell> sudo chown -R mysql:mysql /var/lib/mysql
shell> sudo systemctl start mysql
```

8.5 执行时间点恢复

一旦恢复完整的备份后，你需要恢复二进制日志以获得时间点（point-in-time）恢复。备份集提供截止到备份可用时的二进制日志坐标。

正如第 7 章的 7.7 节所述，应该根据 mysqldump 中指定的--dump-slave 或--master-data 选项从备份所在的服务器中选择二进制日志备份。

8.5.1 如何操作

我们来详细介绍一下操作细节，这里有很多要学习的地方。

mysqldump 或 mysqlpump

根据你传给 mysqldump/mysqlpump 的选项，二进制日志信息被作为 CHANGE MASTER TO 命令存储在 SQL 文件中。

1. 如果你用了--master-data，则应使用从服务器的二进制日志：

```
shell> head -30 /backups/dump.sql
-- MySQL dump 10.13  Distrib 8.0.3-rc, for Linux (x86_64)
--
-- Host: localhost Database:
-- ------------------------------------------------------
-- Server version 8.0.3-rc-log
/*!40101 SET @OLD_CHARACTER_SET_CLIENT=@@CHARACTER_SET_CLIENT */;
/*!40101 SET @OLD_CHARACTER_SET_RESULTS=@@CHARACTER_SET_RESULTS */;
/*!40101 SET @OLD_COLLATION_CONNECTION=@@COLLATION_CONNECTION */;
/*!40101 SET NAMES utf8 */;
/*!40103 SET
@OLD_TIME_ZONE=@@TIME_ZONE */;
/*!40103 SET TIME_ZONE='+00:00' */;
/*!50606 SET
@OLD_INNODB_STATS_AUTO_RECALC=@@INNODB_STATS_AUTO_RECALC */;
/*!50606 SET GLOBAL INNODB_STATS_AUTO_RECALC=OFF */;
/*!40014 SET @OLD_UNIQUE_CHECKS=@@UNIQUE_CHECKS, UNIQUE_CHECKS=0
*/;
```

```
/*!40014 SET @OLD_FOREIGN_KEY_CHECKS=@@FOREIGN_KEY_CHECKS,
FOREIGN_KEY_CHECKS=0 */;
/*!40101 SET @OLD_SQL_MODE=@@SQL_MODE,
SQL_MODE='NO_AUTO_VALUE_ON_ZERO' */;
/*!40111 SET @OLD_SQL_NOTES=@@SQL_NOTES, SQL_NOTES=0 */;
--
-- Position to start replication or point-in-time recovery from
--
CHANGE MASTER TO MASTER_LOG_FILE='server1.000008',
MASTER_LOG_POS=154;
```

在这种情况下，应该从位于从服务器，位置为 154 处的 server1.000008 文件开始恢复。

```
shell> mysqlbinlog --start-position=154 --disable-log-bin /backups/binlogs/server1.000008 | mysql -u<user> -p -h <host> -f
```

2. 如果你用了--dump-slave，则应该使用主服务器上的二进制日志：

```
--
-- Position to start replication or point-in-time recovery from
(the master of this slave)
--
CHANGE MASTER TO MASTER_LOG_FILE='centos7-bin.000001',
MASTER_LOG_POS=463;
```

在这种情况下，应该从位于主服务器，位置为 463 处的 centos7-bin.000001 文件开始恢复。

```
shell> mysqlbinlog --start-position=463
--disable-log-bin /backups/binlogs/centos7-bin.000001
| mysql -u<user> -p -h <host> -f
```

mydumper

二进制日志信息可以从元数据中获取：

```
shell> sudo cat /backups/metadata
Started dump at: 2017-08-26 06:26:19
SHOW MASTER STATUS:
 Log: server1.000012
 Pos: 154
</span> GTID:
SHOW SLAVE STATUS:
 Host: 35.186.158.188
 Log: centos7-bin.000001
 Pos: 463
 GTID:
Finished dump at: 2017-08-26 06:26:24
```

如果你已经从从服务器中获取二进制日志备份，则应从位置为 154（SHOW MASTER STATUS）的 server1.000012 文件开始恢复：

```
shell> mysqlbinlog --start-position=154 --disable-log-bin
/backups/binlogs/server1.000012 | mysql -u<user> -p -h <host> -f
```

如果你有来自主服务器的二进制日志备份，则应从位置为 463（SHOW SLAVE STATUS）的 centos7-bin.000001 文件开始恢复：

```
shell> mysqlbinlog --start-position=463 --disable-log-bin
/backups/binlogs/centos7-bin.000001 | mysql -u<user> -p -h <host> -f
```

第 9 章 复制

在本章中，我们将介绍以下方法：

- 准备复制
- 设置主主复制
- 设置多源复制
- 设置复制筛选器
- 将从数据库由主从复制切换到链式复制
- 将从数据库由链式复制切换到主从复制
- 设置延迟复制
- 设置 GTID 复制
- 设置半同步复制

9.1 引言

正如第 6 章中所述，复制（replication）功能可以将一个 MySQL 数据库服务器（主库）中的数据复制到一个或多个 MySQL 数据库服务器（从库）。默认情况下，复制是异步的；从库不需要永久连接以接收来自主库的更新。你可以将其配置为复制所有数据库、复制指定的数据库，甚至还可以配置为复制数据库中指定的表。

在本章中，我们将学习如何设置传统复制，复制指定的数据库和表，并设置多源复制，链复制、延迟复制和半同步复制。

简单说来，复制的原理如下：在服务器（**主库**）上执行的所有 DDL 和 DML 语句都会被记录到二进制日志中，这些日志由连接到它的服务器（称为**从库**）提取。它们只是被复制到从库，并被保存为中继日志。这个过程由一个称为 **IO 线程**的线程负责。还有一个线程叫作 **SQL 线程**，它按顺序执行中继日志中的语句。

下面这篇博客文章非常清楚地解释了复制的工作原理：

```
https://www.percona.com/blog/2013/01/09/how-does-mysql-replication-really-work/
```

复制有如下优点（摘自 `https://dev.mysql.com/doc/refman/8.0/en/replication.html`）。

- 水平解决方案：将负载分散到多个从库以提高性能。在此环境中，所有的写入和更新都必须在主库上进行。但是，读操作可能发生在一个或多个从库上。该模式可以提高写入的性能（因为主库专门用于更新），同时对于不断增加的从库也能显著加快其读取速度。
- 数据安全性：因为数据被复制到从库，而且从库可以暂停复制过程，所以可以在从库上运行备份服务而不会损坏相应的主库数据。
- 分析：在主库上可以实时创建数据，而对信息的分析可以在从库上进行，不会影响主库的性能。
- 远程数据分发：你可以使用复制为远程服务器站点创建本地数据的副本，无须永久访问主库。

9.2 准备复制

复制有许多拓扑形式。其中一些是传统的主从复制、链式复制、主主复制、多源复制，等等。

传统复制涉及单个主库和多个从库，如下图所示。

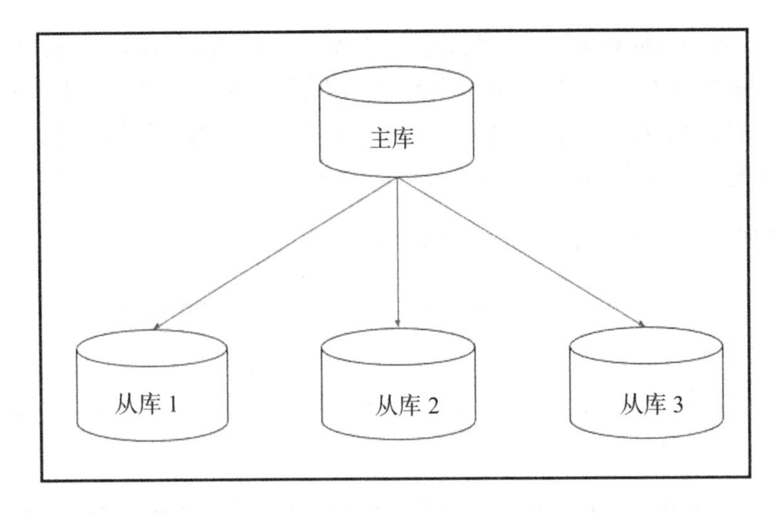

链式复制意味着一台服务器从另一台复制，而另一台服务器又从另一台复制。中间的服务器称为中继主库（主库→中继主库→从库），如下图所示。

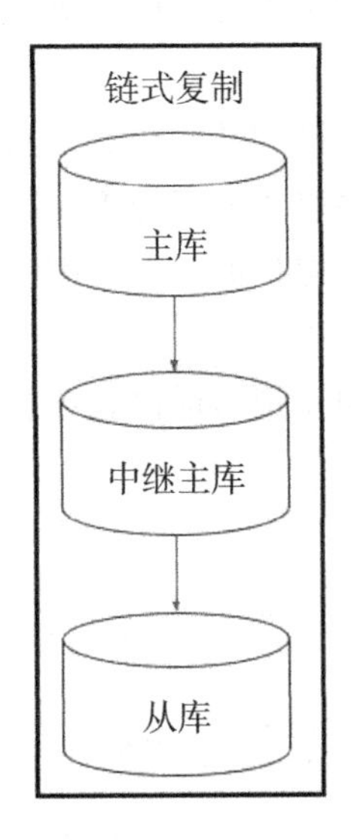

如果想要在两个数据中心之间设置复制，一般就会使用这种方式。主库（the primary master）及其从库将位于一个数据中心内。辅助主库（中继主库）从另一个数据中心的主库进行复制。另一个数据中心的所有从库都从辅助主库（the secondary master）复制。

主主复制：在此拓扑中，两个主库互相之间都可以接受写入和复制，如下图所示。

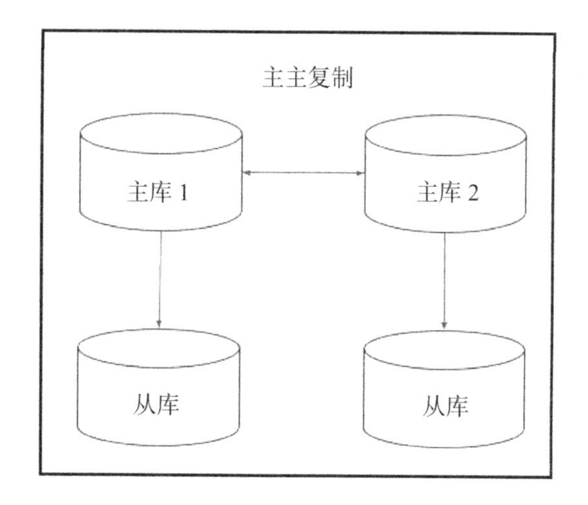

多源复制：在这种拓扑中，一个从库将从多个主库而非从一个主库复制，如下图所示。

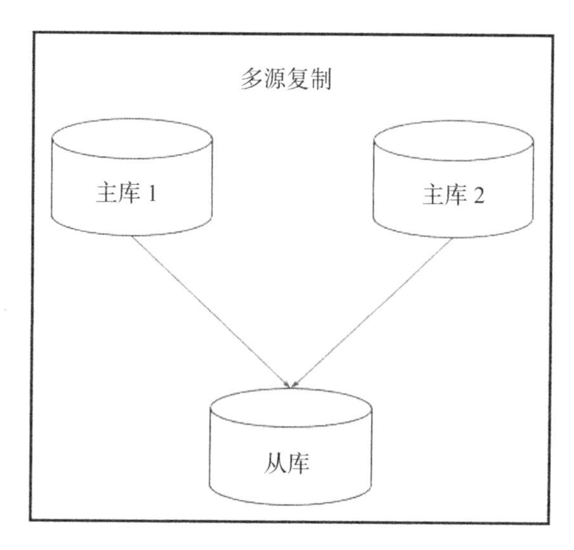

如果要设置链式复制，则可以按照与此处提到的相同步骤操作，用中继主库替换主库即可。

9.2.1 如何操作

本节解释如何设置单个从库。可以应用相同的原则来设置链式复制。通常，我们在创建另一个从库时都是从从库中获取备份副本的。

大致方案如下：

1. 在主库上启用二进制日志记录。
2. 在主库上创建一个复制用户。
3. 在从库上设置唯一的 server_id。
4. 从主库中取得备份。
5. 恢复从库上的备份。
6. 执行 CHANGE MASTER TO 命令。
7. 开始复制。

具体的步骤如下。

1. **在主库上，**启用二进制日志记录并设置 SERVER_ID。请参阅第 6 章，了解如何启用二进制日志记录。

2. **在主库上，**创建一个复制用户。从库使用此账户连接到主库：

```
mysql> GRANT REPLICATION SLAVE ON *.* TO 'binlog_user'@'%' IDENTIFIED
BY 'binlog_P@ss12';
Query OK, 0 rows affected, 1 warning (0.00 sec)
```

3. **在从库上，**设置唯一的 SERVER_ID 选项（它应该与你在主库上设置的不同）：

```
mysql> SET @@GLOBAL.SERVER_ID = 32;
```

4. **在从库上，**通过远程连接从主库进行备份。可以使用 mysqldump 或 mydumper。不能使用 mysqlpump，因为二进制日志的位置不一致。

mysqldump:

```
shell> mysqldump -h <master_host> -u backup_user --password=<pass>
--all-databases --routines --events --single-transaction --master-
data > dump.sql
```

从另一个从库进行备份时，必须设置-slave-dump 选项。

mydumper:

```
shell> mydumper -h <master_host> -u backup_user --password=<pass> -
-use-savepoints --trx-consistency-only --kill-long-queries --
outputdir /backups
```

5. 在从库上，待备份完成后恢复此备份。有关恢复备份的方法，请参阅第 8 章。

```
mysqldump:
  shell> mysql -u <user> -p -f < dump.sql
mydumper:
  shell> myloader --directory=/backups --user=<user> --
  password=<password> --queries-per-transaction=5000 --threads=8 --
  overwrite-tables
```

6. 在从库上，恢复备份后，必须执行以下命令：

```
mysql> CHANGE MASTER TO MASTER_HOST='<master_host>',
MASTER_USER='binlog_user', MASTER_PASSWORD='binlog_P@ss12',
MASTER_LOG_FILE='<log_file_name>', MASTER_LOG_POS=<position>
```

mysqldump: <log_file_name>和<position>包含在备份转储文件中。例如：

```
shell> less dump.sql
--
-- Position to start replication or point-in-time recovery from
(the master of this slave)
--
CHANGE MASTER TO MASTER_LOG_FILE='centos7-bin.000001',
MASTER_LOG_POS=463;
```

mydumper: <log_file_name>和<position>存储在元数据文件中：

```
shell> cat metadata
Started dump at: 2017-08-26 06:26:19
SHOW MASTER STATUS:
    Log: server1.000012
    Pos: 154122
    GTID:
SHOW SLAVE STATUS:
    Host: xx.xxx.xxx.xxx
    Log: centos7-bin.000001
```

```
    Pos: 463223
    GTID:
Finished dump at: 2017-08-26 06:26:24
```

如果你从一个从库或主库进行备份来设置另一个从库，则必须使用来自 SHOW SLAVE STATUS 的位置。如果要设置链式复制，则可以使用 SHOW MASTER STATUS 中的位置。

1. 在从库上，执行 START SLAVE 命令：

```
mysql> START SLAVE;
```

2. 可以通过执行以下命令来检查复制的状态：

```
mysql> SHOW SLAVE STATUS\G
*************************** 1. row ***************************
               Slave_IO_State: Waiting for master to send event
                  Master_Host: xx.xxx.xxx.xxx
                  Master_User: binlog_user
                  Master_Port: 3306
                Connect_Retry: 60
              Master_Log_File: server1-bin.000001
          Read_Master_Log_Pos: 463
               Relay_Log_File: server2-relay-bin.000004
                Relay_Log_Pos: 322
        Relay_Master_Log_File: server1-bin.000001
             Slave_IO_Running: Yes
            Slave_SQL_Running: Yes
              Replicate_Do_DB:
          Replicate_Ignore_DB:
           Replicate_Do_Table:
       Replicate_Ignore_Table:
      Replicate_Wild_Do_Table:
  Replicate_Wild_Ignore_Table:
                   Last_Errno: 0
                   Last_Error:
                 Skip_Counter: 0
          Exec_Master_Log_Pos: 463
              Relay_Log_Space: 1957
```

```
               Until_Condition: None
                Until_Log_File:
                 Until_Log_Pos: 0
            Master_SSL_Allowed: No
            Master_SSL_CA_File:
            Master_SSL_CA_Path:
               Master_SSL_Cert:
             Master_SSL_Cipher:
                Master_SSL_Key:
         Seconds_Behind_Master: 0
 Master_SSL_Verify_Server_Cert: No
                 Last_IO_Errno: 0
                 Last_IO_Error:
                Last_SQL_Errno: 0
                Last_SQL_Error:
   Replicate_Ignore_Server_Ids:
              Master_Server_Id: 32
                   Master_UUID: b52ef45a-7ff4-11e7-9091-42010a940003
              Master_Info_File: /var/lib/mysql/master.info
                     SQL_Delay: 0
           SQL_Remaining_Delay: NULL
       Slave_SQL_Running_State: Slave has read all relay log;
waiting for more updates
            Master_Retry_Count: 86400
                   Master_Bind:
       Last_IO_Error_Timestamp:
      Last_SQL_Error_Timestamp:
                Master_SSL_Crl:
            Master_SSL_Crlpath:
            Retrieved_Gtid_Set:
             Executed_Gtid_Set:
                 Auto_Position: 0
          Replicate_Rewrite_DB:
                  Channel_Name:
            Master_TLS_Version:
1 row in set (0.00 sec)
```

你应该查看 Seconds_Behind_Master 的值，它代表的是复制的延迟情况。如果它的值为 0，则意味着从库与主库同步；如果为非零值，则表示延迟的秒数；如果为 NULL，则表示未复制。

9.3 设置主主复制

这个方法会让很多人感兴趣，因为我们中的很多人都在尝试这样做。现在我们来了解它的细节。

9.3.1 如何操作

假设主库是 master1 和 master2。

具体步骤如下。

1. 按照本章所述的方法，在 master1 和 master2 之间设置复制。

2. 设置 master2 为只读：

```
mysql> SET @@GLOBAL.READ_ONLY=ON;
```

3. 在 master2 上，检查当前二进制日志的坐标。

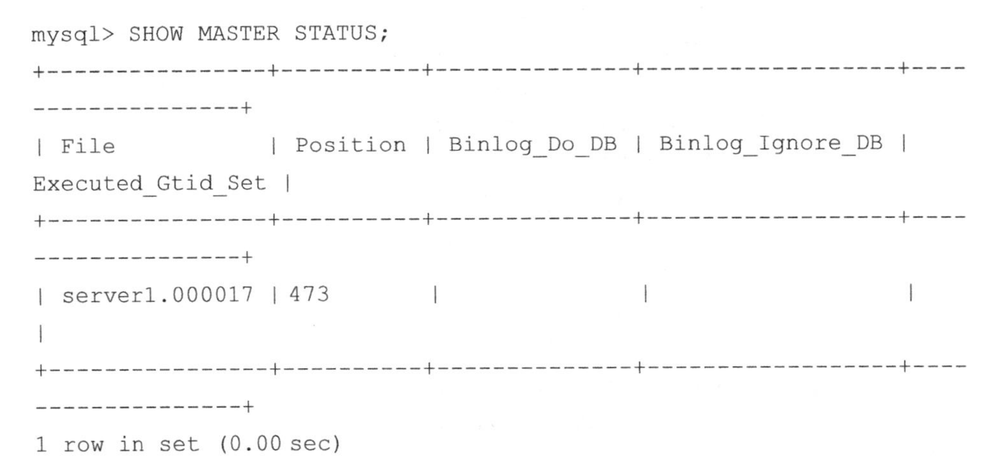

```
mysql> SHOW MASTER STATUS;
+----------------+----------+--------------+------------------+----
---------------+
| File           | Position | Binlog_Do_DB | Binlog_Ignore_DB |
Executed_Gtid_Set |
+----------------+----------+--------------+------------------+----
---------------+
| server1.000017 | 473      |              |                  |
|
+----------------+----------+--------------+------------------+----
---------------+
1 row in set (0.00 sec)
```

根据上面的信息，你可以从位置为 473 的 server1.000017 文件处开始在 master1 上复制。

4. 从第3步中获取的位置开始，在 master1 上执行 CHANGE MASTER TO 命令：

```
mysql> CHANGE MASTER TO MASTER_HOST='<master2_host>',
MASTER_USER='binlog_user', MASTER_PASSWORD='binlog_P@ss12',
MASTER_LOG_FILE='<log_file_name>', MASTER_LOG_POS=<position>
```

5. 在 master1 上开启 Slave 模式：

```
mysql> START SLAVE;
```

6. 最后，你可以设置 master2 为“可读/写”，这样应用程序就可以开始对它写入数据了。

```
mysql> SET @@GLOBAL.READ_ONLY=OFF;
```

9.4　设置多源复制

MySQL 多源复制使从库能够同时接收来自多个源的事务。多源复制可用于将多台服务器备份到单台服务器，合并表分片，以及将多台服务器中的数据整合到单台服务器。多源复制在应用事务时不会执行任何冲突检测或解析，并且如果需要的话，这些任务将留给应用程序来处理。在多源复制拓扑中，从库为每个主库创建一个复制通道，以便从中接收事务。

在本节中，我们将学习如何设置有多个主库的从库。此方法与在复制通道上设置传统复制相同。

9.4.1　如何操作

假设你打算将 server3 设置为 server1 和 server2 的从库。你需要创建从 server1 到 server3 的传统复制通道，并创建另一个从 server2 到 server3 的通道。为保证数据在从库上保持一致，请确保复制不同的数据库，或应用程序可以处理复制时的冲突。

在开始之前，从 server1 上获取备份并在 server3 上进行恢复；同样，从 server2 上获取备份并在 server3 上进行恢复。

1. 在 server3 上，将复制存储库从 FILE 修改为 TABLE。可以运行以下命令来动态地更改它：

```
mysql> STOP SLAVE; // 如果从库还在同步
mysql> SET GLOBAL master_info_repository = 'TABLE';
mysql> SET GLOBAL relay_log_info_repository = 'TABLE';
```

   还要在配置文件中进行一些更改：

```
shell> sudo vi /etc/my.cnf
[mysqld]
master-info-repository=TABLE
relay-log-info-repository=TABLE
```

2. 在 server3 上， 执行 CHANGE MASTER TO 命令，使其通过名为 master-1 的通道成为 server1 的从库。你可以为它取任何名字：

```
mysql> CHANGE MASTER TO MASTER_HOST='server1',
MASTER_USER='binlog_user', MASTER_PORT=3306,
MASTER_PASSWORD='binlog_P@ss12', MASTER_LOG_FILE='server1.000017',
MASTER_LOG_POS=788 FOR CHANNEL 'master-1';
```

3. 在 server3 上，执行 CHANGE MASTER TO 命令，使其通过名为 master-2 的通道成为 server2 的从库：

```
mysql> CHANGE MASTER TO MASTER_HOST='server2',
MASTER_USER='binlog_user', MASTER_PORT=3306,
MASTER_PASSWORD='binlog_P@ss12', MASTER_LOG_FILE='server2.000014',
MASTER_LOG_POS=75438 FOR CHANNEL 'master-2';
```

4. 为每个通道执行 START SLAVE FOR CHANNEL 语句：

```
mysql> START SLAVE FOR CHANNEL 'master-1';
Query OK, 0 rows affected (0.01 sec)
mysql> START SLAVE FOR CHANNEL 'master-2';
Query OK, 0 rows affected (0.00 sec)
```

5. 通过执行 SHOW SLAVE STATUS 语句来验证从库状态：

```
mysql> SHOW SLAVE STATUS\G
```

```
*************************** 1. row ***************************
               Slave_IO_State: Waiting for master to send event
                  Master_Host: server1
                  Master_User: binlog_user
                  Master_Port: 3306
                Connect_Retry: 60
              Master_Log_File: server1.000017
          Read_Master_Log_Pos: 788
               Relay_Log_File: server3-relay-bin-
master@002d1.000002
                Relay_Log_Pos: 318
        Relay_Master_Log_File: server1.000017
             Slave_IO_Running: Yes
            Slave_SQL_Running: Yes
              Replicate_Do_DB:
          Replicate_Ignore_DB:
           Replicate_Do_Table:
       Replicate_Ignore_Table:
      Replicate_Wild_Do_Table:
  Replicate_Wild_Ignore_Table:
                   Last_Errno: 0
                   Last_Error:
                 Skip_Counter: 0
          Exec_Master_Log_Pos: 788
              Relay_Log_Space: 540
              Until_Condition: None
               Until_Log_File:
                Until_Log_Pos: 0
           Master_SSL_Allowed: No
           Master_SSL_CA_File:
           Master_SSL_CA_Path:
              Master_SSL_Cert:
            Master_SSL_Cipher:
               Master_SSL_Key:
        Seconds_Behind_Master: 0
Master_SSL_Verify_Server_Cert: No
                Last_IO_Errno: 0
```

```
                Last_IO_Error:
               Last_SQL_Errno: 0
               Last_SQL_Error:
  Replicate_Ignore_Server_Ids:
             Master_Server_Id: 32
                  Master_UUID: 7cc7fca7-4deb-11e7-a53e-42010a940002
             Master_Info_File: mysql.slave_master_info
                    SQL_Delay: 0
          SQL_Remaining_Delay: NULL
      Slave_SQL_Running_State: Slave has read all relay log;
waiting for more updates
           Master_Retry_Count: 86400
                  Master_Bind:
      Last_IO_Error_Timestamp:
     Last_SQL_Error_Timestamp:
               Master_SSL_Crl:
           Master_SSL_Crlpath:
           Retrieved_Gtid_Set:
            Executed_Gtid_Set:
                Auto_Position: 0
         Replicate_Rewrite_DB:
                 Channel_Name: master-1
           Master_TLS_Version:
*************************** 2. row ***************************
               Slave_IO_State: Waiting for master to send event
                  Master_Host: server2
                  Master_User: binlog_user
                  Master_Port: 3306
                Connect_Retry: 60
              Master_Log_File: server2.000014
          Read_Master_Log_Pos: 75438
               Relay_Log_File: server3-relay-bin-
master@002d2.000002
                Relay_Log_Pos: 322
        Relay_Master_Log_File: server2.000014
             Slave_IO_Running: Yes
            Slave_SQL_Running: Yes
```

```
           Replicate_Do_DB:
       Replicate_Ignore_DB:
        Replicate_Do_Table:
    Replicate_Ignore_Table:
   Replicate_Wild_Do_Table:
Replicate_Wild_Ignore_Table:
                Last_Errno: 0
                Last_Error:
              Skip_Counter: 0
       Exec_Master_Log_Pos: 75438
           Relay_Log_Space: 544
           Until_Condition: None
            Until_Log_File:
             Until_Log_Pos: 0
        Master_SSL_Allowed: No
        Master_SSL_CA_File:
        Master_SSL_CA_Path:
           Master_SSL_Cert:
         Master_SSL_Cipher:
            Master_SSL_Key:
     Seconds_Behind_Master: 0
Master_SSL_Verify_Server_Cert: No
             Last_IO_Errno: 0
             Last_IO_Error:
            Last_SQL_Errno: 0
            Last_SQL_Error:
Replicate_Ignore_Server_Ids:
          Master_Server_Id: 32
               Master_UUID: b52ef45a-7ff4-11e7-9091-42010a940003
          Master_Info_File: mysql.slave_master_info
                 SQL_Delay: 0
       SQL_Remaining_Delay: NULL
   Slave_SQL_Running_State: Slave has read all relay log;
waiting for more updates
        Master_Retry_Count: 86400
               Master_Bind:
   Last_IO_Error_Timestamp:
```

```
        Last_SQL_Error_Timestamp:
                 Master_SSL_Crl:
             Master_SSL_Crlpath:
             Retrieved_Gtid_Set:
              Executed_Gtid_Set:
                  Auto_Position: 0
           Replicate_Rewrite_DB:
                   Channel_Name: master-2
             Master_TLS_Version:
2 rows in set (0.00 sec)
```

6. 要获取特定通道的从库状态，请执行：

```
mysql> SHOW SLAVE STATUS FOR CHANNEL 'master-1' \G
```

7. 这是使用 performance schema 来监控指标的另一种方式：

```
mysql> SELECT * FROM
performance_schema.replication_connection_status\G
*************************** 1. row ***************************
                                      CHANNEL_NAME: master-1
                                        GROUP_NAME:
                                       SOURCE_UUID:
7cc7fca7-4deb-11e7-a53e-42010a940002
                                         THREAD_ID: 36
                                     SERVICE_STATE: ON
                         COUNT_RECEIVED_HEARTBEATS: 73
                          LAST_HEARTBEAT_TIMESTAMP: 2017-09-15
12:42:10.910051
                          RECEIVED_TRANSACTION_SET:
                                 LAST_ERROR_NUMBER: 0
                                LAST_ERROR_MESSAGE:
                              LAST_ERROR_TIMESTAMP: 0000-00-00
00:00:00.000000
                           LAST_QUEUED_TRANSACTION:
 LAST_QUEUED_TRANSACTION_ORIGINAL_COMMIT_TIMESTAMP: 0000-00-00
00:00:00.000000
LAST_QUEUED_TRANSACTION_IMMEDIATE_COMMIT_TIMESTAMP: 0000-00-00
```

```
00:00:00.000000
    LAST_QUEUED_TRANSACTION_START_QUEUE_TIMESTAMP: 0000-00-00
00:00:00.000000
      LAST_QUEUED_TRANSACTION_END_QUEUE_TIMESTAMP: 0000-00-00
00:00:00.000000
                          QUEUEING_TRANSACTION:
   QUEUEING_TRANSACTION_ORIGINAL_COMMIT_TIMESTAMP: 0000-00-00
00:00:00.000000
  QUEUEING_TRANSACTION_IMMEDIATE_COMMIT_TIMESTAMP: 0000-00-00
00:00:00.000000
       QUEUEING_TRANSACTION_START_QUEUE_TIMESTAMP: 0000-00-00
00:00:00.000000
*************************** 2. row ***************************
                                  CHANNEL_NAME: master-2
                                    GROUP_NAME:
                                   SOURCE_UUID:
b52ef45a-7ff4-11e7-9091-42010a940003
                                     THREAD_ID: 38
                                 SERVICE_STATE: ON
                      COUNT_RECEIVED_HEARTBEATS: 73
                       LAST_HEARTBEAT_TIMESTAMP: 2017-09-15
12:42:13.986271
                       RECEIVED_TRANSACTION_SET:
                             LAST_ERROR_NUMBER: 0
                            LAST_ERROR_MESSAGE:
                          LAST_ERROR_TIMESTAMP: 0000-00-00
00:00:00.000000
                        LAST_QUEUED_TRANSACTION:
 LAST_QUEUED_TRANSACTION_ORIGINAL_COMMIT_TIMESTAMP: 0000-00-00
00:00:00.000000
LAST_QUEUED_TRANSACTION_IMMEDIATE_COMMIT_TIMESTAMP: 0000-00-00
00:00:00.000000
    LAST_QUEUED_TRANSACTION_START_QUEUE_TIMESTAMP: 0000-00-00
00:00:00.000000
      LAST_QUEUED_TRANSACTION_END_QUEUE_TIMESTAMP: 0000-00-00
00:00:00.000000
                          QUEUEING_TRANSACTION:
```

```
    QUEUEING_TRANSACTION_ORIGINAL_COMMIT_TIMESTAMP: 0000-00-00
00:00:00.000000
   QUEUEING_TRANSACTION_IMMEDIATE_COMMIT_TIMESTAMP: 0000-00-00
00:00:00.000000
     QUEUEING_TRANSACTION_START_QUEUE_TIMESTAMP: 0000-00-00
00:00:00.000000
2 rows in set (0.00 sec)
```

你可以通过追加 FOR CHANNEL'channel_name'命令来为通道指定所有与从库相关的命令：

```
mysql> STOP SLAVE FOR CHANNEL 'master-1';
mysql> RESET SLAVE FOR CHANNEL 'master-2';
```

9.5 设置复制筛选器

你可以选择要复制哪些表或数据库。在主库上，可以使用--binlog-do-db 和--binlog-ignore-db 选项来选择要记录变更的数据库，以控制二进制日志，详情请参见第 6 章。更好的方法是控制从库。你可以使用--replicate- *选项或通过创建复制筛选器来动态执行或忽略从主库收到的语句。

9.5.1 如何操作

要创建筛选器，你需要执行 CHANGE REPLICATION FILTER 语句。

仅复制数据库

假设你只想复制 db1 和 db2。使用以下语句来创建复制筛选器：

```
mysql> CHANGE REPLICATION FILTER REPLICATE_DO_DB = (db1, db2);
```

请注意，应该在括号内指定要复制的所有数据库。

复制特定的表

可以使用 REPLICATE_DO_TABLE 指定要复制的表：

```
mysql> CHANGE REPLICATION FILTER REPLICATE_DO_TABLE = ('db1.table1');
```

假设你想使用正则表达式来选择表,就可以使用REPLICATE_WILD_DO_TABLE选项:

```
mysql> CHANGE REPLICATION FILTER REPLICATE_WILD_DO_TABLE = ('db1.imp%');
```

你可以使用不同的 IGNORE 选项，并使用正则表达式来指定数据库或表。

忽略数据库

就像可以选择复制某个数据库一样，也可以使用 REPLICATE_IGNORE_DB 指定你不想复制的数据库:

```
mysql> CHANGE REPLICATION FILTER REPLICATE_IGNORE_DB = (db1, db2);
```

忽略特定的表

可以使用REPLICATE_IGNORE_TABLE和REPLICATE_WILD_IGNORE_TABLE选项忽略某些表。REPLICATE_WILD_IGNORE_TABLE 选项允许使用通配符，而REPLICATE_IGNORE_TABLE只接受完整的表名:

```
mysql> CHANGE REPLICATION FILTER REPLICATE_IGNORE_TABLE = ('db1.table1');
mysql> CHANGE REPLICATION FILTER REPLICATE_WILD_IGNORE_TABLE = ('db1.new%',
'db2.new%');
```

也可以通过指定通道名称来为通道设置筛选器:

```
mysql> CHANGE REPLICATION FILTER REPLICATE_DO_DB = (d1) FOR
CHANNEL 'master-1';
```

9.5.2 延伸阅读

有关复制筛选器的更多详细信息，请参阅 https://dev.mysql.com/doc/refman/8.0/en/change-replication-filter.html。如果你使用了多个筛选器，请参阅 https://dev.mysql.com/doc/refman/8.0/en/replication-rules.html 以了解更多关于 MySQL 如何评估筛选器的内容。

9.6 将从库由主从复制切换到链式复制

如果设置了主从复制，则服务器B和C将从服务器A中复制：服务器A→(服务器B、服务器C)。如果希望把服务器C作为服务器B的从库，则必须先在服务器B和服务器C上停止复制，然后使用 START SLAVE UNTIL 命令将它们置于相同的主库日志位置。之后，就可以从服务器B中获取主库日志坐标，并且在服务器C上执行 CHANGE MASTER TO 命令了。

9.6.1 如何操作

1. 在服务器 C 上，停止从库的运行，并记下 Relay_Master_Log_File 和 Exec_Master_Log_Pos 在 SHOW SLAVE STATUS\G 命令中的位置：

```
mysql> STOP SLAVE;
Query OK, 0 rows affected (0.01 sec)

mysql> SHOW SLAVE STATUS\G
*************************** 1. row ***************************
               Slave_IO_State:
                  Master_Host: xx.xxx.xxx.xxx
                  Master_User: binlog_user
                  Master_Port: 3306
                Connect_Retry: 60
              Master_Log_File: server_A-bin.000023
          Read_Master_Log_Pos: 2604
               Relay_Log_File: server_C-relay-bin.000002
                Relay_Log_Pos: 1228
        Relay_Master_Log_File: server_A-bin.000023
~
          Exec_Master_Log_Pos: 2604
              Relay_Log_Space: 1437
              Until_Condition: None
               Until_Log_File:
                Until_Log_Pos: 0
~
1 row in set (0.00 sec)
```

2. 在服务器 B 上，停止从库的运行，并记下 Relay_Master_Log_File 和 Exec_Master_Log_Pos 在 SHOW SLAVE STATUS\G 命令中的日志位置：

```
mysql> STOP SLAVE;
Query OK, 0 rows affected (0.01 sec)

mysql> SHOW SLAVE STATUS\G
*************************** 1. row ***************************
               Slave_IO_State:
                  Master_Host: xx.xxx.xxx.xxx
                  Master_User: binlog_user
                  Master_Port: 3306
                Connect_Retry: 60
              Master_Log_File: server_A-bin.000023
          Read_Master_Log_Pos: 8250241
               Relay_Log_File: server_B-relay-bin.000002
                Relay_Log_Pos: 1228
        Relay_Master_Log_File: server_A-bin.000023
~
          Exec_Master_Log_Pos: 8250241
              Relay_Log_Space: 8248167
              Until_Condition: None
               Until_Log_File:
                Until_Log_Pos:0
~
1 row in set (0.00 sec)
```

3. 将服务器 B 的日志位置与服务器 C 进行比较，找出哪一个是服务器 A 的最新同步。通常，由于先在服务器 C 上停止了从库的运行，服务器 B 的日志位置会更靠前。在本例中，二者的日志位置如下所示。

   服务器 C：(server_A-bin.000023, 2604)

   服务器 B：(server_A-bin.000023, 8250241)

   服务器 B 的日志位置更靠前，所以必须将服务器 C 置于服务器 B 的日志位置。

4. 在服务器 C 上，使用 START SLAVE UNTIL 语句将其同步到服务器 B 的日志位置：

```
mysql> START SLAVE UNTIL MASTER_LOG_FILE='centos7-bin.000023',
MASTER_LOG_POS=8250241;
Query OK, 0 rows affected, 1 warning (0.03 sec)

mysql> SHOW WARNINGS\G
*************************** 1. row ***************************
  Level: Note
   Code: 1278
Message: It is recommended to use --skip-slave-start when doing
step-by-step replication with START SLAVE UNTIL; otherwise, you will
get problems if you get an unexpected slave's mysqld restart 1 row in
set (0.00 sec)
```

5. 在服务器 C 上，检查 SHOW SLAVE STATUS 的输出中的 Exec_Master_Log_Pos 和 Until_Log_Pos（两者应该相同）来等待服务器 C 同步：

```
mysql> SHOW SLAVE STATUS\G
*************************** 1. row ***************************
               Slave_IO_State: Waiting for master to send event
                  Master_Host: xx.xxx.xxx.xxx
                  Master_User: binlog_user
                  Master_Port: 3306
                Connect_Retry: 60
              Master_Log_File: server_A-bin.000023
          Read_Master_Log_Pos: 8250241
               Relay_Log_File: server_C-relay-bin.000003
                Relay_Log_Pos: 8247959
        Relay_Master_Log_File: server_A-bin.000023
             Slave_IO_Running: Yes
            Slave_SQL_Running: No
~
                   Last_Errno: 0
                   Last_Error:
                 Skip_Counter: 0
          Exec_Master_Log_Pos: 8250241
              Relay_Log_Space: 8249242
```

```
              Until_Condition: Master
               Until_Log_File: server_A-bin.000023
                Until_Log_Pos: 8250241
           Master_SSL_Allowed: No
           Master_SSL_CA_File:
           Master_SSL_CA_Path:
              Master_SSL_Cert:
            Master_SSL_Cipher:
               Master_SSL_Key:
        Seconds_Behind_Master: NULL
~
1 row in set (0.00 sec)
```

6. 在服务器 B 上，找出主库状态，启动从库，并确保它在复制：

```
mysql> SHOW MASTER STATUS;
+----------------------+----------+--------------+------------------
+-------------------+
| File                 | Position | Binlog_Do_DB | Binlog_Ignore_DB
| Executed_Gtid_Set |
+----------------------+----------+--------------+------------------
+-------------------+
| server_B-bin.000003  | 36379324 |              |
|                   |
+----------------------+----------+--------------+------------------
+-------------------   +
1 row in set (0.00 sec)
mysql> START SLAVE;
Query OK, 0 rows affected (0.02 sec)

mysql> SHOW SLAVE STATUS\G
*************************** 1. row ***************************
               Slave_IO_State:
                  Master_Host: xx.xxx.xxx.xxx
                  Master_User: binlog_user
                  Master_Port: 3306
                Connect_Retry: 60
              Master_Log_File: server_A-bin.000023
```

```
            Read_Master_Log_Pos: 8250241
                 Relay_Log_File: server_B-relay-bin.000002
                  Relay_Log_Pos: 1228
          Relay_Master_Log_File: server_A-bin.000023
~
            Exec_Master_Log_Pos: 8250241
                Relay_Log_Space: 8248167
                Until_Condition: None
                 Until_Log_File:
                  Until_Log_Pos:0
~
1 row in set (0.00 sec)
```

7. 在服务器 C 上，停止从库的运行，执行 CHANGE MASTER TO 命令，并指向服务器 B。必须使用在前面的步骤中获得的日志位置：

```
mysql> STOP SLAVE;
Query OK, 0 rows affected (0.04 sec)
mysql> CHANGE MASTER TO MASTER_HOST = 'Server B', MASTER_USER =
'binlog_user', MASTER_PASSWORD = 'binlog_P@ss12',
MASTER_LOG_FILE='server_B-bin.000003', MASTER_LOG_POS=36379324;
Query OK, 0 rows affected, 1 warning (0.04 sec)
```

8. 在服务器 C 上，启动复制并验证从库的状态：

```
mysql> START SLAVE;
Query OK, 0 rows affected (0.00 sec)

mysql> SHOW SLAVE STATUS\G
Query OK, 0 rows affected, 1 warning (0.00 sec)

*************************** 1. row ***************************
               Slave_IO_State: Waiting for master to send event
                  Master_Host: xx.xxx.xxx.xx
                  Master_User: binlog_user
                  Master_Port: 3306
                Connect_Retry: 60
              Master_Log_File: server_B-bin.000003
```

```
            Read_Master_Log_Pos: 36380416
                 Relay_Log_File: server_C-relay-bin.000002
                  Relay_Log_Pos: 1413
          Relay_Master_Log_File: server_B-bin.000003
               Slave_IO_Running: Yes
              Slave_SQL_Running: Yes
~
            Exec_Master_Log_Pos: 36380416
                Relay_Log_Space: 1622
~
          Seconds_Behind_Master: 0
Master_SSL_Verify_Server_Cert: No
                  Last_IO_Errno: 0
                  Last_IO_Error:
                 Last_SQL_Errno: 0
                 Last_SQL_Error:
  Replicate_Ignore_Server_Ids:
~
1 row in set (0.00 sec)
```

9.7　将从库由链式复制切换到主从复制

如果你的服务器设置的是链式复制（例如服务器 A→服务器 B→服务器 C），并且希望将服务器 C 作为服务器 A 的直接从库，则必须停止服务器 B 上的复制，让服务器 C 追上 B，然后查找服务器 A 对应于服务器 B 停下来的位置的坐标。使用这些坐标，你就可以在服务器 C 上执行 CHANGE MASTER TO 命令并使其成为服务器 A 的从库。

9.7.1　如何操作

1. **在服务器 B 上，停止从库的运行，并记下主库状态：**

```
mysql> STOP SLAVE;
Query OK, 0 rows affected (0.04 sec)

mysql> SHOW MASTER STATUS;
+--------------------+---------+-------------+-----------------
```

```
+-------------------+
| File                 | Position | Binlog_Do_DB | Binlog_Ignore_DB
| Executed_Gtid_Set |
+---------------------+----------+-------------+-----------------
+-------------------+
| server_B-bin.000003 | 44627878 |              |
|       |
+---------------------+----------+-------------+-----------------
+-------------------+
1 row in set (0.00 sec)
```

2. **在服务器 C 上，确保从库的延迟已被追上。**

   `Relay_Master_Log_File` 和 `Exec_Master_Log_Pos` 应该等于服务器 B 上主库状态的输出。一旦延迟被追上，就停止从库的运行：

```
mysql> SHOW SLAVE STATUS\G
*************************** 1. row ***************************
               Slave_IO_State: Waiting for master to send event
                  Master_Host: 35.186.157.16
                  Master_User: repl
                  Master_Port: 3306
                Connect_Retry: 60
              Master_Log_File: server_B-bin.000003
          Read_Master_Log_Pos: 44627878
               Relay_Log_File: ubuntu2-relay-bin.000002
                Relay_Log_Pos: 8248875
        Relay_Master_Log_File: server_B-bin.000003
             Slave_IO_Running: Yes
            Slave_SQL_Running: Yes
              Replicate_Do_DB:
          Replicate_Ignore_DB:
           Replicate_Do_Table:
       Replicate_Ignore_Table:
      Replicate_Wild_Do_Table:
  Replicate_Wild_Ignore_Table:
                   Last_Errno: 0
                   Last_Error:
```

```
            Skip_Counter: 0
     Exec_Master_Log_Pos: 44627878
         Relay_Log_Space: 8249084
         Until_Condition: None
          Until_Log_File:
           Until_Log_Pos: 0
      Master_SSL_Allowed: No
      Master_SSL_CA_File:
      Master_SSL_CA_Path:
         Master_SSL_Cert:
       Master_SSL_Cipher:
          Master_SSL_Key:
   Seconds_Behind_Master: 0
~
1 row in set (0.00 sec)

mysql> STOP SLAVE;
Query OK, 0 rows affected (0.01 sec)
```

3. 在服务器B上，从 SHOW SLAVE STATUS 输出中获取服务器A的日志坐标（记下 Relay_Master_Log_File 和 Exec_Master_Log_Pos 的值），并启动从库：

```
mysql> SHOW SLAVE STATUS\G
*************************** 1. row ***************************
              Slave_IO_State:
                 Master_Host: xx.xxx.xxx.xxx
                 Master_User: repl
                 Master_Port: 3306
               Connect_Retry: 60
             Master_Log_File: server_A-bin.000023
         Read_Master_Log_Pos: 16497695
              Relay_Log_File: server_B-relay-bin.000004
               Relay_Log_Pos: 8247776
       Relay_Master_Log_File: server_A-bin.000023
            Slave_IO_Running: No
           Slave_SQL_Running: No
             Replicate_Do_DB:
```

```
              Replicate_Ignore_DB:
               Replicate_Do_Table:
           Replicate_Ignore_Table:
          Replicate_Wild_Do_Table:
      Replicate_Wild_Ignore_Table:
                       Last_Errno: 0
                       Last_Error:
                     Skip_Counter: 0
              Exec_Master_Log_Pos: 16497695
                  Relay_Log_Space: 8248152
                  Until_Condition: None
                   Until_Log_File:
                    Until_Log_Pos: 0
               Master_SSL_Allowed: No
               Master_SSL_CA_File:
               Master_SSL_CA_Path:
                  Master_SSL_Cert:
                Master_SSL_Cipher:
                   Master_SSL_Key:
            Seconds_Behind_Master: NULL

mysql> START SLAVE;
Query OK, 0 rows affected (0.01 sec)
```

4. 在服务器 C 上，停止从库的运行，并执行 CHANGE MASTER TO 命令，指向服务器 A。使用在前面的步骤中记下的位置信息（server_A-bin.000023 和 16497695）。最后，启动从库并验证从库的状态：

```
mysql> STOP SLAVE;
Query OK, 0 rows affected (0.07 sec)

mysql> CHANGE MASTER TO MASTER_HOST = 'Server A', MASTER_USER =
'binlog_user', MASTER_PASSWORD = 'binlog_P@ss12',
MASTER_LOG_FILE='server_A-bin.000023', MASTER_LOG_POS=16497695;
Query OK, 0 rows affected, 1 warning (0.02 sec)

mysql> START SLAVE;
```

```
Query OK, 0 rows affected (0.07 sec)
mysql> SHOW SLAVE STATUS\G
*************************** 1. row ***************************
               Slave_IO_State:
                  Master_Host: xx.xxx.xxx.xxx
                  Master_User: binlog_user
                  Master_Port: 3306
                Connect_Retry: 60
              Master_Log_File: server_A-bin.000023
          Read_Master_Log_Pos: 16497695
               Relay_Log_File: server_C-relay-bin.000001
                Relay_Log_Pos: 4
        Relay_Master_Log_File: server_A-bin.000023
             Slave_IO_Running: No
            Slave_SQL_Running: No
~
                 Skip_Counter: 0
          Exec_Master_Log_Pos: 16497695
              Relay_Log_Space: 154
              Until_Condition: None
               Until_Log_File:
                Until_Log_Pos: 0
           Master_SSL_Allowed: No
           Master_SSL_CA_File:
           Master_SSL_CA_Path:
              Master_SSL_Cert:
            Master_SSL_Cipher:
               Master_SSL_Key:
        Seconds_Behind_Master: 0
~
1 row in set (0.00 sec)
```

9.8　设置延迟复制

有时，你需要用一个延迟的从库进行灾难恢复。假设在主库上执行了一条灾难性语句（例如 DROP DATABASE 命令）。你必须使用备份中的时间点恢复（*point-in-time recovery*）

来恢复数据库。这将导致长时间停机，停机的具体时长取决于数据库的大小。为了避免出现这种情况，你可以使用一个延迟的从库（delayed slave），该从库总是比主库延迟一段时间（这个时间是可以配置的）。如果发生了灾难，并且该延迟的从库没有执行这条灾难性语句，则可以先停止从库的运行，再启动从库，一直运行到这条灾难性语句的位置，这样该语句就不会被执行。最后，把该从库提升为主库。

除了在 `CHANGE MASTER TO` 命令中指定 `MASTER_DELAY` 之外，该过程与设置传统复制的过程完全相同。

如何度量延迟

在 MySQL 8.0 之前的版本中，延迟是基于 `Seconds_Behind_Master` 的值来度量的。在 MySQL 8.0 中，延迟是基于写入二进制日志的 `original_commit_timestamp` 和 `immediate_commit_timestamp` 的值来度量的。

`original_commit_timestamp` 是自事务写入（提交）到原主库的二进制日志时以来的微秒数。

`immediate_commit_timestamp` 是自事务写入（提交）到直接主库的二进制日志时以来的微秒数。

9.8.1 如何操作

1. 停止从库的运行：

```
mysql> STOP SLAVE;
Query OK, 0 rows affected (0.06 sec)
```

2. 执行 `CHANGE MASTER TO MASTER_DELAY =` 命令，并启动从库。假设你想要 1 小时的延迟，可以设置 `MASTER_DELAY` 为 `3600` 秒：

```
mysql> CHANGE MASTER TO MASTER_DELAY = 3600;
Query OK, 0 rows affected (0.04 sec)

mysql> START SLAVE;
Query OK, 0 rows affected (0.00 sec)
```

3. 在从库状态中检查以下内容。

   `SQL_Delay`：从库必须延迟于主库的秒数。

   `SQL_Remaining_Delay`：延迟还剩余的秒数。当保持延迟时，这个值是NULL。

   `Slave_SQL_Running_State`：SQL线程的状态。

```
mysql> SHOW SLAVE STATUS\G
*************************** 1. row ***************************
               Slave_IO_State: Waiting for master to send event
                  Master_Host: 35.186.158.188
                  Master_User: repl
                  Master_Port: 3306
                Connect_Retry: 60
              Master_Log_File: server_A-bin.000023
          Read_Master_Log_Pos: 24745149
               Relay_Log_File: server_B-relay-bin.000002
                Relay_Log_Pos: 322
        Relay_Master_Log_File: server_A-bin.000023
             Slave_IO_Running: Yes
            Slave_SQL_Running: Yes
~
                   Last_Errno: 0
                   Last_Error:
                 Skip_Counter: 0
          Exec_Master_Log_Pos: 16497695
              Relay_Log_Space: 8247985
              Until_Condition: None
               Until_Log_File:
                Until_Log_Pos: 0
~
        Seconds_Behind_Master: 52
Master_SSL_Verify_Server_Cert: No
                Last_IO_Errno: 0
                Last_IO_Error:
               Last_SQL_Errno: 0
               Last_SQL_Error:
~
                    SQL_Delay: 3600
```

```
            SQL_Remaining_Delay: 3549
        Slave_SQL_Running_State: Waiting until MASTER_DELAY seconds
after master executed event
             Master_Retry_Count: 86400
                    Master_Bind:
        Last_IO_Error_Timestamp:
       Last_SQL_Error_Timestamp:
~
1 row in set (0.00 sec)
```

请注意，延迟消除后，`Seconds_Behind_Master` 将显示为 0。

9.9 设置 GTID 复制

全局事务标识符（Global Transaction Identifier，GTID）是在程序中创建的唯一标识符，并与主库上提交的每个事务相关联。此标识符是唯一的，不仅在其主库上，在给定的复制设置中的所有数据库上，它都是唯一的。所有事务和所有 GTID 之间都是一对一的映射关系。

GTID 用一对坐标表示，用冒号（:）分隔：

```
GTID = source_id:transaction_id
```

`source_id` 是主库的标识。通常，服务器的 `server_uuid` 选项就代表此标识。`transaction_id` 是一个序列号，由在该服务器上提交事务的顺序决定。例如，提交的第一个事务，`transaction_id` 为 1；在同一个主库上提交的第 10 个事务的 `transaction_id` 为 10。

正如你在前述方法中所见，必须指明二进制日志文件，以及作为复制起点的位置。如果将一个从库的主库切换到另一个主库，尤其是在故障转移期间，就必须从新主库获取位置以同步该从库，这个过程可能会很麻烦。为了避免这些问题，可以使用基于 GTID 的复制，其中 MySQL 使用 GTID 自动检测二进制日志的位置。

9.9.1 如何操作

如果已在服务器之间设置过复制，请按照下列步骤操作。

1. 在 my.cnf 中启用 GTID:

```
shell> sudo vi /etc/my.cnf
[mysqld]
gtid_mode=ON
enforce-gtid-consistency=true
skip_slave_start
```

2. 将主库设置为只读，并确保所有从库都能与主库同步。这一点非常重要，因为主库和从库之间不应该有任何数据不一致:

```
On master
mysql> SET @@global.read_only = ON;

On Slaves (if replication is already setup)
mysql> SHOW SLAVE STATUS\G
```

3. 重新启动所有的从库使 GTID 生效。由于在配置文件中给出了 skip_slave_start，所以只有在执行了 START SLAVE 命令之后，从库才会启动。如果启动从库，它将因为下面的错误而启动失败: The replication receiver thread cannot start because the master has GTID_MODE = OFF and this server has GTID_MODE = ON:

```
shell> sudo systemctl restart mysql
```

4. 重新启动主库。当重新启动主库时，它将以读/写模式开始运行，并开始接受以 GTID 模式写入:

```
shell> sudo systemctl restart mysql
```

5. 执行 CHANGE MASTER TO 命令来设置 GTID 复制:

```
mysql> CHANGE MASTER TO MASTER_HOST = <master_host>, MASTER_PORT =
<port>, MASTER_USER = 'binlog_user', MASTER_PASSWORD =
'binlog_P@ss12', MASTER_AUTO_POSITION = 1;
```

你可能观察到这里没有给出二进制日志文件和日志位置，而是给出了 MASTER_AUTO_POSITION，它会自动找到执行的 GTID。

6. 在所有从库上执行 START SLAVE：

```
mysql> START SLAVE;
```

7. 确认从库正在复制：

```
mysql> SHOW SLAVE STATUS\G
*************************** 1. row ***************************
               Slave_IO_State: Waiting for master to send event
                  Master_Host: xx.xxx.xxx.xxx
                  Master_User: binlog_user
                  Master_Port: 3306
                Connect_Retry: 60
              Master_Log_File: server1-bin.000002
          Read_Master_Log_Pos: 345
               Relay_Log_File: server2-relay-bin.000002
                Relay_Log_Pos: 562
        Relay_Master_Log_File: server1-bin.000002
             Slave_IO_Running: Yes
            Slave_SQL_Running: Yes
              Replicate_Do_DB:
          Replicate_Ignore_DB:
           Replicate_Do_Table:
       Replicate_Ignore_Table:
      Replicate_Wild_Do_Table:
  Replicate_Wild_Ignore_Table:
                   Last_Errno: 0
                   Last_Error:
                 Skip_Counter: 0
          Exec_Master_Log_Pos: 345
              Relay_Log_Space: 770
              Until_Condition: None
               Until_Log_File:
                Until_Log_Pos: 0
           Master_SSL_Allowed: No
           Master_SSL_CA_File:
           Master_SSL_CA_Path:
              Master_SSL_Cert:
```

```
             Master_SSL_Cipher:
                Master_SSL_Key:
         Seconds_Behind_Master: 0
 Master_SSL_Verify_Server_Cert: No
                 Last_IO_Errno: 0
                 Last_IO_Error:
                Last_SQL_Errno: 0
                Last_SQL_Error:
   Replicate_Ignore_Server_Ids:
              Master_Server_Id: 32
                   Master_UUID: b52ef45a-7ff4-11e7-9091-42010a940003
              Master_Info_File: /var/lib/mysql/master.info
                     SQL_Delay: 0
           SQL_Remaining_Delay: NULL
       Slave_SQL_Running_State: Slave has read all relay log;
waiting for more updates
            Master_Retry_Count: 86400
                   Master_Bind:
       Last_IO_Error_Timestamp:
      Last_SQL_Error_Timestamp:
                Master_SSL_Crl:
            Master_SSL_Crlpath:
            Retrieved_Gtid_Set:
b52ef45a-7ff4-11e7-9091-42010a940003:1
             Executed_Gtid_Set:
b52ef45a-7ff4-11e7-9091-42010a940003:1
                 Auto_Position: 1
          Replicate_Rewrite_DB:
                  Channel_Name:
            Master_TLS_Version:
1 row in set (0.00 sec)
```

要了解有关 GTID 的更多信息，请参阅 https://dev.mysql.com/doc/refman/5.6/en/replication-gtids-concepts.html。

9.10 设置半同步复制

默认情况下，复制是异步的。主库不知道写入操作是否已经到达从库。如果主库和从库之间存在延迟，并且主库崩溃，尚未到达从库的那些数据就会丢失。为了解决这种问题，你可以使用半同步复制。

在半同步复制中，主库会一直等待，直到至少有一个从库接收到写入的数据。默认情况下，rpl_semi_sync_master_wait_point 的值是 AFTER_SYNC，这意味着主库将事务同步到二进制日志，再由从库读取使用。

之后，从库向主库发送确认消息，然后主库提交事务并将结果返回给客户端。所以，写入操作能到达中继日志就足够了，从库不需要提交这个事务。你可以将变量 rpl_semi_sync_master_wait_point 更改为 AFTER_COMMIT 来改变此行为。在这种情况下，主库将事务提交给存储引擎，但不会将结果返回给客户端。一旦事务在从库上提交，主库就会收到对事务的确认消息，然后将结果返回给客户端。

如果你希望在更多的从库上确认事务，则可以增加动态变量 rpl_semi_sync_master_ wait_for_slave_count 的值。你还可以设置主库必须等待多少毫秒才能通过动态变量 rpl_semi_sync_master_timeout 获取从库的确认，其默认值是 10 秒。

在完全同步复制中，主库会一直等待，直到所有从库都提交了事务。要实现这一点，你必须使用 Galera Cluster。

9.10.1 如何操作

简单来说，你要在想做半同步复制的主库和所有从库上安装并启用半同步插件。你必须重新启动从库 IO 线程才能使这个变动生效。可以根据你的网络情况和应用程序调整 rpl_semi_sync_master_timeout 的值。可以把它设置为 1 秒。

1. 在主库上，安装 rpl_semi_sync_master 插件：

```
mysql> INSTALL PLUGIN rpl_semi_sync_master SONAME
'semisync_master.so';
Query OK, 0 rows affected (0.86 sec)
```

确认插件已激活：

```
mysql> SELECT PLUGIN_NAME, PLUGIN_STATUS FROM
INFORMATION_SCHEMA.PLUGINS WHERE PLUGIN_NAME LIKE '%semi%';
+----------------------+---------------+
| PLUGIN_NAME          | PLUGIN_STATUS |
+----------------------+---------------+
| rpl_semi_sync_master | ACTIVE        |
+----------------------+---------------+
1 row in set (0.01 sec)
```

2. 在主库上，启用半同步复制并调整超时（例如调整为 1 秒）：

```
mysql> SET @@GLOBAL.rpl_semi_sync_master_enabled=1;
Query OK, 0 rows affected (0.00 sec)

mysql> SHOW VARIABLES LIKE 'rpl_semi_sync_master_enabled';
+------------------------------+-------+
| Variable_name                | Value |
+------------------------------+-------+
| rpl_semi_sync_master_enabled | ON    |
+------------------------------+-------+
1 row in set (0.00 sec)

mysql> SET @@GLOBAL.rpl_semi_sync_master_timeout=1000;
Query OK, 0 rows affected (0.00 sec)

mysql> SHOW VARIABLES LIKE 'rpl_semi_sync_master_timeout';
+------------------------------+-------+
| Variable_name                | Value |
+------------------------------+-------+
| rpl_semi_sync_master_timeout | 1000  |
+------------------------------+-------+
1 row in set (0.00 sec)
```

3. 在从库上，安装 rpl_semi_sync_slave 插件：

```
mysql> INSTALL PLUGIN rpl_semi_sync_slave SONAME
'semisync_slave.so';
Query OK, 0 rows affected (0.22 sec)
```

```
mysql> SELECT PLUGIN_NAME, PLUGIN_STATUS FROM
INFORMATION_SCHEMA.PLUGINS WHERE PLUGIN_NAME LIKE '%semi%';
+---------------------+---------------+
| PLUGIN_NAME         |PLUGIN_STATUS  |
+---------------------+---------------+
| rpl_semi_sync_slave | ACTIVE        |
+---------------------+---------------+
1 row in set (0.08 sec)
```

4. 在从库上，启用半同步复制，并重新启动从库IO线程：

```
mysql> SET GLOBAL rpl_semi_sync_slave_enabled = 1;
Query OK, 0 rows affected (0.00 sec)

mysql> STOP SLAVE IO_THREAD;
Query OK, 0 rows affected (0.02 sec)

mysql> START SLAVE IO_THREAD;
Query OK, 0 rows affected (0.00 sec)
```

5. 你可以通过以下方式监控半同步复制的状态。
   要查找以半同步连接到主库的客户端数量，请执行：

```
mysql> SHOW STATUS LIKE 'Rpl_semi_sync_master_clients';
+------------------------------+-------+
| Variable_name                | Value |
+------------------------------+-------+
| Rpl_semi_sync_master_clients | 1     |
+------------------------------+-------+
1 row in set (0.01 sec)
```

当发生超时且从库赶上来时，主库在异步和半同步复制之间切换。要检查主库在使用的复制类型，请查看 Rpl_semi_sync_master_status 的状态（on 表示半同步，off 表示异步）：

```
mysql> SHOW STATUS LIKE 'Rpl_semi_sync_master_status';
+-----------------------------+-------+
| Variable_name               | Value |
```

```
+-----------------------------+-------+
| Rpl_semi_sync_master_status | ON    |
+-----------------------------+-------+
1 row in set (0.00 sec)
```

可以使用此方法来验证半同步复制。

1. 停止从库的运行：

```
mysql> STOP SLAVE;
Query OK, 0 rows affected (0.01 sec)
```

2. 在主库上随便执行一条语句：

```
mysql> USE employees;
Database changed
mysql> DROP TABLE IF EXISTS employees_test;
Query OK, 0 rows affected, 1 warning (0.00 sec)
```

你会注意到主库已切换到异步复制，因为即使在 1 秒之后（rpl_semi_sync_master_timeout 的值为 1 秒），主库也没有收到从库的任何确认信息：

```
mysql> SHOW STATUS LIKE 'Rpl_semi_sync_master_status';
+-----------------------------+-------+
| Variable_name               | Value |
+-----------------------------+-------+
| Rpl_semi_sync_master_status | ON    |
+-----------------------------+-------+
1row in set (0.00 sec)

mysql> DROP TABLE IF EXISTS employees_test;
Query OK, 0 rows affected (1.02 sec)
mysql> SHOW STATUS LIKE 'Rpl_semi_sync_master_status';
+-----------------------------+-------+
| Variable_name               | Value |
+-----------------------------+-------+
| Rpl_semi_sync_master_status |OFF    |
+-----------------------------+-------+
1 row in set (0.01 sec
```

3. 启动从库：

```
mysql> START SLAVE;
Query OK, 0 rows affected (0.02 sec)
```

4. 在主库上，你会注意到主库已切换回半同步复制：

```
mysql> SHOW STATUS LIKE 'Rpl_semi_sync_master_status';
+-----------------------------+-------+
| Variable_name               | Value |
+-----------------------------+-------+
| Rpl_semi_sync_master_status |ON     |
+-----------------------------+-------+
1 row in set (0.00 sec)
```

第 10 章 表维护

在本章中，我们将介绍以下内容：

- 安装 Percona 工具包
- 修改表结构
- 在数据库之间移动表
- 使用一种在线模式更改工具来修改表
- 归档表
- 克隆表
- 为表分区
- 分区截断和选择
- 分区管理
- 分区信息
- 有效地管理生存时间和软删除行

10.1 引言

维护数据库的一项关键工作就是管理表。通常，你需要修改一个很大的表或克隆（clone）一个表。在本章中，我们将学习如何管理大表（big table）。由于 MySQL 不支持某些操作，因此我们会使用一些开源的第三方工具。本章还介绍了第三方工具的安装和使用。

10.2 安装 Percona 工具包

Percona 工具包是一个高级开源命令行工具集，它是由 Percona 开发和使用的，用于执行各种不易手动执行的困难或复杂的任务。本节将介绍如何安装这个工具包，在后面几节将学习如何使用它。

10.2.1 如何操作

我们来看如何在各种操作系统上安装 Percona 工具包。

在 Debian/Ubuntu 系统上

1. 下载软件包：

```
shell> wget https://repo.percona.com/apt/percona-
release_0.1-4.$(lsb_release -sc)_all.deb
```

2. 安装软件包：

```
shell> sudo dpkg -i percona-release_0.1-4.$(lsb_release -
sc)_all.deb
```

3. 更新本地包列表：

```
shell> sudo apt-get update
```

4. 确保 Percona 软件包可用：

```
shell> apt-cache search percona
```

你应该会看到类似于下面的输出：

```
percona-xtrabackup-dbg - Debug symbols for Percona XtraBackup
percona-xtrabackup-test - Test suite for Percona XtraBackup
percona-xtradb-cluster-client - Percona XtraDB Cluster database
client
percona-xtradb-cluster-server - Percona XtraDB Cluster database
server
percona-xtradb-cluster-testsuite - Percona XtraDB Cluster database
regression test suite
```

```
percona-xtradb-cluster-testsuite-5.5 - Percona Server database test
suite
...
```

5. 安装 percona-toolkit 包：

```
shell> sudo apt-get install percona-toolkit
```

如果你不想安装软件包，也可以直接执行下面的安装：

```
shell> wget
https://www.percona.com/downloads/percona-toolkit/3.0.4/binary/debian/xenia
l/x86_64/percona-toolkit_3.0.4-1.xenial_amd64.deb
shell> sudo dpkg -i percona-toolkit_3.0.4-1.yakkety_amd64.deb;
shell> sudo apt-get install -f
```

在 CentOS/Red Hat/Fedora 系统上

1. 安装软件包：

```
shell> sudo yum install http://www.percona.com/downloads/percona-
release/redhat/0.1-4/percona-release-0.1-4.noarch.rpm
```

如果安装成功，则会看到如下信息：

```
Installed:
  percona-release.noarch 0:0.1-4
Complete!
```

2. 确保 Percona 软件包可用：

```
shell> sudo yum list | grep percona
```

你应该会看到类似于下面的输出：

```
percona-release.noarch                  0.1-4
@/percona-release-0.1-4.noarch
Percona-Server-55-debuginfo.x86_64   5.5.54-rel38.7.el7
percona-release-x86_64
Percona-Server-56-debuginfo.x86_64   5.6.35-rel81.0.el7
percona-release-x86_64
```

```
Percona-Server-57-debuginfo.x86_64    5.7.17-13.1.el7
percona-release-x86_64
...
```

3. 安装 Percona 工具包：

```
shell> sudo yum install percona-toolkit
```

如果不想安装软件包，可以直接使用 YUM 进行安装：

```
shell> sudo yum install
https://www.percona.com/downloads/percona-toolkit/3.0.4/binary/redhat/7/x86
_64/percona-toolkit-3.0.4-1.el7.x86_64.rpm
```

10.3 修改表结构

ALTER TABLE 语句用于改变表的结构。例如，可以添加或删除列、创建或销毁索引、更改现有列的类型，或者对列或这个表重命名。

在执行某些修改（alter）操作时（如更改列的数据类型、添加 SPATIAL INDEX、删除主键、转换字符集、添加/删除加密等），对表的 DML 操作会被阻塞。如果表很大，则需要花费更多的时间来执行 alter 操作，并且在此期间应用程序无法访问表，这是无法接受的。在这种情况下，采用 pt-online-schema 更改（change）就很有用，它允许使用 DML 语句。

修改（alter）表的操作有以下两种算法。

- In-place（默认）：不需要复制整个表的数据。
- Copy：将数据复制到一个临时的磁盘文件中并重新命名。

只有特定的 alter 操作可以在本地完成。在线 DDL 操作的性能很大程度上取决于该操作是否能就地执行，或者需要复制和重建整个表。具体信息请参考 https://dev.mysql.com/doc/refman/8.0/en/innodb-create-index-overview.html#innodb-online-ddl-summary-grid，以查看哪些操作可以就地执行，以及为了避免表复制操作有哪些要求。

可以阅读 *How Copy Algorithm Works*（“复制算法是如何工作的”，摘自 MySQL 文档，

https://dev.mysql.com/doc/refman/8.0/en/altertable.html）。

没有就地（*in-place*）执行的 ALTER TABLE 操作会创建原始表的临时副本。MySQL 等待修改表的其他操作完成后，然后继续运行。它将修改合并到这个副本中，删除原始表，并重命名新的表。当 ALTER TABLE 正在执行时，原始表可供其他会话读。在 ALTER TABLE 操作开始之后，对表开始做的更新和写操作将停止，直到新表准备就绪，然后这些更新和写操作会被自动重定向到这个没有发生过任何更新失败的新表。原始表的临时副本在新表的数据库目录中创建。ALTER TABLE 操作可以改变原始表的数据库目录，该操作将表重命名，放到不同的数据库。

要了解 DDL 操作是在原地执行还是在表的副本中执行，请查看命令执行完以后显示的 rows affected 的值。

- 更改列的默认值（超快，完全不影响表的数据），输出将是这样的：
  ```
  Query OK, 0 rows affected (0.07 sec)
  ```
- 添加索引（需要一点时间，但如果显示 0 rows affected，则表示这个表未被复制），输出将是这样的：
  ```
  Query OK, 0 rows affected (21.42 sec)
  ```
- 更改列的数据类型（要花费大量时间，并且需要重新构建表的所有行），输出将是这样的：
  ```
  Query OK, 1671168 rows affected (1 min 35.54 sec)
  ```

更改列的数据类型时需要重新构建表中的所有行，更改 VARCHAR 大小则例外，它可以使用在线 ALTER TABLE 命令来实现。请参考 10.5 节的示例，这一节展示了如何使用 pt-online-schema 来修改列属性。

10.3.1　如何操作

如果要向 employees 表添加一个新列，可以执行 ADD COLUMN 语句：

```
mysql> ALTER TABLE employees ADD COLUMN address varchar(100);
Query OK, 0 rows affected (5.10 sec)
Records: 0  Duplicates: 0  Warnings: 0
```

你可以看到受影响的行数为 0，这意味着该表没有被复制，并且操作已经就地完成。

如果想要增加 varchar 列的长度，可以执行 MODIFY COLUMN 语句：

```
mysql> ALTER TABLE employees MODIFY COLUMN address VARCHAR(255);
Query OK, 0 rows affected (0.01 sec)
Records: 0  Duplicates: 0  Warnings: 0
```

如果你认为 varchar(255) 不足以存储地址，并且希望将其更改为 tinytext，那么可以使用 MODIFY COLUMN 语句。但是，在本例中，由于你正在修改一个列的数据类型，所以应该修改现有表中的所有行，这需要执行复制表的操作，而且 DML 会被阻塞：

```
mysql> ALTER TABLE employees MODIFY COLUMN address tinytext;
Query OK, 300025 rows affected (4.36 sec)
Records: 300025  Duplicates: 0  Warnings: 0
```

你将注意到受影响的行数是 300025，这就是这个表的大小。

你还可以做许多其他操作，比如重命名一个列、更改默认值、对列位置重新排序等，请参考 MySQL 文档（https://dev.mysql.com/doc/refman/8.0/en/innodb-create-index-overview.html），获取更多详细信息。

添加一个虚拟生成的列不过是一个元数据发生变化，几乎是瞬时完成的：

```
mysql> ALTER TABLE employees ADD COLUMN full_name VARCHAR(40) AS
(CONCAT('first_name', ' ', 'last_name'));
Query OK, 0 rows affected (0.09 sec)
Records: 0  Duplicates: 0  Warnings: 0
```

但是，添加一个 STORED GENERATED 列并修改 VIRTUAL GENERATED 列，则不是在线上完成的：

```
mysql> ALTER TABLE employees MODIFY COLUMN full_name VARCHAR(40) AS
(CONCAT(first_name, '-', last_name)) VIRTUAL;
Query OK, 300026 rows affected (4.37 sec)
Records: 300026  Duplicates: 0  Warnings: 0
```

10.4 在数据库之间移动表

你可以通过执行 RENAME TABLE 语句重命名一个表。

为了让后面的例子能正常运行，先创建示例表和数据库。

```
mysql> CREATE DATABASE prod;
mysql> CREATE TABLE prod.audit_log (id int NOT NULL, msg varchar(64));
mysql> CREATE DATABASE archive;
```

10.4.1　如何操作

例如，如果你想把 audit_log 表重命名为 audit_log_archive_2018，可以执行以下操作：

```
mysql> USE prod;
Database changed

mysql> RENAME TABLE audit_log TO audit_log_archive_2018;
Query OK, 0 rows affected (0.07 sec)
```

如果希望将这张表从一个数据库移到另一个数据库，可以使用点记法指定数据库的名称。例如，如果你希望将 audit_log 表从名为 prod 的数据库移到名为 archive 的数据库，则可以执行以下操作：

```
mysql> USE prod
Reading table information for completion of table and column names
You can turn off this feature to get a quicker startup with -A

mysql> SHOW TABLES;
+------------------------+
| Tables_in_prod         |
+------------------------+
| audit_log_archive_2018 |
+------------------------+
1 row in set (0.00 sec)

mysql> RENAME TABLE audit_log_archive_2018 TO archive.audit_log;
Query OK, 0 rows affected (0.03 sec)

mysql> SHOW TABLES;
Empty set (0.00 sec)
```

```
mysql> USE archive
Reading table information for completion of table and column names
You can turn off this feature to get a quicker startup with -A

Database changed
mysql> SHOW TABLES;
+-------------------+
| Tables_in_archive |
+-------------------+
| audit_log         |
+-------------------+
1 row in set (0.00 sec)
```

10.5 使用在线模式更改工具修改表

在本节中，你将了解 Percona 的 pt-online-schema-change（pt-osc）工具，该工具用于在 DML 未阻塞的情况下执行 ALTER TABLE 操作。

pt-osc 附在 Percona 工具包中。在本章前面已经介绍了 Percona 工具包的安装。

10.5.1 如何运行[1]

pt-online-schema-change 会先创建表的空副本，然后根据需要对其进行修改，再将原始表中的行复制到这个新表中。复制完成后，它将移除原来的表并将其替换为新的表。默认情况下，它也会删除（drop）原始表。

数据的复制是以小块数据为单位执行的，这些数据块的大小可以调整，以便能在指定时间内完成复制。在复制期间对原始表数据的任何修改都将反映在新表中，因为 pt-online-schema-change 会在原始表上创建触发器以更新新表中的相应行。触发器的使用意味着，如果表中已经定义了触发器，pt-online-schema-change 将不起作用。

1 本节内容摘自 https://www.percona.com/doc/percona-toolkit/LATEST/pt-online-schema-change.html。

当 pt-online-schema-change 将数据复制到新表中时，它使用原子操作 RENAME TABLE 来同时重命名原始表和新表。完成此操作后，它将删除原始表。

外键会使这个工具的操作复杂化，并引入额外的风险。当外键指向表时，原子重命名原始表和新表的技术不起作用。在模式更改完成后，pt-online-schema-change 必须更新外键来引用新表。它支持两种实现方法。你可以阅读 --alter-foreign-keys-method 的文档，从中找到更多信息。

10.5.2 如何操作

修改列数据类型可以像如下这般操作：

```
shell> pt-online-schema-change D=employees,t=employees,h=localhost -u root
--ask-pass --alter="MODIFY COLUMN address VARCHAR(100)" --alter-foreign-
keys-method=auto --execute
Enter MySQL password:
No slaves found.  See --recursion-method if host server1 has slaves.
Not checking slave lag because no slaves were found and --check-slave-lag
was not specified.
Operation, tries, wait:
  analyze_table, 10, 1
  copy_rows, 10, 0.25
  create_triggers, 10, 1
  drop_triggers, 10, 1
  swap_tables, 10, 1
  update_foreign_keys, 10, 1
Child tables:
  `employees`.`dept_emp` (approx. 331143 rows)
  `employees`.`titles` (approx. 442605 rows)
  `employees`.`salaries` (approx. 2838426 rows)
  `employees`.`dept_manager` (approx. 24 rows)
Will automatically choose the method to update foreign keys.
Altering `employees`.`employees`...
Creating new table...
Created new table employees._employees_new OK.
Altering new table...
Altered `employees`.`_employees_new` OK.
```

```
2017-09-24T09:56:49 Creating triggers...
2017-09-24T09:56:49 Created triggers OK.
2017-09-24T09:56:49 Copying approximately 299478 rows...
2017-09-24T09:56:56 Copied rows OK.
2017-09-24T09:56:56 Max rows for the rebuild_constraints method: 88074
Determining the method to update foreign keys...
2017-09-24T09:56:56  `employees`.`dept_emp`: too many rows: 331143; must
use drop_swap
2017-09-24T09:56:56 Drop-swapping tables...
2017-09-24T09:56:56 Analyzing new table...
2017-09-24T09:56:56 Dropped and swapped tables OK.
Not dropping old table because --no-drop-old-table was specified.
2017-09-24T09:56:56 Dropping triggers...
2017-09-24T09:56:56 Dropped triggers OK.
Successfully altered `employees`.`employees`.
```

你会注意到，`pt-online-schema-change` 创建了一个列数据类型为改后的类型的新表，在表中创建了触发器，将行复制到新表，最后，重命名这张新表。

如果你想修改（alter）`salaries` 表，而它已经有触发器，那么你需要指定 `--preserver-triggers` 选项。否则，最后将出现错误：

`The table `employees`.`salaries` has triggers but --preserve-triggers was not specified.`

```
shell> pt-online-schema-change D=employees,t=salaries,h=localhost -u user -
-ask-pass --alter="MODIFY COLUMN salary int" --alter-foreign-keys-
method=auto --execute --no-drop-old-table --preserve-triggers
No slaves found.  See --recursion-method if host server1 has slaves.
Not checking slave lag because no slaves were found and --check-slave-lag
was not specified.

Operation, tries, wait:
  analyze_table, 10, 1
  copy_rows, 10, 0.25
  create_triggers, 10, 1
  drop_triggers, 10, 1
  swap_tables, 10, 1
```

```
  update_foreign_keys, 10, 1
No foreign keys reference `employees`.`salaries`; ignoring --alter-foreign-
keys-method.
Altering `employees`.`salaries`...
Creating new table...
Created new table employees._salaries_new OK.
Altering new table...
Altered `employees`.`_salaries_new` OK.
2017-09-24T11:11:58 Creating triggers...
2017-09-24T11:11:58 Created triggers OK.
2017-09-24T11:11:58 Copying approximately 2838045 rows...
2017-09-24T11:12:20 Copied rows OK.
2017-09-24T11:12:20 Adding original triggers to new table.
2017-09-24T11:12:21 Analyzing new table...
2017-09-24T11:12:21 Swapping tables...
2017-09-24T11:12:21 Swapped original and new tables OK.
Not dropping old table because --no-drop-old-table was specified.
2017-09-24T11:12:21 Dropping triggers...
2017-09-24T11:12:21 Dropped triggers OK.
Successfully altered `employees`.`salaries`
```

如果此数据库服务器有从库，那么 pt-online-schema-change 在将现有表复制到新表时可能会使从库产生延时。为了避免这种情况，可以指定--check-slave-lag（默认是启用的）；它会暂停数据的复制，直到此复制的延时小于--max-lag（默认值为 1 秒）。你可以通过传递--max-lag 选项来指定--max-lag 的值。

如果想确保从库的延时不会超过 10 秒，可以设置--max-lag=10：

```
shell> pt-online-schema-change D=employees,t=employees,h=localhost -u user
--ask-pass --alter="MODIFY COLUMN address VARCHAR(100)" --alter-foreign-
keys-method=auto --execute --preserve-triggers --max-lag=10
Enter MySQL password:
Found 1 slaves:
server2 -> xx.xxx.xxx.xx:socket
Will check slave lag on:
server2 -> xx.xxx.xxx.xx:socket
Operation, tries, wait:
```

```
  analyze_table, 10, 1
  copy_rows, 10, 0.25
  create_triggers, 10, 1
  drop_triggers, 10, 1
  swap_tables, 10, 1
  update_foreign_keys, 10, 1
Child tables:
  `employees`.`dept_emp` (approx. 331143 rows)
  `employees`.`titles` (approx. 442605 rows)
  `employees`.`salaries` (approx. 2838426 rows)
  `employees`.`dept_manager` (approx. 24 rows)
Will automatically choose the method to update foreign keys.
Altering `employees`.`employees`...
Creating new table...
Created new table employees._employees_new OK.
Waiting forever for new table `employees`.`_employees_new` to replicate to
ubuntu...
Altering new table...
Altered `employees`.`_employees_new` OK.
2017-09-24T12:00:58 Creating triggers...
2017-09-24T12:00:58 Created triggers OK.
2017-09-24T12:00:58 Copying approximately 299342 rows...
2017-09-24T12:01:05 Copied rows OK.
2017-09-24T12:01:05 Max rows for the rebuild_constraints method: 86446
Determining the method to update foreign keys...
2017-09-24T12:01:05  `employees`.`dept_emp`: too many rows: 331143; must
use drop_swap
2017-09-24T12:01:05 Skipping triggers creation since --no-swap-tables was
specified along with --drop-new-table
2017-09-24T12:01:05 Drop-swapping tables...
2017-09-24T12:01:05 Analyzing new table...
2017-09-24T12:01:05 Dropped and swapped tables OK.
Not dropping old table because --no-drop-old-table was specified.
2017-09-24T12:01:05 Dropping triggers...
2017-09-24T12:01:05 Dropped triggers OK.
Successfully altered `employees`.`employees`.
```

更多有关的详细信息和选项，请参阅 Percona 文档，网址为 https://www.percona.com/doc/percona-toolkit/LATEST/pt-online-schema-change.html。

> pt-online-schema-change 仅在有主键或唯一键时才起作用，否则其执行将失败，并显示以下错误：
>
> The new table `employees`.`_employees_new` does not have a PRIMARY KEY or a unique index which is required for the DELETE trigger.

因此，如果该表没有唯一键，则不能使用 pt-online-schema-change。

10.6　归档表

有时，你不想保留旧数据并希望删除它。如果你想删除一个月前最后一次访问的所有行，如果表很小（< 10,000 行），则可以直接使用以下方法：

```
DELETE FROM <TABLE> WHERE last_accessed<DATE_ADD(NOW(), INTERVAL -1 MONTH)
```

如果表很大,会怎样？你应该知道 InnoDB 创建了一个 UNDO 日志来恢复失效的事务。因此，所有被删除的行都被保存在 UNDO 日志空间中，以便在 DELETE 语句的执行中止时恢复它们。不幸的是，如果 DELETE 语句在执行时被中止，InnoDB 将从 UNDO 日志空间将行复制到表中，这可能使表无法访问。

为了克服这种弊端，可以限制删除的行数然后提交事务，循环做这个操作，直到删除所有不需要的行。

以下是一个伪代码示例：

```
WHILE count<=0:
    DELETE FROM <TABLE> WHERE last_accessed<DATE_ADD(NOW(), INTERVAL -1
MONTH) LIMIT 10000;
    count=SELECT COUNT(*) FROM <TABLE> WHERE last_accessed<DATE_ADD(NOW(),
INTERVAL -1 MONTH);
```

如果 last_accessed 上没有索引，则会锁定该表。在这种情况下，需要找到要删除的行的主键，并且基于主键删除这些行。

下面是伪代码（假设 id 是主键）：

```
WHILE count<=0:
    SELECT id FROM <TABLE> WHERE last_accessed < DATE_ADD(NOW(), INTERVAL
-1 MONTH) LIMIT 10000;
    DELETE FROM <TABLE> WHERE id IN ('ids from above statement');
    count=SELECT COUNT(*) FROM <TABLE> WHERE
last_accessed<DATE_ADD(NOW(), INTERVAL -1 MONTH);
```

你可以使用 Percona 的 pt-archiver 工具，而不是编写删除行的代码，前者做的事情在本质上与后者是一样的，并且提供了许多其他选项，比如将行保存到另一个表或文件中，对加载和复制延迟进行精细的控制，等等。

10.6.1 如何操作

在 pt-archiver 中有许多选项，我们将从简单的清除操作（purge）开始。

清除数据

如果你想要删除 employees 表中 hire_date 超过 30 年的所有行，可以执行以下操作：

```
shell> pt-archiver --source h=localhost,D=employees,t=employees -u <user> -
p<pass> --where="hire_date<DATE_ADD(NOW(), INTERVAL -30 YEAR)" --no-check-
charset --limit 10000 --commit-each
```

你可以通过--source 选项传递主机名、数据库名和表名，可以使用--limit 选项限制在批处理操作中删除的行数。

如果指定--progress，其输出就是一个标题行，外加时不时的状态输出。状态输出中的每一行都会列出当前的日期和时间、pt-archiver 已经运行了多少秒，以及它已经归档了多少行。

如果指定--statistics，pt-archiver 会输出定时时间（timing）和其他信息，以帮助确定归档过程的哪一部分花的时间最多。

如果指定--check-slave-lag，pt-archiver 将暂停归档，直到从库的延时少于--max-lag。

归档数据

如果你想将删除后剩余的行保存到一个单独的表或文件中，可以指定--dest 选项。

假设你想将 employees 数据库中 employees 表的所有行移到 employees_archive 表，则可以执行以下操作：

```
shell> pt-archiver --source h=localhost,D=employees,t=employees --dest
h=localhost,D=employees_archive -u <user> -p<pass> --where="1=1" --no-
check-charset --limit 10000 --commit-each
```

如果指定--where="1=1"，它将复制所有的行。

复制数据

如果你想将数据从一个表复制到另一个表，可以使用 mysqldump 或 mysqlpump 来备份某些行，然后将它们加载到目标表中。你还可以使用 pt-archive 来复制数据。如果指定--no-delete 选项，pt-archiver 将不会从源表中删除行：

```
shell> pt-archiver --source h=localhost,D=employees,t=employees --dest
h=localhost,D=employees_archive -u <user> -p<pass> --where="1=1" --no-
check-charset --limit 10000 --commit-each --no-delete
```

10.6.2　延伸阅读

有关 pt-archiver 的更多细节和选项，请参考 https://www.percona.com/doc/percona-toolkit/LATEST/pt-archiver.html。

10.7　克隆表

如果你想克隆一个表，有许多方法可以选择。

10.7.1 如何操作

1. 使用 INSERT INTO SELECT 语句：

```
mysql> CREATE TABLE employees_clone LIKE employees;
mysql> INSERT INTO employees_clone SELECT * FROM employees;
```

请注意，如果有任何生成的列，上述语句将不起作用。在这种情况下，应该给出完整的 insert 语句，但要排除生成的列。

```
mysql> INSERT INTO employees_clone SELECT * FROM employees;
ERROR 3105 (HY000): The value specified for generated column
'hire_date_year' in table 'employees_clone' is not allowed.
mysql> INSERT INTO employees_clone(emp_no, birth_date, first_name,
last_name, gender, hire_date) SELECT emp_no, birth_date, first_name,
last_name, gender, hire_date FROM employees;
Query OK, 300024 rows affected (3.21 sec)
Records: 300024  Duplicates: 0  Warnings: 0
```

但是，在大表上执行上面的语句是非常慢且危险的。请记住，如果语句执行失败，为了恢复表的状态，InnoDB 会将所有的行保存在 UNDO 日志中。

2. 使用 mysqldump 或 mysqlpump，并对单个表进行备份，在目标位置恢复它。如果表很大，这个操作可能需要很长时间。

3. 使用 Innobackupex 对特定的表进行备份，并将数据文件恢复到目标位置。

4. 在使用 pt-archiver 时带上--no-delete 的选项，它会将所需的行或所有行复制到目标表。

你还可以使用可传输的表空间来克隆表，我们将在第 11 章的 11.5 节中详细讲解。

10.7.2 为表分区

你可以使用分区将单个表的各部分分散到一个文件系统中。用户所选择的划分数据的规则被称为分区函数（partitioning function），它可以是模量（modulu），与一组范围或值列表、一个内部哈希函数或一个线性哈希函数简单匹配。

表的不同行可以被分配给不同的物理分区，称为水平分区。MySQL 不支持垂直分区，在垂直分区中，表的不同列被分配给不同的物理分区。

对一个表做分区有很多种方法。

- RANGE：这种类型的分区根据落在给定范围内的列值，将行分配给分区。
- LIST：类似于按 RANGE 分区，不同的是其分区是基于与一组离散值匹配的列来选择的。
- HASH：在这种类型的分区操作中，一个分区是根据用户定义的表达式返回的值来选择的，该表达式对插入到表的行中的列值进行操作。HASH 函数可以包含任何在 MySQL 中具有非负整数值的有效表达式。
- KEY：这种类型的分区类似于 HASH 分区，只是它仅提供一个或多个列，而且 MySQL 服务器提供自己的哈希函数。这些列可以包含除整数值以外的其他值，因为 MySQL 提供的哈希函数保证不管列数据是什么类型，结果都为整数。

上述每一个分区类型都有一个扩展。RANGE 的扩展为 RANGE COLUMNS，LIST 的扩展为 LIST COLUMNS，HASH 的扩展为 LINEAR HASH，KEY 的扩展为 LINEAR KEY。

对于 [LINEAR] KEY、RANGE COLUMNS 和 LIST COLUMNS 分区，分区表达式包含一个或多个列的列表。

在 RANGE、LIST 和[LINEAR] HASH 分区中，分区的列的值被传递给分区函数，该函数返回一个整数值，表示该特定记录应该被存储在第几个分区。这个函数的返回值必须为既非常数也非随机数。

数据库分区的一个非常常见的用途是按日期分隔数据。

有关分区的优点和其他细节信息，请参阅 https://dev.mysql.com/doc/refman/8.0/en/partitioning-overview.html。

请注意，分区仅适用于 InnoDB 表，并且外键不能与分区结合使用。

10.7.3　如何操作

你可以在创建表时指定分区，也可以通过执行 ALTER TABLE 命令来指定分区。分区列应该是表中所有唯一键的一部分。

如果基于 created_at 列定义了分区，并且 id 是主键，则应该将 create_at 列作为 PRIMARY KEY 的一部分包含在内，即（id，created_at）。

以下示例假定该表的外键没有被引用。

如果希望在 MySQL 8.0 中基于范围或时间间隔实施分区计划，有两种选择：

- 按 RANGE 对表进行分区。对于分区表达式，应用在 DATE、TIME 或 DATETIME 列上运行的函数，并返回一个整数值。
- 使用 DATE 或 DATETIME 列作为分区列，通过 RANGE COLUMNS 对表进行分区。

RANGE 分区

如果想要根据 emp_no 对 employees 表分区，并希望在一个分区中保留 100,000 名员工，可以这样创建分区：

```
mysql> CREATE TABLE `employees` (
  `emp_no` int(11) NOT NULL,
  `birth_date` date NOT NULL,
  `first_name` varchar(14) NOT NULL,
  `last_name` varchar(16) NOT NULL,
  `gender` enum('M','F') NOT NULL,
  `hire_date` date NOT NULL,
  `address` varchar(100) DEFAULT NULL,
  PRIMARY KEY (`emp_no`),
  KEY `name` (`first_name`,`last_name`)
) ENGINE=InnoDB DEFAULT CHARSET=utf8mb4
PARTITION BY RANGE (emp_no)
(PARTITION p0 VALUES LESS THAN (100000) ENGINE = InnoDB,
 PARTITION p1 VALUES LESS THAN (200000) ENGINE = InnoDB,
 PARTITION p2 VALUES LESS THAN (300000) ENGINE = InnoDB,
 PARTITION p3 VALUES LESS THAN (400000) ENGINE = InnoDB,
 PARTITION p4 VALUES LESS THAN (500000) ENGINE = InnoDB);
```

因此，所有 emp_no 小于 100,000 的员工将被划入分区 p0，所有 emp_no 小于 200,000 和大于 100,000 的员工将被划入分区 p1，依此类推。

如果员工编号大于 500,000，因为没有为这些编号定义分区，所以这时插入操作将会失败，并报错。为了避免这种情况，必须定期检查并添加分区或创建一个 MAXVALUE 分区，以捕获所有类似这种情况的异常：

```
mysql> CREATE TABLE `employees` (
  `emp_no` int(11) NOT NULL,
  `birth_date` date NOT NULL,
  `first_name` varchar(14) NOT NULL,
  `last_name` varchar(16) NOT NULL,
  `gender` enum('M','F') NOT NULL,
  `hire_date` date NOT NULL,
  `address` varchar(100) DEFAULT NULL,
  PRIMARY KEY (`emp_no`),
  KEY `name` (`first_name`,`last_name`)
) ENGINE=InnoDB DEFAULT CHARSET=utf8mb4
PARTITION BY RANGE (emp_no)
(PARTITION p0 VALUES LESS THAN (100000) ENGINE = InnoDB,
 PARTITION p1 VALUES LESS THAN (200000) ENGINE = InnoDB,
 PARTITION p2 VALUES LESS THAN (300000) ENGINE = InnoDB,
 PARTITION p3 VALUES LESS THAN (400000) ENGINE = InnoDB,
 PARTITION p4 VALUES LESS THAN (500000) ENGINE = InnoDB,
 PARTITION pmax VALUES LESS THAN MAXVALUE ENGINE = InnoDB
);
```

如果你想基于 hire_date 分区，可以使用 YEAR(hire_date) 函数作为分区表达式：

```
mysql> CREATE TABLE `employees` (
  `emp_no` int(11) NOT NULL,
  `birth_date` date NOT NULL,
  `first_name` varchar(14) NOT NULL,
  `last_name` varchar(16) NOT NULL,
  `gender` enum('M','F') NOT NULL,
  `hire_date` date NOT NULL,
  `address` varchar(100) DEFAULT NULL,
  PRIMARY KEY (`emp_no`,`hire_date`),
  KEY `name` (`first_name`,`last_name`)
) ENGINE=InnoDB DEFAULT CHARSET=utf8mb4
PARTITION BY RANGE (YEAR(hire_date))
```

```
(PARTITION p1980 VALUES LESS THAN (1980) ENGINE = InnoDB,
 PARTITION p1990 VALUES LESS THAN (1990) ENGINE = InnoDB,
 PARTITION p2000 VALUES LESS THAN (2000) ENGINE = InnoDB,
 PARTITION p2010 VALUES LESS THAN (2010) ENGINE = InnoDB,
 PARTITION p2020 VALUES LESS THAN (2020) ENGINE = InnoDB,
 PARTITION pmax VALUES LESS THAN MAXVALUE ENGINE = InnoDB
);
```

MySQL 中的分区被广泛应用于 date、datetime 或 timestamp 列。如果你想在数据库中存储一些事件，并且所有的查询都基于一个时间范围，则可以像这样使用分区。

分区函数 to_days()返回自 0000-01-01 以来的天数，这是一个整数值：

```
mysql> CREATE TABLE `event_history` (
  `event_id` int(11) NOT NULL,
  `event_name` varchar(10) NOT NULL,
  `created_at` datetime NOT NULL,
  `last_updated` timestamp DEFAULT CURRENT_TIMESTAMP ON UPDATE
CURRENT_TIMESTAMP,
  `event_type` varchar(10) NOT NULL,
  `msg` tinytext NOT NULL,
  PRIMARY KEY (`event_id`,`created_at`)
) ENGINE=InnoDB DEFAULT CHARSET=utf8mb4
PARTITION BY RANGE (to_days(created_at))
(PARTITION p20170930 VALUES LESS THAN (736967) ENGINE = InnoDB,
PARTITION p20171001 VALUES LESS THAN (736968) ENGINE = InnoDB,
PARTITION p20171002 VALUES LESS THAN (736969) ENGINE = InnoDB,
PARTITION p20171003 VALUES LESS THAN (736970) ENGINE = InnoDB,
PARTITION p20171004 VALUES LESS THAN (736971) ENGINE = InnoDB,
PARTITION p20171005 VALUES LESS THAN (736972) ENGINE = InnoDB,
PARTITION p20171006 VALUES LESS THAN (736973) ENGINE = InnoDB,
PARTITION p20171007 VALUES LESS THAN (736974) ENGINE = InnoDB,
PARTITION p20171008 VALUES LESS THAN (736975) ENGINE = InnoDB,
PARTITION p20171009 VALUES LESS THAN (736976) ENGINE = InnoDB,
PARTITION p20171010 VALUES LESS THAN (736977) ENGINE = InnoDB,
PARTITION p20171011 VALUES LESS THAN (736978) ENGINE = InnoDB,
PARTITION p20171012 VALUES LESS THAN (736979) ENGINE = InnoDB,
PARTITION p20171013 VALUES LESS THAN (736980) ENGINE = InnoDB,
```

```
PARTITION p20171014 VALUES LESS THAN (736981) ENGINE = InnoDB,
PARTITION p20171015 VALUES LESS THAN (736982) ENGINE = InnoDB,
PARTITION pmax VALUES LESS THAN MAXVALUE ENGINE = InnoDB
);
```

如果希望将现有的一个表转换为分区的表，如果分区键不是主键的一部分，则需要删除（drop）主键，并将分区键作为主键和所有唯一键的一部分添加进来。否则，你将收到一条报错的消息“ERROR 1503 (HY000): A PRIMARY KEY must include all columns in the table's partitioning function..”。你可以这样做：

```
mysql> ALTER TABLE employees DROP PRIMARY KEY, ADD PRIMARY
  KEY(emp_no,hire_date);
  Query OK, 0 rows affected (0.11 sec)
  Records: 0  Duplicates: 0  Warnings: 0

mysql> ALTER TABLE employees PARTITION BY RANGE (YEAR(hire_date))
        (PARTITION p1980 VALUES LESS THAN (1980) ENGINE = InnoDB,
        PARTITION p1990 VALUES LESS THAN (1990) ENGINE = InnoDB,
        PARTITION p2000 VALUES LESS THAN (2000) ENGINE = InnoDB,
        PARTITION p2010 VALUES LESS THAN (2010) ENGINE = InnoDB,
        PARTITION p2020 VALUES LESS THAN (2020) ENGINE = InnoDB,
        PARTITION pmax VALUES LESS THAN MAXVALUE ENGINE = InnoDB
      );
Query OK, 300025 rows affected (4.71 sec)
Records: 300025  Duplicates: 0  Warnings: 0
```

有关 RANGE 分区的更多细节，请参阅 https://dev.mysql.com/doc/refman/8.0/en/partitionrange.html。

删除分区

如果希望删除分区，可以执行 REMOVE PARTITIONING 语句：

```
mysql> ALTER TABLE employees REMOVE PARTITIONING;
Query OK, 0 rows affected (0.09 sec)
Records: 0  Duplicates: 0  Warnings: 0
```

RANGE COLUMNS 分区

RANGE COLUMNS 分区类似于 RANGE 分区，但是它允许使用基于多个列值的范围来定义分区。此外，你可以使用非整数类型的列来定义范围。RANGE COLUMNS 分区与 RANGE 分区在以下几方面有很大的区别：

- RANGE COLUMNS 不接受表达式，只接受列的名称。
- RANGE COLUMNS 接受一个或多个列的列表。
- RANGE COLUMNS 的分区列不限于整数列；字符串、DATE 和 DATETIME 列也可以用作分区列。

在 RANGE COLUMNS 中可以直接使用 hire_date 列，而不是使用 to_days()或 year()函数。

```
mysql> ALTER TABLE employees
    PARTITION BY RANGE COLUMNS (hire_date)
    (PARTITION p0 VALUES LESS THAN ('1970-01-01'),
     PARTITION p1 VALUES LESS THAN ('1980-01-01'),
     PARTITION p2 VALUES LESS THAN ('1990-01-01'),
     PARTITION p3 VALUES LESS THAN ('2000-01-01'),
     PARTITION p4 VALUES LESS THAN ('2010-01-01'),
     PARTITION p5 VALUES LESS THAN (MAXVALUE)
    );
Query OK, 300025 rows affected (4.71 sec)
Records: 300025  Duplicates: 0  Warnings: 0
```

或者你可以根据员工的姓氏（last_name）来划分员工。这么做不能保证员工在各个分区之间均匀分布：

```
mysql> ALTER TABLE employees
PARTITION BY RANGE COLUMNS (last_name)
    (PARTITION p0 VALUES LESS THAN ('b'),
     PARTITION p1 VALUES LESS THAN ('f'),
     PARTITION p2 VALUES LESS THAN ('l'),
     PARTITION p3 VALUES LESS THAN ('q'),
     PARTITION p4 VALUES LESS THAN ('u'),
     PARTITION p5 VALUES LESS THAN ('z')
```

```
);
Query OK, 300025 rows affected (4.71 sec)
Records: 300025  Duplicates: 0  Warnings: 0
```

使用 RANGE COLUMNS，可以在分区函数中放置多个列：

```
mysql> CREATE TABLE range_columns_example (
    a INT,
    b INT,
    c INT,
    d INT,
    e INT,
    PRIMARY KEY(a, b, c)
)
PARTITION BY RANGE COLUMNS(a,b,c) (
    PARTITION p0 VALUES LESS THAN (0,25,50),
    PARTITION p1 VALUES LESS THAN (10,50,100),
    PARTITION p2 VALUES LESS THAN (10,100,200),
    PARTITION p3 VALUES LESS THAN (MAXVALUE,MAXVALUE,MAXVALUE)
);
```

如果插入值 a = 10，b = 20，c = 100，d = 100，e = 100，则它将被分配到 p1 分区。在通过 RANGE COLUMNS 设计表分区时，你可以利用 mysql 客户端比较所需的元组来测试连续分区的定义，如下所示：

```
mysql> SELECT (10,20,100) < (0,25,50) p0, (10,20,100) < (10,50,100) p1,
(10,20,100) < (10,100,200) p2;
+----+----+----+
| p0 | p1 | p2 |
+----+----+----+
| 0 | 1| 1     |
+----+----+----+
1 row in set (0.00 sec)
```

在本例中，插入语句将被分配到 p1 分区。

LIST 和 LIST COLUMNS 分区

LIST 分区与 RANGE 分区类似，其每个分区都是根据一组值列表中的一个列值的成员来定义和选择的，而不是在一组连续的值范围内进行。

你需要通过 PARTITION BY LIST（<expr>）来定义它，其中 expr 是一个列值或基于列值的表达式，并返回一个整数值。

分区定义包含 VALUES IN (<value_list>)，其中 value_list 是一个用逗号分隔的整数列表，而不是 VALUES LESS THAN (<value>)。

如果希望使用除整数以外的数据类型，可以使用 LIST COLUMNS。

与 RANGE 分区的情况不同，这里没有像 MAXVALUE 之类的 catch-all，分区表达式期望的所有值都应该包含在 PARTITION 表达式中。

假设有一张带有邮政编码和城市信息的客户表。例如，如果你想在分区中按照特定的邮政编码来划分客户，可以使用 LIST 分区：

```
mysql> CREATE TABLE customer (
customer_id INT,
zipcode INT,
city varchar(100),
PRIMARY KEY (customer_id, zipcode)
)
PARTITION BY LIST(zipcode) (
  PARTITION pnorth VALUES IN (560030, 560007, 560051, 560084),
  PARTITION peast VALUES IN (560040, 560008, 560061, 560085),
  PARTITION pwest VALUES IN (560050, 560009, 560062, 560086),
  PARTITION pcentral VALUES IN (560060, 560010, 560063, 560087)
);
```

如果希望直接使用列而不是整数，则可以使用 LIST COLUMNS 来分区：

```
mysql> CREATE TABLE customer (
customer_id INT,
zipcode INT,
city varchar(100),
PRIMARY KEY (customer_id, city)
```

```
)
PARTITION BY LIST COLUMNS(city) (
   PARTITION pnorth VALUES IN ('city1','city2','city3'),
   PARTITION peast VALUES IN ('city4','city5','city6'),
   PARTITION pwest VALUES IN ('city7','city8','city9'),
   PARTITION pcentral VALUES IN ('city10','city11','city12')
);
```

HASH 和 LINEAR HASH 分区

HASH分区主要是为了确保数据均匀地分布在数量预先确定的一组分区中。使用RANGE 或 LIST 分区的话，必须明确指定应该将给定的列值或列值集合存储在哪一个分区中；而如果使用 HASH 分区，这个决定将由你来做，你只需要根据要进行哈希的列值指定一个列值或表达式，以及分区表要分为多少个分区即可。

如果你希望员工在分区中均匀地分布，可以指定分区的数量，并根据 YEAR(hire_date)进行 HASH 分区，而不是根据 YEAR(hire_date)进行 RANGE 分区。当使用 PARTITION BY HASH 时，存储引擎会根据该表达式结果的模来确定要使用哪一个分区。

例如，如果 hire_date 是 1987-11-28，YEAR（hire_date）将是 1987，MOD（1987,8）的结果是 3，所以这一行将被分到第三个分区：

```
mysql> CREATE TABLE `employees` (
  `emp_no` int(11) NOT NULL,
  `birth_date` date NOT NULL,
  `first_name` varchar(14) NOT NULL,
  `last_name` varchar(16) NOT NULL,
  `gender` enum('M','F') NOT NULL,
  `hire_date` date NOT NULL,
  `address` varchar(100) DEFAULT NULL,
  PRIMARY KEY (`emp_no`,`hire_date`),
  KEY `name` (`first_name`,`last_name`)
) ENGINE=InnoDB DEFAULT CHARSET=utf8mb4
PARTITION BY HASH(YEAR(hire_date))
PARTITIONS 8;
```

效率最高的哈希函数是对单个列进行操作的函数，其值与该列值同步地增加或减少。

在 LINEAR HASH 分区中，可以使用相同的语法，只不过要添加 LINEAR 关键字。MySQL 不使用 MODULUS 操作，而是使用 2 的幂算法来确定分区。更多有关的详细信息，请参阅 https://dev.mysql.com/doc/refman/8.0/en/partitioning-linear-hash.html。

```
mysql> CREATE TABLE `employees` (
  `emp_no` int(11) NOT NULL,
  `birth_date` date NOT NULL,
  `first_name` varchar(14) NOT NULL,
  `last_name` varchar(16) NOT NULL,
  `gender` enum('M','F') NOT NULL,
  `hire_date` date NOT NULL,
  `address` varchar(100) DEFAULT NULL,
  PRIMARY KEY (`emp_no`,`hire_date`),
  KEY `name` (`first_name`,`last_name`)
) ENGINE=InnoDB DEFAULT CHARSET=utf8mb4
PARTITION BY LINEAR HASH(YEAR(hire_date))
PARTITIONS 8;
```

KEY 和 LINEAR KEY 分区

KEY 分区与 HASH 分区类似，不同之处在于，HASH 分区使用用户定义的表达式，KEY 分区的哈希函数由 MySQL 服务器提供。这个内部哈希函数采用的是与 PASSWORD() 函数相同的算法。

KEY 仅包含零个或几个列名称的列表。如果表有主键的话，则用作 KEY 分区的任何列都必须是主键的一部分或全部。如果没有列名可以被指定为分区键，则使用表的主键（如果有的话）：

```
mysql> CREATE TABLE `employees` (
  `emp_no` int(11) NOT NULL,
  `birth_date` date NOT NULL,
  `first_name` varchar(14) NOT NULL,
  `last_name` varchar(16) NOT NULL,
  `gender` enum('M','F') NOT NULL,
```

```
  `hire_date` date NOT NULL,
  `address` varchar(100) DEFAULT NULL,
PRIMARY KEY (`emp_no`,`hire_date`),
KEY `name` (`first_name`,`last_name`)
) ENGINE=InnoDB DEFAULT CHARSET=utf8mb4
PARTITION BY KEY()
PARTITIONS 8;
```

子分区

你可以将每个分区进一步划分为一个分区表，称为**子分区**或**复合分区**：

```
mysql> CREATE TABLE `employees` (
  `emp_no` int(11) NOT NULL,
  `birth_date` date NOT NULL,
  `first_name` varchar(14) NOT NULL,
  `last_name` varchar(16) NOT NULL,
  `gender` enum('M','F') NOT NULL,
  `hire_date` date NOT NULL,
  `address` varchar(100) DEFAULT NULL,
  PRIMARY KEY (`emp_no`,`hire_date`),
  KEY `name` (`first_name`,`last_name`)
) ENGINE=InnoDB DEFAULT CHARSET=utf8mb4
PARTITION BY RANGE( YEAR(hire_date) )
  SUBPARTITION BY HASH(emp_no)
    SUBPARTITIONS 4 (
        PARTITION p0 VALUES LESS THAN (1990),
        PARTITION p1 VALUES LESS THAN (2000),
        PARTITION p2 VALUES LESS THAN (2010),
        PARTITION p3 VALUES LESS THAN (2020),
        PARTITION p4 VALUES LESS THAN MAXVALUE
    );
```

10.8 分区修剪和指定

MySQL不扫描没有匹配值的分区，这是自动的操作，称为分区修剪(partition pruning)。对给定的值，MySQL优化器会计算分区表达式，以确定哪个分区包含该值，并且只扫描这个分区。

SELECT、DELETE 和 UPDATE 语句支持分区修剪。INSERT 语句目前不能被裁剪。

你还可以显式地指定用于匹配 WHERE 条件的行的分区和子分区。

10.8.1 如何操作

分区修剪只适用于查询语句，但查询语句和许多 DML 语句都支持显示地选择分区。

分区修剪

以 employees 表为例，该表基于 emp_no 做了分区：

```
mysql> CREATE TABLE `employees` (
  `emp_no` int(11) NOT NULL,
  `birth_date` date NOT NULL,
  `first_name` varchar(14) NOT NULL,
  `last_name` varchar(16) NOT NULL,
  `gender` enum('M','F') NOT NULL,
  `hire_date` date NOT NULL,
  `address` varchar(100) DEFAULT NULL,
  PRIMARY KEY (`emp_no`,`hire_date`),
  KEY `name` (`first_name`,`last_name`)
) ENGINE=InnoDB DEFAULT CHARSET=utf8mb4
PARTITION BY RANGE (YEAR(hire_date))
(PARTITION p1980 VALUES LESS THAN (1980) ENGINE = InnoDB,
 PARTITION p1990 VALUES LESS THAN (1990) ENGINE = InnoDB,
 PARTITION p2000 VALUES LESS THAN (2000) ENGINE = InnoDB,
 PARTITION p2010 VALUES LESS THAN (2010) ENGINE = InnoDB,
 PARTITION p2020 VALUES LESS THAN (2020) ENGINE = InnoDB,
 PARTITION pmax VALUES LESS THAN MAXVALUE ENGINE = InnoDB
);
```

假设执行以下 SELECT 查询：

```
mysql> SELECT last_name,birth_date FROM employees WHERE
hire_date='1999-02-01' AND first_name='Mariangiola';
```

MySQL 优化器检测到在查询中使用了分区列，会自动确定要扫描的分区。

在这个查询中，它首先计算 YEAR('1999-02-01')，即 1999 年，然后扫描 p2000 分区而不是整个表。这就大大减少了查询时间。

如果给定的范围是 hire_date>='1999-02-01'，而不是 hire_date='1999-02-01'，则 MySQL 会对分区 p2000、p2010、p2020 和 pmax 进行扫描。

如果在 WHERE 子句中没有给出表达式 hire_date='1999-02-01'，那么 MySQL 必须扫描整个表。

要想知道优化器扫描的是哪个分区，可以执行查询的 explain 计划（将在第 13 章的 13.1 节中解释）：

```
mysql> EXPLAIN SELECT last_name,birth_date FROM employees WHERE
hire_date='1999-02-01' AND first_name='Mariangiola'\G
*************************** 1. row ***************************
           id: 1
  select_type: SIMPLE
        table: employees
   partitions: p2000
         type: ref
possible_keys: name
          key: name
      key_len: 58
          ref: const
         rows: 120
     filtered: 10.00
        Extra: Using index condition

mysql> EXPLAIN SELECT last_name,birth_date FROM employees WHERE
hire_date>='1999-02-01' AND first_name='Mariangiola'\G
*************************** 1. row ***************************
           id: 1
  select_type: SIMPLE
        table: employees
   partitions: p2000,p2010,p2020,pmax
         type: ref
```

```
possible_keys: name
          key: name
      key_len: 58
          ref: const
         rows: 121
     filtered: 33.33
        Extra: Using index condition
1 row in set, 1 warning (0.00 sec)
```

指定分区

分区修剪是基于 WHERE 子句的自动选择。你可以在查询中显式地指定要扫描的分区。这些查询可以是 SELECT、DELETE、INSERT、REPLACE、UPDATE、LOAD DATA 和 LOAD XML。PARTITION 选项用于从给定的表中选择分区，你应该在所有其他选项之前、表名（包括所有表别名）之后，指定关键字 PARTITION<分区名>。例如：

```
mysql> SELECT emp_no,hire_date FROM employees PARTITION (p1990) LIMIT 10;
+--------+------------+
| emp_no | hire_date  |
+--------+------------+
| 413688 | 1989-12-10 |
| 242368 | 1989-08-06 |
| 283280 | 1985-11-22 |
| 405098 | 1985-11-16 |
|  30404 | 1985-07-17 |
| 419259 | 1988-03-21 |
| 466254 | 1986-11-28 |
| 428971 | 1986-12-13 |
|  94467 | 1987-01-28 |
| 259555 | 1987-07-30 |
+--------+------------+
10 rows in set (0.00 sec)
```

同样，我们可以在 DELETE 中使用 WHERE 子句指定分区：

```
mysql> DELETE FROM employees PARTITION (p1980, p1990) WHERE first_name LIKE
'j%';
Query OK, 7001 rows affected (0.12 sec)
```

10.9 管理分区

当涉及管理分区时，最重要的事情就是在基于时间的 RANGE 分区中事先添加足够数量的分区。如果不这样做，就会导致插入错误；或者，如果定义了 MAXVALUE 分区，所有的插入操作就都会在 MAXVALUE 分区执行。例如，在没有 pmax 分区的情况下使用 event_history 表：

```
mysql> CREATE TABLE `event_history` (
  `event_id` int(11) NOT NULL,
 `event_name` date NOT NULL,
 `created_at` datetime NOT NULL,
 `last_updated` timestamp DEFAULT CURRENT_TIMESTAMP ON UPDATE
CURRENT_TIMESTAMP,
  `event_type` varchar(10) NOT NULL,
  `msg` tinytext NOT NULL,
  PRIMARY KEY (`event_id`,`created_at`)
) ENGINE=InnoDB DEFAULT CHARSET=utf8mb4
PARTITION BY RANGE (to_days(created_at))
(PARTITION p20170930 VALUES LESS THAN (736967) ENGINE = InnoDB,
PARTITION p20171001 VALUES LESS THAN (736968) ENGINE = InnoDB,
PARTITION p20171002 VALUES LESS THAN (736969) ENGINE = InnoDB,
PARTITION p20171003 VALUES LESS THAN (736970) ENGINE = InnoDB,
PARTITION p20171004 VALUES LESS THAN (736971) ENGINE = InnoDB,
PARTITION p20171005 VALUES LESS THAN (736972) ENGINE = InnoDB,
PARTITION p20171006 VALUES LESS THAN (736973) ENGINE = InnoDB,
PARTITION p20171007 VALUES LESS THAN (736974) ENGINE = InnoDB,
PARTITION p20171008 VALUES LESS THAN (736975) ENGINE = InnoDB,
PARTITION p20171009 VALUES LESS THAN (736976) ENGINE = InnoDB,
PARTITION p20171010 VALUES LESS THAN (736977) ENGINE = InnoDB,
PARTITION p20171011 VALUES LESS THAN (736978) ENGINE = InnoDB,
PARTITION p20171012 VALUES LESS THAN (736979) ENGINE = InnoDB,
PARTITION p20171013 VALUES LESS THAN (736980) ENGINE = InnoDB,
PARTITION p20171014 VALUES LESS THAN (736981) ENGINE = InnoDB,
PARTITION p20171015 VALUES LESS THAN (736982) ENGINE = InnoDB
);
```

该表在 2017 年 10 月 15 日前接受 INSERTS；在此日期之后，INSERTS 操作都会失败。

另一个重要的事情是，如果数据超出保留日期（cross retention），则删除它。

10.9.1 如何操作

要执行这些操作，需要执行 ALTER 命令。

添加分区

要添加新分区，请执行 ADD PARTITION（<PARTITION DEFINITION>）语句：

```
mysql> ALTER TABLE event_history ADD PARTITION (
PARTITION p20171016 VALUES LESS THAN (736983) ENGINE = InnoDB,
PARTITION p20171017 VALUES LESS THAN (736984) ENGINE = InnoDB
);
```

此语句会在很短的时间内锁定整个表。

重组分区

如果存在 MAXVALUE 分区，则不能在 MAXVALUE 之后添加分区。在这种情况下，你需要将 MAXVALUE 分区用 REORGANIZE MAXVALUE 语句分为两个分区：

```
mysql> ALTER TABLE event_history REORGANIZE PARTITION pmax INTO (PARTITION
p20171016 VALUES LESS THAN (736983) ENGINE = InnoDB,
PARTITION pmax VALUES LESS THAN MAXVALUE ENGINE = InnoDB);
```

记住，在重新组织分区时，MySQL 得不得移动大量的数据，并且在此期间将锁定表。

你还可以将多个分区重组为一个分区：

```
mysql> ALTER TABLE event_history REORGANIZE PARTITION
p20171001,p20171002,p20171003,p20171004,p20171005,p20171006,p20171007
INTO (PARTITION p2017_oct_week1 VALUES LESS THAN (736974));
```

删除分区

如果数据已超出保留日期，则可以删除（DROP）整个分区，与传统的 DELETE FROM TABLE 语句相比，这种操作是超级快的，在高效存档数据时非常有用。

如果 p20170930 已超出保留日期，则可以使用 ALTER TABLE...DROP PARTITION

语句删除分区。

```
mysql> ALTER TABLE event_history DROP PARTITION p20170930;
Query OK, 0 rows affected (0.02 sec)
Records: 0  Duplicates: 0  Warnings: 0
```

删除（DROP）分区会从表中删除指定分区。

TRUNCATE 分区

如果你希望在表中保留 PARTITION DEFINITION 且仅删除数据，则可以执行 TRUNCATE PARTITION 命令：

```
mysql> ALTER TABLE event_history TRUNCATE PARTITION p20171001;
Query OK, 0 rows affected (0.08 sec)
```

管理 HASH 和 KEY 分区

在 HASH 和 KEY 分区上执行的操作完全不同。你只能减少或增加分区的数量。

假设 employees 表是基于 HASH 进行分区的：

```
mysql> CREATE TABLE `employees` (
  `emp_no` int(11) NOT NULL,
  `birth_date` date NOT NULL,
  `first_name` varchar(14) NOT NULL,
  `last_name` varchar(16) NOT NULL,
  `gender` enum('M','F') NOT NULL,
  `hire_date` date NOT NULL,
  `address` varchar(100) DEFAULT NULL,
  PRIMARY KEY (`emp_no`,`hire_date`),
  KEY `name` (`first_name`,`last_name`)
) ENGINE=InnoDB DEFAULT CHARSET=utf8mb4
PARTITION BY HASH(YEAR(hire_date))
PARTITIONS 8;
```

要将分区数从 8 减少到 6，可以执行 COALESCE PARTITION 语句，并指定要减少的分区数，即 8 – 6 = 2：

```
mysql> ALTER TABLE employees COALESCE PARTITION 2;
Query OK, 0 rows affected (0.31 sec)
Records: 0  Duplicates: 0  Warnings: 0
```

要将分区数从 6 增加到 16，可以执行 ADD PARTITION 语句并指定要增加的分区数，即 16 – 6 = 10：

```
mysql> ALTER TABLE employees ADD PARTITION PARTITIONS 10;
Query OK, 0 rows affected (5.11 sec)
Records: 0  Duplicates: 0  Warnings: 0
```

其他操作

对于特定的分区，还可以执行其他操作，例如 REBUILD、OPTIMIZE、ANALYZE 和 REPAIR 语句。举一个例子：

```
mysql> ALTER TABLE event_history REPAIR PARTITION p20171009, p20171010;
```

10.10 分区信息

本节讨论如何获取现有分区的信息，这些信息可以通过多种方式获得。

10.10.1 如何操作

让我们了解一下细节。

使用 SHOW CREATE TABLE

要知道一张表是否已分区，可以执行 SHOW CREATE TABLE\G 语句，该语句会列出表的定义和表的分区。例如：

```
mysql> SHOW CREATE TABLE employees \G
*************************** 1. row ***************************
       Table: employees
Create Table: CREATE TABLE `employees` (
  `emp_no` int(11) NOT NULL,
  `birth_date` date NOT NULL,
  `first_name` varchar(14) NOT NULL,
```

```
  `last_name` varchar(16) NOT NULL,
  `gender` enum('M','F') NOT NULL,
  `hire_date` date NOT NULL,
  `address` varchar(100) DEFAULT NULL,
  PRIMARY KEY (`emp_no`,`hire_date`),
  KEY `name` (`first_name`,`last_name`)
) ENGINE=InnoDB DEFAULT CHARSET=utf8mb4
/*!50100 PARTITION BY RANGE (YEAR(hire_date))
(PARTITION p1980 VALUES LESS THAN (1980) ENGINE = InnoDB,
 PARTITION p1990 VALUES LESS THAN (1990) ENGINE = InnoDB,
 PARTITION p2000 VALUES LESS THAN (2000) ENGINE = InnoDB,
 PARTITION p2010 VALUES LESS THAN (2010) ENGINE = InnoDB,
 PARTITION p2020 VALUES LESS THAN (2020) ENGINE = InnoDB,
 PARTITION pmax VALUES LESS THAN MAXVALUE ENGINE = InnoDB) */
```

使用 SHOW TABLE STATUS

你可以执行 SHOW TABLE STATUS 命令，并在输出中查看 Create_options：

```
mysql> SHOW TABLE STATUS LIKE 'employees'\G
*************************** 1. row ***************************
           Name: employees
         Engine: InnoDB
        Version: 10
     Row_format: Dynamic
           Rows: NULL
 Avg_row_length: NULL
    Data_length: NULL
Max_data_length: NULL
   Index_length: NULL
      Data_free: NULL
 Auto_increment: NULL
    Create_time: 2017-10-01 05:01:53
    Update_time: NULL
     Check_time: NULL
      Collation: utf8mb4_0900_ai_ci
       Checksum: NULL
 Create_options: partitioned
```

```
        Comment:
1 row in set (0.00 sec)
```

使用 EXPLAIN

explain 计划会显示一条查询所扫描的所有分区。如果运行 EXPLAIN SELECT * FROM <table>，则会列出所有分区。例如：

```
mysql> EXPLAIN SELECT * FROM employees\G
*************************** 1. row ***************************
           id: 1
  select_type: SIMPLE
        table: employees
   partitions: p1980,p1990,p2000,p2010,p2020,pmax
         type: ALL
possible_keys: NULL
          key: NULL
      key_len: NULL
          ref: NULL
         rows: 292695
     filtered: 100.00
        Extra: NULL
1 row in set, 1 warning (0.00 sec)
```

查询 INFORMATION_SCHEMA.PARTITIONS 表

与前面所有的方法相比，INFORMATION_SCHEMA.PARTITIONS 提供了关于分区的更多信息：

```
mysql> SHOW CREATE TABLE INFORMATION_SCHEMA.PARTITIONS\G
*************************** 1. row ***************************
       Table: PARTITIONS
Create Table: CREATE TEMPORARY TABLE `PARTITIONS` (
  `TABLE_CATALOG` varchar(512) NOT NULL DEFAULT '',
  `TABLE_SCHEMA` varchar(64) NOT NULL DEFAULT '',
  `TABLE_NAME` varchar(64) NOT NULL DEFAULT '',
  `PARTITION_NAME` varchar(64) DEFAULT NULL,
  `SUBPARTITION_NAME` varchar(64) DEFAULT NULL,
```

```
  `PARTITION_ORDINAL_POSITION` bigint(21) unsigned DEFAULT NULL,
  `SUBPARTITION_ORDINAL_POSITION` bigint(21) unsigned DEFAULT NULL,
  `PARTITION_METHOD` varchar(18) DEFAULT NULL,
  `SUBPARTITION_METHOD` varchar(12) DEFAULT NULL,
  `PARTITION_EXPRESSION` longtext,
  `SUBPARTITION_EXPRESSION` longtext,
  `PARTITION_DESCRIPTION` longtext,
  `TABLE_ROWS` bigint(21) unsigned NOT NULL DEFAULT '0',
  `AVG_ROW_LENGTH` bigint(21) unsigned NOT NULL DEFAULT '0',
  `DATA_LENGTH` bigint(21) unsigned NOT NULL DEFAULT '0',
  `MAX_DATA_LENGTH` bigint(21) unsigned DEFAULT NULL,
  `INDEX_LENGTH` bigint(21) unsigned NOT NULL DEFAULT '0',
  `DATA_FREE` bigint(21) unsigned NOT NULL DEFAULT '0',
  `CREATE_TIME` datetime DEFAULT NULL,
  `UPDATE_TIME` datetime DEFAULT NULL,
  `CHECK_TIME` datetime DEFAULT NULL,
  `CHECKSUM` bigint(21) unsigned DEFAULT NULL,
  `PARTITION_COMMENT` varchar(80) NOT NULL DEFAULT '',
  `NODEGROUP` varchar(12) NOT NULL DEFAULT '',
  `TABLESPACE_NAME` varchar(64) DEFAULT NULL
) ENGINE=InnoDB DEFAULT CHARSET=utf8
1 row in set (0.00 sec)
```

要了解有关表的分区的更多详细信息，可以通过 TABLE_SCHEMA 指定数据库名称并通过 TABLE_NAME 指定表名来查询 INFORMATION_SCHEMA.PARTITIONS 表。例如：

```
mysql> SELECT PARTITION_NAME FROM INFORMATION_SCHEMA.PARTITIONS WHERE
TABLE_SCHEMA='employees' AND TABLE_NAME='employees';
+----------------+
| PARTITION_NAME |
+----------------+
| p1980          |
| p1990          |
| p2000          |
| p2010          |
| p2020          |
| pmax           |
```

```
+-----------------+
6 rows in set (0.00 sec)
```

你可以获知该分区的 PARTITION_METHOD、PARTITION_EXPRESSION、PARTITION_DESCRIPTION 和 TABLE_ROWS 等详细信息：

```
mysql> SELECT * FROM INFORMATION_SCHEMA.PARTITIONS WHERE
TABLE_SCHEMA='employees' AND TABLE_NAME='employees' AND
PARTITION_NAME='p1990'\G
*************************** 1. row ***************************
                TABLE_CATALOG: def
                 TABLE_SCHEMA: employees
                   TABLE_NAME: employees
               PARTITION_NAME: p1990
            SUBPARTITION_NAME: NULL
   PARTITION_ORDINAL_POSITION: 2
SUBPARTITION_ORDINAL_POSITION: NULL
             PARTITION_METHOD: RANGE
          SUBPARTITION_METHOD: NULL
         PARTITION_EXPRESSION: YEAR(hire_date)
      SUBPARTITION_EXPRESSION: NULL
        PARTITION_DESCRIPTION: 1990
                   TABLE_ROWS: 157588
               AVG_ROW_LENGTH: 56
                  DATA_LENGTH: 8929280
              MAX_DATA_LENGTH: NULL
                 INDEX_LENGTH: 8929280
                    DATA_FREE: 0
                  CREATE_TIME: NULL
                  UPDATE_TIME: NULL
                   CHECK_TIME: NULL
                     CHECKSUM: NULL
            PARTITION_COMMENT:
                    NODEGROUP: default
              TABLESPACE_NAME: NULL
1 row in set (0.00 sec)
```

更多有关的详细信息，请参阅 https://dev.mysql.com/doc/refman/8.0/en/partitions-table.html。

10.11　有效地管理生存时间和软删除行

RANGE COLUMNS 在管理生存时间和软删除行方面非常有用。假设你有一个应用程序指定了行的失效时间（即在一定时间之后要删除此行），而到期时间是变化的。

假设该应用程序可以执行以下几类插入操作：

- 插入持久化数据
- 插入有到期时间的数据

如果到期时间是一致的，比如插入的所有行将在一定时间后被删除，我们可以使用 RANGE 分区。但是，如果到期时间有分别，比如有些行将在一周内删除，有些在一个月内删除，有些在一年内删除，还有些不会过期，则无法创建分区。在这种情况下，可以使用 RANGE COLUMNS 分区，下面将说明其中的原因。

10.11.1　如何运行

我们引入一个名为 soft_delete 的列，它将由触发器设置。该 soft_delete 列将成为 RANGE 列分区的一部分。

分区有点像是 soft_delete 与到期时间的函数。soft_delete 和到期时间合起来控制某一行应该被分到哪个分区。soft_delete 列决定了行被保留到哪个分区。如果到期时间为 0，则触发器将 soft_delete 值设置为 0，将此行放入 no_retention 分区，并且如果到期时间的值超出分区范围，则触发器将 soft_delete 值设置为 1，并将该行放入 long_retention 分区。如果到期时间的值在分区范围内，则触发器将 soft_delete 值设置为 2。根据到期时间的值，该行将被放入相应的分区。

总之，soft_delete 的值有以下几种：

- 0——如果到期时间的值是 0。
- 1——如果到期时间距离时间戳超过 7 天。
- 2——如果到期时间距离时间戳小于或等于 7 天。

对应地，我们创建如下数量的分区：

- 1 个 `no_retention` 分区（`soft_delete = 0`）。
- 1 个 `long_retention` 分区（`soft_delete = 1`）。
- 7 个按天的分区（`soft_delete = 2`）。

10.11.2 如何操作

你可以像这样创建一张表：

```
mysql> CREATE TABLE `customer_data` (
  `id` int(11) NOT NULL AUTO_INCREMENT,
  `msg` text,
  `timestamp` bigint(20) NOT NULL DEFAULT '0',
  `expires` bigint(20) NOT NULL DEFAULT '0',
  `soft_delete` tinyint(3) unsigned NOT NULL DEFAULT '1',
  PRIMARY KEY (`id`,`expires`,`soft_delete`)
) ENGINE=InnoDB DEFAULT CHARSET=utf8
/*!50500 PARTITION BY RANGE COLUMNS(soft_delete,expires)
(PARTITION no_retention VALUES LESS THAN (0,MAXVALUE) ENGINE = InnoDB,
 PARTITION long_retention VALUES LESS THAN (1,MAXVALUE) ENGINE = InnoDB,
 PARTITION pd20171017 VALUES LESS THAN (2,1508198400000) ENGINE = InnoDB,
 PARTITION pd20171018 VALUES LESS THAN (2,1508284800000) ENGINE = InnoDB,
 PARTITION pd20171019 VALUES LESS THAN (2,1508371200000) ENGINE = InnoDB,
 PARTITION pd20171020 VALUES LESS THAN (2,1508457600000) ENGINE = InnoDB,
 PARTITION pd20171021 VALUES LESS THAN (2,1508544000000) ENGINE = InnoDB,
 PARTITION pd20171022 VALUES LESS THAN (2,1508630400000) ENGINE = InnoDB,
 PARTITION pd20171023 VALUES LESS THAN (2,1508716800000) ENGINE = InnoDB,
 PARTITION pd20171024 VALUES LESS THAN (2,1508803200000) ENGINE = InnoDB,
 PARTITION pd20171025 VALUES LESS THAN (2,1508869800000) ENGINE = InnoDB,
   PARTITION pd20171026 VALUES LESS THAN (2,1508956200000) ENGINE = InnoDB)
*/;
```

我们创建 10 个分区，每个分区代表一天，而有些分区经常会为空，就把它们作为缓冲分区。这样，我们就可以将 10 个分区分成 7+3 两个部分，7 是指 7 个工作分区，3 是指 3 个缓冲分区。

```
mysql> DROP TRIGGER IF EXISTS customer_data_insert;
DELIMITER $$
CREATE TRIGGER customer_data_insert
BEFORE INSERT
   ON customer_data FOR EACH ROW
BEGIN
    SET NEW.soft_delete = (IF((NEW.expires =
0),0,IF((ROUND((((((NEW.expires - NEW.timestamp) / 1000) / 60) / 60) /
24),0) <= 7),2,1)));
END;
$$
DELIMITER ;

mysql> DROP TRIGGER IF EXISTS customer_data_update;
DELIMITER $$
CREATE TRIGGER customer_data_update
BEFORE UPDATE
    ON customer_data FOR EACH ROW
BEGIN
    SET NEW.soft_delete = (IF((NEW.expires =
0),0,IF((ROUND((((((NEW.expires - NEW.timestamp) / 1000) / 60) / 60) /
24),0) <= 7),2,1)));
END;
$$
DELIMITER ;
```

- 假设客户端插入的行的时间戳为 1508265000(2017-10-17 18:30:00)，到期时间为 1508351400(2017-10-18 18:30:00)，那么 soft_delete 的值将会是 2，这一行将被分到 pd20171019 分区。

```
mysql> INSERT INTO customer_data(id, msg, timestamp, expires)
VALUES(1,'test',1508265000000,1508351400000);
Query OK, 1 row affected (0.05 sec)

mysql> SELECT * FROM customer_data PARTITION (pd20171019);
+----+------+---------------+---------------+-------------+
| id | msg  | timestamp     | expires       | soft_delete |
```

```
+----+------+---------------+---------------+-------------+
|  1 | test | 1508265000000 | 1508351400000 |           2 |
+----+------+---------------+---------------+-------------+
1 row in set (0.00 sec)
```

- 假设客户端未设置到期时间，那么到期时间这一列将为 0，这将导致 soft_delete 也为 0，那么这一行将被分到 no_retention 分区。

```
mysql> INSERT INTO customer_data(id, msg, timestamp, expires)
VALUES(2,'non_expiry_row',1508265000000,0);
Query OK, 1 row affected (0.07 sec)

mysql> SELECT * FROM customer_data PARTITION (no_retention);
+----+----------------+---------------+---------+-------------+
| id | msg            | timestamp     | expires | soft_delete |
+----+----------------+---------------+---------+-------------+
|  2 | non_expiry_row | 1508265000000 |       0 |           0 |
+----+----------------+---------------+---------+-------------+
1 row in set (0.00 sec)
```

- 假设客户端想设置到期时间（比如，2017-10-19 06:30:00），可以更新到期时间这一列，将该行从 no_retention 分区转移到相应的分区（对性能有一定的影响，因为该行必须跨分区移动）。

```
mysql> UPDATE customer_data SET expires=1508394600000 WHERE id=2;
Query OK, 1 row affected (0.06 sec)
Rows matched: 1  Changed: 1  Warnings: 0

mysql> SELECT * FROM customer_data PARTITION (no_retention);
Empty set (0.00 sec)

mysql> SELECT * FROM customer_data PARTITION (pd20171020);
+----+----------------+---------------+---------------+-----------
--+
| id | msg            | timestamp     | expires       | soft_delete |
+----+----------------+---------------+---------------+-----------
--+
|  2 | non_expiry_row | 1508265000000 | 1508394600000 |           2 |
```

```
+----+---------------+--------------+--------------+------------
--+
1 row in set (0.00 sec)
```

- 假设客户端设置的到期时间超出了我们的分区范围，那么这一行将自动进入 long_retention 分区。

```
mysql> INSERT INTO customer_data(id, msg, timestamp, expires)
VALUES(3,'long_expiry',1507852800000,1608025600000);

mysql> SELECT * FROM customer_data PARTITION (long_retention);
+----+-------------+--------------+--------------+-------------+
| id | msg         | timestamp    | expires      | soft_delete |
+----+-------------+--------------+--------------+-------------+
|  3 | long_expiry | 1507852800000 | 1608025600000 |          1 |
+----+-------------+--------------+--------------+-------------+
1 row in set (0.00 sec)
```

跨分区的行移动很慢，如果你更新了 soft_delete，那么该行将从默认分区移到其他分区，这个过程将会很慢。

扩展逻辑

我们可以扩展逻辑并增加 soft_delete 的值以容纳更多类型的分区。将 soft_delete 可能的值增加到以下 4 个：

- 0——如果到期时间的值为 0。
- 3——如果到期时间距离时间戳小于或等于 7 天。
- 2——如果到期时间距离时间戳小于或等于 60 天。
- 1——如果到期时间大于 60 天。

soft_delete 列将是分区的一部分。

- 如果 soft_delete 的值为 0，则创建一个 no_retention 分区。
- 如果 soft_delete 的值为 1，则创建一个 long_retention 分区。
- 如果 soft_delete 的值为 2，则每周创建一个分区。
- 如果 soft_delete 的值为 3，则每天创建一个分区。

分区表结构示例

我们将创建一个按天和按周的混合分区，分别存储到期时间距离时间戳小于或等于 7 天的数据，以及到期时间距离时间戳小于或等于 60 天的数据。每一类分区都会有冗余，冗余的部分就作为缓冲区。。

```
mysql> DROP TRIGGER IF EXISTS customer_data_insert;
DELIMITER $$
CREATE TRIGGER customer_data_insert
BEFORE INSERT
    ON customer_data FOR EACH ROW
BEGIN
    SET NEW.soft_delete = (IF((NEW.expires =
0),0,IF((ROUND((((((NEW.expires - NEW.timestamp) / 1000) / 60) / 60) /24),0)
<= 7),3,IF((ROUND((((((NEW.expires - NEW.timestamp) / 1000) / 60) / 60) /
24),0) <= 42),2,1))));
END;
$$
DELIMITER ;

mysql> DROP TRIGGER IF EXISTS customer_data_update;
DELIMITER $$
CREATE TRIGGER customer_data_update
BEFORE INSERT
   ON customer_data FOR EACH ROW
BEGIN
    SET NEW.soft_delete = (IF((NEW.expires =
     0),0,IF((ROUND((((((NEW.expires - NEW.timestamp) / 1000) / 60) / 60)
/24),0) <= 7),3,IF((ROUND((((((NEW.expires - NEW.timestamp) / 1000) / 60)
/60) / 24),0) <= 42),2,1))));
END;
$$
DELIMITER ;

mysql> CREATE TABLE `customer_data` (
  `id` int(11) NOT NULL AUTO_INCREMENT,
  `msg` text,
```

```
  `timestamp` bigint(20) NOT NULL DEFAULT '0',
  `expires` bigint(20) NOT NULL DEFAULT '0',
  `soft_delete` tinyint(3) unsigned NOT NULL DEFAULT '1',
  PRIMARY KEY (`id`,`expires`,`soft_delete`)
) ENGINE=InnoDB AUTO_INCREMENT=609585360 DEFAULT CHARSET=utf8
/*!50500 PARTITION BY RANGE  COLUMNS(`soft_delete`,`expires`)
(
 PARTITION no_retention VALUES LESS THAN (0,MAXVALUE) ENGINE = InnoDB,
 PARTITION long_retention VALUES LESS THAN (1,MAXVALUE) ENGINE = InnoDB,
 PARTITION pw20171022 VALUES LESS THAN (2,1508630400000) ENGINE = InnoDB,
 PARTITION pw20171029 VALUES LESS THAN (2,1509235200000) ENGINE = InnoDB,
 PARTITION pw20171105 VALUES LESS THAN (2,1509840000000) ENGINE = InnoDB,
 PARTITION pw20171112 VALUES LESS THAN (2,1510444800000) ENGINE = InnoDB,
 PARTITION pw20171119 VALUES LESS THAN (2,1511049600000) ENGINE = InnoDB,
 PARTITION pw20171126 VALUES LESS THAN (2,1511654400000) ENGINE = InnoDB,
 PARTITION pw20171203 VALUES LESS THAN (2,1512259200000) ENGINE = InnoDB,
 -- buffer partition which will be 67 days away and will be always empty so
that we can split
 PARTITION pw20171210 VALUES LESS THAN (2,1512864000000) ENGINE = InnoDB,
 PARTITION pd20171016 VALUES LESS THAN (3,1508112000000) ENGINE = InnoDB,
 PARTITION pd20171017 VALUES LESS THAN (3,1508198400000) ENGINE = InnoDB,
 PARTITION pd20171018 VALUES LESS THAN (3,1508284800000) ENGINE = InnoDB,
 PARTITION pd20171019 VALUES LESS THAN (3,1508371200000) ENGINE = InnoDB,
 PARTITION pd20171020 VALUES LESS THAN (3,1508457600000) ENGINE = InnoDB,
 PARTITION pd20171021 VALUES LESS THAN (3,1508544000000) ENGINE = InnoDB,
 PARTITION pd20171022 VALUES LESS THAN (3,1508630400000) ENGINE = InnoDB,
 PARTITION pd20171023 VALUES LESS THAN (3,1508716800000) ENGINE = InnoDB,
 PARTITION pd20171024 VALUES LESS THAN (3,1508803200000) ENGINE = InnoDB
) */;
```

管理分区

你可以在 Linux 中创建 CRON 或者在 MySQL 中创建 EVENT 以管理分区。当临近保留日期（期限）时，分区管理工具应该将缓冲区分区重新组织成一个可用分区和一个缓冲区分区，并删除超出保留日期的分区。

以前面提到的 customer_data 表为例：

在 2017 年 12 月 3 日，必须将分区 pw20171210 分为 pw20171210 和 pw20171217。

在 2017 年 10 月 17 日，必须将分区 pd20171024 分为 pd20171024 和 pd20171025。

只有在没有数据（或数据非常少）的情况下，分割（重新组织）分区会非常快（如果没有查询锁定这个表的话，只需要几毫秒）。因此，我们应该在数据进入分区之前对分区进行重新组织，以保持该分区为空。

第 11 章 管理表空间

在本章中，我们将会介绍如下内容：

- 更改 InnoDB REDO 日志文件的数量和大小
- 调整 InnoDB 系统的表空间的大小
- 在数据目录之外创建独立表空间（file-per-table）
- 将独立表空间复制到另一个实例中
- 管理 UNDO 表空间
- 管理通用表空间
- 压缩 InnoDB 表

11.1 引言

在开始学习本章之前，你应该了解一点 InnoDB 的基础知识。

本节以下内容均引自 MySQL 文档。

系统表空间（共享表空间）

“InnoDB 系统表空间包含 InnoDB 数据字典（与 InnoDB 相关的对象的元数据），它是 doublewrite buffer、change buffer 和 UNDO 日志的存储区域。系统表空间还包含在系统表空间中创建的表以及所有用户创建的表的索引数据。系统表空间被认为是共享的表空间，因为它由多个表共享。

系统表空间用一个或多个数据文件表示。默认情况下，将在 MySQL 数据目录中创建一个名为 ibdata1 的系统数据文件。系统数据文件的大小和数量由 innodb_data_file_path 启动项控制。”

独立表空间

每个独立表空间都是一个单表表空间，它是在自己的数据文件中创建的，而不是在系统表空间中创建的。当启用 innodb_file_per_table 选项时，将在独立表空间中创建表；否则，将在系统表空间中创建 InnoDB 表。每个独立表空间由一个.ibd 数据文件表示，该文件默认是在数据库目录中创建的。

独立表空间支持 DYNAMIC 和 COMPRESSED 的行格式，这些格式支持对可变长度的数据和表压缩的跨页存储等特性。

要了解独立表空间的优缺点，请参考 https:// dev.mysql.com/doc/refman/8.0/en/innodb-multiple-tablespaces.html 和 https://dev.mysql.com/doc/refman/8.0/en/innodb-parameters.html#sysvar_innodb_file_per_table。

通用表空间

通用表空间是使用语法 CREATE TABLESPACE 创建的共享 InnoDB 表空间。通用表空间可以在 MySQL 数据目录之外创建，可以容纳多张表，并支持所有行格式的表。

UNDO 表空间

UNDO（撤销）日志是与单个事务关联的 UNDO 日志记录的集合。一条 UNDO 日志记录包含如何将最近的事务更改还原为聚簇索引记录的信息。如果另一个事务需要查看原始数据（作为一致读操作的一部分），则从 UNDO 日志记录中检索未修改的数据。UNDO 日志存在于 UNDO 日志段中，这些日志段包含在回滚段中。回滚段驻留在系统表空间、临时表空间和 UNDO 表空间中。

UNDO 表空间由一个或多个包含 UNDO 日志的文件组成。InnoDB 使用的 UNDO 表空间的数量由 innodb_undo_tablespaces 配置项定义。

这些日志用于回滚事务和多版本并发控制。

数据字典

数据字典（data dictionary）是元数据，它跟踪数据库对象，如表、索引和表列。对于 MySQL 8.0 中引入的 MySQL 数据字典而言，元数据的物理位置位于 MySQL 数据库目录中的 InnoDB 独立表空间文件中。对于 InnoDB 数据字典，元数据的物理位置位于 InnoDB 系统表空间中。

MySQL 数据字典

MySQL 服务器集成了一个事务数据字典，它存储关于数据库对象的信息。在之前的 MySQL 版本中，字典数据存储在元数据文件、非事务性表和存储引擎特定的数据字典中。

在之前的 MySQL 版本中，字典数据部分地存储在元数据文件中。基于文件的元数据存储存在的问题有：文件扫描的开销大，易受到文件系统相关的 bug 的影响，处理复制以及从崩溃中恢复的失败状态要编写复杂代码，同时它还缺乏可扩展性，很难为新特性和关系型对象添加元数据。

MySQL 数据字典的优点有：

- 以集中式数据字典模式统一存储字典数据，更简单。
- 删除了基于文件的元数据存储。
- 对字典数据的事务性、崩溃时的安全存储。
- 对字典对象的统一和集中缓存。
- 对一些 `INFORMATION_SCHEMA` 表的实现进行简化和改进。
- 原子 DDL。

下面列出从 MySQL 中删除的元数据文件。除非另有说明，以前存储在元数据文件中的数据现在均存储在数据字典表中。

- `.frm` 文件：用于表定义的表元数据文件。
- `.par` 文件：分区定义文件。在 MySQL 5.7 中，InnoDB 停止使用`.definition` 定义分区文件，引入了对 InnoDB 表的本地分区支持。
- `.trn` 文件：触发器命名空间文件。
- `.trg` 文件：触发器参数文件。
- `.isl` 文件：InnoDB 的符号链接文件，其中包含了在 MySQL 数据目录之外创建的每个表空间文件的位置。

- db.opt 文件：数据库配置文件。这些文件（每个数据库目录有一个 db.opt 文件），包含数据库默认字符集的属性。

MySQL 数据字典的局限性如下：

- 不支持在数据目录下手动创建数据库目录（例如，使用 mkdir）。手动创建的数据库目录不被 MySQL 服务器识别。
- 不支持通过复制和移动 MyISAM 数据文件来移动存储在 MyISAM 表中的数据。使用此方法移动的表不会被服务器发现。
- 不支持使用复制的数据文件对单个 MyISAM 表进行简单的备份和恢复。
- 由于要写入存储引擎、UNDO 日志和 REDO（重做）日志，而不是写入 .frm 文件，因此 DDL 操作需要更长的时间。

字典数据的事务性存储

数据字典模式将字典数据存储在事务性（InnoDB）表中。数据字典表与非数据字典系统表都位于 MySQL 数据库中。

数据字典表是在MySQL数据目录中名为mysql.ibd的一个InnoDB表空间中创建的。mysql.ibd 表空间文件必须驻留在 MySQL 数据目录中，它的名称不能被另一个表空间修改或使用。以前，这些表是在 MySQL 数据库目录各自的表空间文件中创建的。

11.2 更改 InnoDB REDO 日志文件的数量或大小

ib_logfile0 和 ib_logfile1 是在数据目录中创建的默认的 InnoDB REDO 日志文件，每个文件大小为 48 MB。如果你希望更改 REDO 日志文件的大小，只需在配置文件中进行修改，并重新启动 MySQL。

在以前的 MySQL 版本中，必须安全关闭 MySQL 服务器，删除 REDO 日志文件，更改配置文件，然后启动 MySQL 服务器。

在 MySQL 8 中，InnoDB 会检测到 innodb_log_file_size 与 REDO 日志文件大小不同。它记下一个日志检查点，关闭并删除旧的日志文件，按照要求的大小创建新的日志文件，并打开它们。

11.2.1 如何操作

1. 检查当前文件的大小：

```
shell> sudo ls -lhtr /var/lib/mysql/ib_logfile*
-rw-r-----. 1 mysql mysql 48M Oct  7 10:16
/var/lib/mysql/ib_logfile1
-rw-r-----. 1 mysql mysql 48M Oct  7 10:18
/var/lib/mysql/ib_logfile0
```

2. 关闭 MySQL 服务器，确保它在关闭时没有出现报错：

```
shell> sudo systemctl stop mysqld
```

3. 修改配置文件：

```
shell> sudo vi /etc/my.cnf
[mysqld]
innodb_log_file_size=128M
innodb_log_files_in_group=4
```

4. 启动 MySQL 服务器：

```
shell> sudo systemctl start mysqld
```

5. 你可以验证 MySQL 在日志文件中做了什么：

```
shell> sudo less /var/log/mysqld.log
2017-10-07T11:09:35.111926Z 1 [Warning] InnoDB: Resizing redo log
from 2*3072 to 4*8192 pages, LSN=249633608
2017-10-07T11:09:35.213717Z 1 [Warning] InnoDB: Starting to delete
and rewrite log files.
2017-10-07T11:09:35.224724Z 1 [Note] InnoDB: Setting log file
./ib_logfile101 size to 128 MB
2017-10-07T11:09:35.225531Z 1 [Note] InnoDB: Progress in MB:
100
2017-10-07T11:09:38.924955Z 1 [Note] InnoDB: Setting log file
./ib_logfile1 size to 128 MB
2017-10-07T11:09:38.925173Z 1 [Note] InnoDB: Progress in MB:
100
```

```
2017-10-07T11:09:42.516065Z 1 [Note] InnoDB: Setting log file
./ib_logfile2 size to 128 MB
2017-10-07T11:09:42.516309Z 1 [Note] InnoDB: Progress in MB:
100
2017-10-07T11:09:46.098023Z 1 [Note] InnoDB: Setting log file
./ib_logfile3 size to 128 MB
2017-10-07T11:09:46.098246Z 1 [Note] InnoDB: Progress in MB:
100
2017-10-07T11:09:49.715400Z 1 [Note] InnoDB: Renaming log file
./ib_logfile101 to ./ib_logfile0
2017-10-07T11:09:49.715497Z 1 [Warning] InnoDB: New log files
created, LSN=249633608
```

6. 你还可以看到创建的新日志文件：

```
shell> sudo ls -lhtr /var/lib/mysql/ib_logfile*
-rw-r-----. 1 mysql mysql 128M Oct  7 11:09
/var/lib/mysql/ib_logfile1
-rw-r-----. 1 mysql mysql 128M Oct  7 11:09
/var/lib/mysql/ib_logfile2
-rw-r-----. 1 mysql mysql 128M Oct  7 11:09
/var/lib/mysql/ib_logfile3
-rw-r-----. 1 mysql mysql 128M Oct  7 11:09
/var/lib/mysql/ib_logfile0
```

11.3 调整 InnoDB 系统的表空间大小

数据目录中的 ibdata1 文件是默认的系统表空间。你可以使用 innodb_data_file_path 和 innodb_data_home_dir 配置项来配置 ibdata1。innodb_data_file_path 选项用于配置 InnoDB 系统表空间数据文件。它的值应该是一个或多个数据文件规范的列表。如果指定了两个或多个数据文件，则用分号（;）分隔。

如果你想要表空间中包含一个名为 ibdata1 的、固定大小为 50 MB 的数据文件，并且在数据目录中有一个名为 ibdata2 的 50 MB 的自动扩展文件，则可以这样配置：

```
shell> sudo vi /etc/my.cnf
[mysqld]
```

```
innodb_data_file_path=ibdata1:50M;ibdata2:50M:autoextend
```

如果 ibdata 文件变得很大，特别是当 innodb_file_per_table 没有被启用且磁盘已满时，你可能需要在另一个磁盘上再添加一个数据文件。

11.3.1　如何操作

调整 InnoDB 系统表空间的大小是你可能希望详细了解的一个主题。下面来讲解它的细节。

增加 InnoDB 系统表空间

假设 innodb_data_file_path 是 ibdata1:50m:autoextend，其大小已达到 76 MB，而你的磁盘仅为 100 MB，那么可以再加一个磁盘，并在新磁盘上配置，添加另一个表空间。

1. 停止 MySQL 服务器的运行：

```
shell> sudo systemctl stop mysql
```

2. 检查已存在的 ibdata1 文件的大小：

```
shell> sudo ls -lhtr /var/lib/mysql/ibdata1
-rw-r----- 1 mysql mysql 76M Oct  6 13:33 /var/lib/mysql/ibdata1
```

3. 挂载新磁盘，假设它挂载在/var/lib/mysql_extend 上，将所有权归属更改为 mysql，确保此文件尚未创建。如果你使用的是 AppArmour 或 SELinux，请确保正确设置了别名或环境：

```
shell> sudo chown mysql:mysql /var/lib/mysql_extend
shell> sudo chmod 750 /var/lib/mysql_extend
shell> sudo ls -lhtr /var/lib/mysql_extend
```

4. 打开 my.cnf 并添加以下内容：

```
shell> sudo vi /etc/my.cnf
[mysqld]
innodb_data_home_dir=
```

```
innodb_data_file_path =
ibdata1:76M;/var/lib/mysql_extend/ibdata2:50M:autoextend
```

由于现有 ibdata1 的大小为 76 MB，所以你必须选择最大值至少为 76 MB。下一个 ibdata 文件将在挂载在/var/lib/mysql_extend/上的新磁盘中创建。应该指定 innodb_data_home_dir 选项；否则，mysqld 就会查看另一条路径，并出现错误。

```
2017-10-07T06:30:00.658039Z 1 [ERROR] InnoDB: Operating system
error number 2 in a file operation.
2017-10-07T06:30:00.658084Z 1 [ERROR] InnoDB: The error means the
system cannot find the path specified.
2017-10-07T06:30:00.658088Z 1 [ERROR] InnoDB: If you are installing
InnoDB, remember that you must create directories yourself, InnoDB
does not create them.
2017-10-07T06:30:00.658092Z 1 [ERROR] InnoDB: File
.//var/lib/mysql_extend/ibdata2: 'create' returned OS error 71.
Cannot continue operation
```

5. 重启 MySQL 服务器：

```
shell> sudo systemctl start mysql
```

6. 验证新文件。既然已经指定它的大小为 50 MB，那么这个文件的初始大小将是 50 MB：

```
shell> sudo ls -lhtr /var/lib/mysql_extend/
total 50M
-rw-r-----. 1 mysql mysql 50M Oct  7 07:38 ibdata2
mysql> SHOW VARIABLES LIKE 'innodb_data_file_path';
+-----------------------+--------------------------------------------
----------------+
| Variable_name         | Value
|
+-----------------------+--------------------------------------------
----------------+
| innodb_data_file_path |
ibdata1:12M;/var/lib/mysql_extend/ibdata2:50M:autoextend |
+-----------------------+--------------------------------------------
----------------+
```

```
1 row in set (0.00 sec)
```

缩小 InnoDB 系统表空间

如果你使用的不是 innodb_file_per_table，那么所有的表数据都存储在系统表空间中。如果你删除（drop）了一个表，那么该空间不会被回收。你可以缩小系统表空间并回收磁盘空间。这需要较长的停机时间，因此建议的做法是在一个切换后没有负载的从库上操作，然后提升它为主库。

可以通过查询 INFORMATION_SCHEMA 表来查看可用空间：

```
mysql> SELECT SUM(data_free)/1024/1024 FROM INFORMATION_SCHEMA.TABLES;
+--------------------------+
| sum(data_free)/1024/1024 |
+--------------------------+
|           6.00000000     |
+--------------------------+
1 row in set (0.00 sec)
```

1. 停止写入数据库。如果它是主库， mysql> SET @@GLOBAL.READ_ONLY=1；如果它是从库，则停止复制并保存二进制日志。

```
mysql> STOP SLAVE;
mysql> SHOW SLAVE STATUS\G
```

2. 使用 mysqldump 或 mydumper 进行完整备份，不包括 sys 数据库。

```
shell> mydumper -u root --password=<password> --trx-consistency-
only --kill-long-queries --long-query-guard 500 --regex '^(?!sys)'
--outputdir /backups
```

3. 停止 MySQL 服务器的运行。

```
shell> sudo systemctl stop mysql
```

4. 删除所有的*.ibd、*.ib_log 和 ibdata 文件。如果只使用了 InnoDB 表，则可以删除数据目录和存储系统表空间的所有位置（innodb_data_file_path）。

```
shell> sudo rm -rf /var/lib/mysql/ib* /var/lib/mysql/<database
directories>
```

```
shell> sudo rm -rf /var/lib/mysql_extend/*
```

5. 初始化数据目录。

```
shell> sudo mysqld --initialize --datadir=/var/lib/mysql
shell> chown -R  mysql:mysql  /var/lib/mysql/
shell> chown -R  mysql:mysql  /var/lib/mysql_extend/
```

6. 获取临时密码。

```
shell> sudo grep "temporary password is generated"
/var/log/mysql/error.log | tail -1
2017-10-07T09:33:31.966223Z 4 [Note] A temporary password is
generated for root@localhost: lI-qerr5agpa
```

7. 启动 MySQL 并更改密码。

```
shell> sudo systemctl start mysqld
shell> mysql -u root -plI-qerr5agpa

mysql> ALTER USER 'root'@'localhost' IDENTIFIED BY 'xxxx';
Query OK, 0 rows affected (0.01 sec)
```

8. 恢复备份。使用临时密码连接到 MySQL。

```
shell> /opt/mydumper/myloader --directory=/backups/ --queries-per-
transaction=50000 --threads=6 --user=root --password=xxxx  --
overwrite-tables
```

9. 如果它是一个主库，则通过语句 `mysql> SET @@GLOBAL.READ_ONLY =0;`来启用写。如果它是一个从库，则通过执行 `CHANGE MASTER TO COMMAND` 和 `START SLAVE` 来恢复复制。

11.4 在数据目录之外创建独立表空间

上一节介绍了如何在另一个磁盘中创建系统表空间。本节将介绍如何在另一个磁盘中创建单个表空间。

11.4.1　如何操作

你可以挂载一个具有特定性能或容量特性的新磁盘到一个目录，如快速 SSD 或大容量 HDD，并配置 InnoDB 来使用它。在目标目录中，MySQL 创建一个与数据库名称相应的子目录，并在该目录中为新表创建一个 `.ibd` 文件。请记住，不能使用带 `DATA DIRECTORY` 子句的 `ALTER TABLE` 语句。

1. 安装新磁盘并更改权限。如果你使用的是 AppArmour 或 SELinux，请确保正确设置了别名或环境：

```
shell> sudo chown -R mysql:mysql /var/lib/mysql_fast_storage
shell> sudo chmod 750 /var/lib/mysql_fast_storage
```

2. 创建一个新表：

```
mysql> CREATE TABLE event_tracker (
event_id INT UNSIGNED AUTO_INCREMENT PRIMARY KEY,
event_name varchar(10),
ts timestamp NOT NULL,
event_type varchar(10)
)
TABLESPACE = innodb_file_per_table
DATA DIRECTORY = '/var/lib/mysql_fast_storage';
```

3. 检查在新设备中创建的 `.ibd` 文件：

```
shell> sudo ls -lhtr  /var/lib/mysql_fast_storage/employees/
total 128K
-rw-r-----. 1 mysql mysql 128K Oct  7 13:48 event_tracker.i
```

11.5　将独立表空间复制到另一个实例

复制表空间文件（`.ibd` 文件）是移动数据的最快方式，比通过 `mysqldump` 或 `mydumper` 导出和导入数据还快。移动以后数据是立即可用的，而不必重新插入和重建索引。有很多原因可以解释为什么需要将一个 InnoDB 文件/表空间复制到一个不同的实例：

- 为了运行报表而不会给生产服务器增加额外的负载。

- 为了在新的从服务器上为一个表设置相同的数据。
- 为了在出现问题或错误后恢复备份的表或分区。
- 为了把繁忙的表放在 SSD 设备上，或在大容量的 HDD 设备上建大型表。

11.5.1 如何操作

大致的步骤是这样的：在目标库上创建表，并使用相同的表定义，并在目标库上执行 DISCARD TABLESPACE 命令。在源库上执行 FLUSH TABLES FOR EXPORT ，这将确保对指定表的更改被刷新到磁盘，因此可以在运行实例时生成二进制表的副本。在执行该语句之后，表被锁定，并且不接受任何写入操作；当然，可以进行读操作。你可以将该表的.ibd 文件复制到目标库，在源库上执行 UNLOCK，最后执行 IMPORT TABLESPACE 命令，该命令接受复制的.ibd 文件。

例如，你希望将测试数据库的 events_history 表从一个服务器（源库）复制到另一个服务器（目标库）中。

如果没有 event_history 的话，就先创建这个表，并且插入一些行以便演示：

```
mysql> USE test;
mysql> CREATE TABLE IF NOT EXISTS `event_history`(
  `event_id` int(11) NOT NULL,
  `event_name` varchar(10) DEFAULT NULL,
  `created_at` datetime NOT NULL,
  `last_updated` timestamp NULL DEFAULT CURRENT_TIMESTAMP ON UPDATE
CURRENT_TIMESTAMP,
  `event_type` varchar(10) NOT NULL,
  `msg` tinytext NOT NULL,
  PRIMARY KEY (`event_id`,`created_at`)
) ENGINE=InnoDB DEFAULT CHARSET=utf8mb4
PARTITION BY RANGE (to_days(`created_at`))
(PARTITION 2017_oct_week1 VALUES LESS THAN (736974) ENGINE = InnoDB,
PARTITION p20171008 VALUES LESS THAN (736975) ENGINE = InnoDB,
PARTITION p20171009 VALUES LESS THAN (736976) ENGINE = InnoDB,
PARTITION p20171010 VALUES LESS THAN (736977) ENGINE = InnoDB,
PARTITION p20171011 VALUES LESS THAN (736978) ENGINE = InnoDB,
PARTITION p20171012 VALUES LESS THAN (736979) ENGINE = InnoDB,
```

```
PARTITION p20171013 VALUES LESS THAN (736980) ENGINE = InnoDB,
PARTITION p20171014 VALUES LESS THAN (736981) ENGINE = InnoDB,
PARTITION p20171015 VALUES LESS THAN (736982) ENGINE = InnoDB,
PARTITION p20171016 VALUES LESS THAN (736983) ENGINE = InnoDB,
PARTITION p20171017 VALUES LESS THAN (736984) ENGINE = InnoDB);
mysql> INSERT INTO event_history VALUES
(1,'test','2017-10-07','2017-10-08','click','test_message'),
(2,'test','2017-10-08','2017-10-08','click','test_message'),
(3,'test','2017-10-09','2017-10-09','click','test_message'),
(4,'test','2017-10-10','2017-10-10','click','test_message'),
(5,'test','2017-10-11','2017-10-11','click','test_message'),
(6,'test','2017-10-12','2017-10-12','click','test_message'),
(7,'test','2017-10-13','2017-10-13','click','test_message'),
(8,'test','2017-10-14','2017-10-14','click','test_message');
Query OK, 8 rows affected (0.01 sec)
Records: 8  Duplicates: 0  Warnings: 0
```

复制全部的表

1. 在目标库上，创建与源库上定义相同的表：

```
mysql> USE test;
mysql> CREATE TABLE IF NOT EXISTS `event_history`(
  `event_id` int(11) NOT NULL,
  `event_name` varchar(10) DEFAULT NULL,
  `created_at` datetime NOT NULL,
  `last_updated` timestamp NULL DEFAULT CURRENT_TIMESTAMP ON UPDATE
CURRENT_TIMESTAMP,
  `event_type` varchar(10) NOT NULL,
  `msg` tinytext NOT NULL,
  PRIMARY KEY (`event_id`,`created_at`)
) ENGINE=InnoDB DEFAULT CHARSET=utf8mb4
PARTITION BY RANGE (to_days(`created_at`))
(PARTITION 2017_oct_week1 VALUES LESS THAN (736974) ENGINE =
InnoDB,
PARTITION p20171008 VALUES LESS THAN (736975) ENGINE = InnoDB,
PARTITION p20171009 VALUES LESS THAN (736976) ENGINE = InnoDB,
PARTITION p20171010 VALUES LESS THAN (736977) ENGINE = InnoDB,
```

```
PARTITION p20171011 VALUES LESS THAN (736978) ENGINE = InnoDB,
PARTITION p20171012 VALUES LESS THAN (736979) ENGINE = InnoDB,
PARTITION p20171013 VALUES LESS THAN (736980) ENGINE = InnoDB,
PARTITION p20171014 VALUES LESS THAN (736981) ENGINE = InnoDB,
PARTITION p20171015 VALUES LESS THAN (736982) ENGINE = InnoDB,
PARTITION p20171016 VALUES LESS THAN (736983) ENGINE = InnoDB,
PARTITION p20171017 VALUES LESS THAN (736984) ENGINE = InnoDB);
```

2. **在目标库上**，Discard 表空间：

```
mysql> ALTER TABLE event_history DISCARD TABLESPACE;
Query OK, 0 rows affected (0.05 sec)
```

3. **在源库上**，执行 `Flush TABLES FOR EXPORT`：

```
mysql> FLUSH TABLES event_history FOR EXPORT;
Query OK, 0 rows affected (0.00 sec)
```

4. **在源库上**，从数据目录中将所有与表相关的文件（`.ibd`、`.cfg`）复制到目标库的数据目录：

```
shell> sudo scp -i /home/mysql/.ssh/id_rsa
/var/lib/mysql/test/event_history#P#*
mysql@xx.xxx.xxx.xxx:/var/lib/mysql/test/
```

5. **在源库上**，解锁表格，以便进行写操作：

```
mysql> UNLOCK TABLES;
Query OK, 0 rows affected (0.00 sec)
```

6. **在目标库上**，确保这些文件的所有权被设置为 `mysql`：

```
shell> sudo ls -lhtr /var/lib/mysql/test
total 1.4M
-rw-r----- 1 mysql mysql 128K Oct  7 17:17
event_history#P#p20171017.ibd
-rw-r----- 1 mysql mysql 128K Oct  7 17:17
event_history#P#p20171016.ibd
-rw-r----- 1 mysql mysql 128K Oct  7 17:17
event_history#P#p20171015.ibd
```

```
-rw-r----- 1 mysql mysql 128K Oct  7 17:17
event_history#P#p20171014.ibd
-rw-r----- 1 mysql mysql 128K Oct  7 17:17
event_history#P#p20171013.ibd
-rw-r----- 1 mysql mysql 128K Oct  7 17:17
event_history#P#p20171012.ibd
-rw-r----- 1 mysql mysql 128K Oct  7 17:17
event_history#P#p20171011.ibd
-rw-r----- 1 mysql mysql 128K Oct  7 17:17
event_history#P#p20171010.ibd
-rw-r----- 1 mysql mysql 128K Oct  7 17:17
event_history#P#p20171009.ibd
-rw-r----- 1 mysql mysql 128K Oct  7 17:17
event_history#P#p20171008.ibd
-rw-r----- 1 mysql mysql 128K Oct  7 17:17
event_history#P#2017_oct_week1.ibd
```

7. 在目标库上，导入表空间。只要表的定义相同，就可以忽略这些警告。如果已经复制了.cfg 文件，则不会出现警告：

```
mysql> ALTER TABLE event_history IMPORT TABLESPACE;
Query OK, 0 rows affected, 12 warnings (0.31 sec)
```

8. 在目标库上，验证数据：

```
mysql> SELECT * FROM event_history;
+----------+------------+---------------------+--------------------
-+------------+--------------+
| event_id | event_name | created_at          | last_updated
| event_type | msg          |
+----------+------------+---------------------+--------------------
-+------------+--------------+
|        1 | test       | 2017-10-07 00:00:00 | 2017-10-08 00:00:00
| click      | test_message |
|        2 | test       | 2017-10-08 00:00:00 | 2017-10-08 00:00:00
| click      | test_message |
| 3        | test       | 2017-10-09 00:00:00 | 2017-10-09 00:00:00
| click      | test_message |
```

```
| 4          | test          | 2017-10-10 00:00:00 | 2017-10-10 00:00:00
| click       | test_message |
| 5          | test          | 2017-10-11 00:00:00 | 2017-10-11 00:00:00
| click       | test_message |
| 6          | test          | 2017-10-12 00:00:00 | 2017-10-12 00:00:00
| click       | test_message |
| 7          | test          | 2017-10-13 00:00:00 | 2017-10-13 00:00:00
| click       | test_message |
| 8 | test | 2017-10-14 00:00:00 | 2017-10-14 00:00:00
| click       | test_message |
+----------+-----------+-------------------+-------------------
-+-----------+-------------+
8 rows in set (0.00 sec)
```

如果你是在一个生产系统上执行这些操作，为了减少停机时间，可以在本地复制文件，这会非常快。立即执行 `UNLOCK TABLES`，然后将文件复制到目标库。如果你无法忍受停机时间，可以使用 Percona XtraBackup，备份单个表，并应用 UNDO 日志（该日志生成 `.ibd` 文件）。你可以将它们复制到目标库并导入进来。

复制表的各个分区

你在源库上添加了 `events_history` 表的一个新分区，并且希望仅将这个新分区复制到目标库。为了便于理解，我们先在 `events_history` 表上创建新的分区并插入几行数据：

```
mysql> ALTER TABLE event_history ADD PARTITION
 (PARTITION p20171018 VALUES LESS THAN (736985) ENGINE = InnoDB,
 PARTITION p20171019 VALUES LESS THAN (736986) ENGINE = InnoDB);
Query OK, 0 rows affected (0.06 sec)
Records: 0  Duplicates: 0  Warnings: 0

mysql> INSERT INTO event_history VALUES
(9,'test','2017-10-17','2017-10-17','click','test_message'),(10,'test','201
7-10-18','2017-10-18','click','test_message');
Query OK, 1 row affected (0.01 sec)

mysql> SELECT * FROM event_history PARTITION (p20171018,p20171019);
```

```
+----------+-----------+-------------------+--------------------+---
------+-------------+
| event_id | event_name | created_at          | last_updated         |
event_type | msg          |
+----------+-----------+-------------------+--------------------+---
------+-------------+
|        9 | test       | 2017-10-17 00:00:00 | 2017-10-17 00:00:00 | click
| test_message |
|       10 | test       | 2017-10-18 00:00:00 | 2017-10-18 00:00:00 | click
| test_message |
+----------+-----------+-------------------+--------------------+--
------+-------------+
2 rows in set (0.00 sec)
```

假设你希望将新建的分区复制到目标库。

1. 在目标库上，创建分区：

```
mysql> ALTER TABLE event_history ADD PARTITION
(PARTITION p20171018 VALUES LESS THAN (736985) ENGINE = InnoDB,
 PARTITION p20171019 VALUES LESS THAN (736986) ENGINE = InnoDB);
Query OK, 0 rows affected (0.05 sec)
Records: 0  Duplicates: 0  Warnings: 0
```

2. 在目标库上，仅 Discard 你想要导入的分区：

```
mysql> ALTER TABLE event_history DISCARD PARTITION p20171018,
p20171019 TABLESPACE;
 Query OK, 0 rows affected (0.06 sec)
```

3. 在源库上，执行 `FLUSH TABLE FOR EXPORT`：

```
mysql> FLUSH TABLES event_history FOR EXPORT;
Query OK, 0 rows affected (0.01 sec)
```

4. 在源库上，将分区的 `.ibd` 文件复制到目标库：

```
shell> sudo scp -i /home/mysql/.ssh/id_rsa \
/var/lib/mysql/test/event_history#P#p20171018.ibd \
/var/lib/mysql/test/event_history#P#p20171019.ibd \
```

```
mysql@35.198.210.229:/var/lib/mysql/test/
event_history#P#p20171018.ibd                          100%
128KB 128.0KB/s   00:00   event_history#P#p20171019.ibd
100%  128KB 128.0KB/s   00:00
```

5. 在目标库上，确保所需分区的`.ibd`文件已被复制过来了，并将其所有者设置为`mysql`：

```
shell> sudo ls -lhtr
/var/lib/mysql/test/event_history#P#p20171018.ibd
-rw-r----- 1 mysql mysql 128K Oct  7 17:54
/var/lib/mysql/test/event_history#P#p20171018.ibd

shell> sudo ls -lhtr
/var/lib/mysql/test/event_history#P#p20171019.ibd
-rw-r----- 1 mysql mysql 128K Oct  7 17:54
/var/lib/mysql/test/event_history#P#p20171019.ibd
```

6. 在目标库上，执行`IMPORT PARTITION TABLESPACE`：

```
mysql> ALTER TABLE event_history IMPORT PARTITION p20171018,
p20171019 TABLESPACE;
Query OK, 0 rows affected, 2 warnings (0.10 sec)
```

只要表的定义相同，就可以忽略这些警告。如果你已经复制了`.cfg`文件，就不会有警告：

```
mysql> SHOW WARNINGS;
+---------+------+-----------------------------------------------------
------------------------------------------------------------------------
------------------------------------------+
| Level   | Code | Message
|
+---------+------+-----------------------------------------------------
------------------------------------------------------------------------
------------------------------------------+
| Warning | 1810 | InnoDB: IO Read error: (2, No such file or
directory) Error opening './test/event_history#P#p20171018.cfg',
will attempt to import without schema verification |
```

```
| Warning | 1810 | InnoDB: IO Read error: (2, No such file or
directory) Error opening './test/event_history#P#p20171019.cfg',
will attempt to import without schema verification |
+---------+------+--------------------------------------------------
--------------------------------------------------------------------
---------------------------------------------+
2 rows in set (0.00 sec)
```

7. 在目标库上，验证数据：

```
mysql> SELECT * FROM event_history PARTITION (p20171018,p20171019);
+----------+------------+---------------------+--------------------
-+------------+-------------+
| event_id | event_name | created_at          | last_updated
| event_type | msg         |
+----------+------------+---------------------+--------------------
-+------------+-------------+
| 9        | test       | 2017-10-17 00:00:00 | 2017-10-17 00:00:00
| click      | test_message    |
| 10       | test       | 2017-10-18 00:00:00 | 2017-10-18 00:00:00
| click      | test_message    |
+----------+------------+---------------------+--------------------
-+------------+-------------+
2 rows in set (0.00 sec)
```

11.5.2　延伸阅读

可参考文档 https://dev.mysql.com/doc/refman/8.0/en/tablespace-copying.html 以了解更多关于此过程的局限性。

11.6　管理 UNDO 表空间

你可以通过动态变量 innodb_max_undo_log_size（默认值为 1 GB）来管理 UNDO 表空间的大小，以及通过 innodb_undo_tablespaces（默认值为 2 GB，从 MySQL 8.0.2 开始，该变量为动态变量）来调整 UNDO 表空间的数量。

默认情况下，innodb_undo_log_truncate 是被启用的。超过 innodb_max_

undo_log_size 定义的阈值的表空间被标记为被截断。只有 UNDO 表空间才能被截断。不支持在系统表空间中删除驻留的 UNDO 日志。要实现截断，必须至少有两个 UNDO 表空间。

11.6.1 如何操作

验证 UNDO 日志的大小：

```
shell> sudo ls -lhtr /var/lib/mysql/undo_00*
-rw-r-----. 1 mysql mysql 19M Oct  7 17:43 /var/lib/mysql/undo_002
-rw-r-----. 1 mysql mysql 16M Oct  7 17:43 /var/lib/mysql/undo_001
```

假设你希望减小超过 15 MB 的文件的体积。请记住，只有一个 UNDO 表空间可以被截断。为了避免每次都截断同一个 UNDO 表空间，将以循环方式选择 UNDO 表空间。在 UNDO 表空间中的所有回滚段被释放后，执行截断操作，此 UNDO 表空间就被截断为初始大小。UNDO 表空间文件的初始大小为 10 MB：

1. 确保启用了 innodb_undo_log_truncate：

```
mysql> SELECT @@GLOBAL.innodb_undo_log_truncate;
+-----------------------------------+
| @@GLOBAL.innodb_undo_log_truncate |
+-----------------------------------+
|                                 1 |
+-----------------------------------+
1 row in set (0.00 sec)
```

2. 将 innodb_max_undo_log_size 设置为 15 MB：

```
mysql> SELECT @@GLOBAL.innodb_max_undo_log_size;
+-----------------------------------+
| @@GLOBAL.innodb_max_undo_log_size |
+-----------------------------------+
|                        1073741824 |
+-----------------------------------+
1 row in set (0.00 sec)
mysql> SET @@GLOBAL.innodb_max_undo_log_size=15*1024*1024;
Query OK, 0 rows affected (0.00 sec)
```

```
mysql> SELECT @@GLOBAL.innodb_max_undo_log_size;
+-----------------------------------+
| @@GLOBAL.innodb_max_undo_log_size |
+-----------------------------------+
|                          15728640 |
+-----------------------------------+
1 row in set (0.00 sec)
```

3. 在回滚段被释放之前，UNDO 表空间不能被截断。正常情况下，清除系统（purge system）每被调用 128 次就会释放一次回滚段。为了加快 UNDO 表空间的截断，可以使用 innodb_purge_rseg_truncate_frequency 选项临时增加清除系统释放回滚段的频率：

```
mysql> SELECT @@GLOBAL.innodb_purge_rseg_truncate_frequency;
+-----------------------------------------------+
| @@GLOBAL.innodb_purge_rseg_truncate_frequency |
+-----------------------------------------------+
|                                           128 |
+-----------------------------------------------+
1 row in set (0.00 sec)

mysql> SET @@GLOBAL.innodb_purge_rseg_truncate_frequency=1;
Query OK, 0 rows affected (0.00 sec)

mysql> SELECT @@GLOBAL.innodb_purge_rseg_truncate_frequency;
+-----------------------------------------------+
| @@GLOBAL.innodb_purge_rseg_truncate_frequency |
+-----------------------------------------------+
|                                             1 |
+-----------------------------------------------+
1 row in set (0.00 sec)
```

4. 通常在繁忙的系统中，至少有一个清除操作已经被初始化，而且应该已经启动截断。如果你是在自己的机器上练习，可以通过创建一个大的事务来启动清除操作：

```
mysql> BEGIN;
Query OK, 0 rows affected (0.00 sec)
```

```
mysql> DELETE FROM employees;
Query OK, 300025 rows affected (16.23 sec)

mysql> ROLLBACK;
Query OK, 0 rows affected (2.38 sec)
```

5. 在删除的过程中，可以看到 UNDO 日志文件在增长：

```
shell> sudo ls -lhtr /var/lib/mysql/undo_00*
-rw-r-----. 1 mysql mysql 19M Oct  7 17:43 /var/lib/mysql/undo_002
-rw-r-----. 1 mysql mysql 16M Oct  7 17:43 /var/lib/mysql/undo_001

shell> sudo ls -lhtr /var/lib/mysql/undo_00*
-rw-r-----. 1 mysql mysql 10M Oct  8 04:52 /var/lib/mysql/undo_001
-rw-r-----. 1 mysql mysql 27M Oct  8 04:52 /var/lib/mysql/undo_002

shell> sudo ls -lhtr /var/lib/mysql/undo_00*
-rw-r-----. 1 mysql mysql 10M Oct  8 04:52 /var/lib/mysql/undo_001
-rw-r-----. 1 mysql mysql 28M Oct  8 04:52 /var/lib/mysql/undo_002

shell> sudo ls -lhtr /var/lib/mysql/undo_00*
-rw-r-----. 1 mysql mysql 10M Oct  8 04:52 /var/lib/mysql/undo_001
-rw-r-----. 1 mysql mysql 29M Oct  8 04:52 /var/lib/mysql/undo_002

shell> sudo ls -lhtr /var/lib/mysql/undo_00*
-rw-r-----. 1 mysql mysql 10M Oct  8 04:52 /var/lib/mysql/undo_001
-rw-r-----. 1 mysql mysql 29M Oct  8 04:52 /var/lib/mysql/undo_002
```

你可以看到 undo_001 被截断为 10 MB，而 undo_002 还在增长，它为 DELETE 语句删除的行提供了空间。

6. 一段时间之后，你可以注意到 unod_002 也被截断为 10 MB：

```
shell> sudo ls -lhtr /var/lib/mysql/undo_00*
-rw-r-----. 1 mysql mysql 10M Oct  8 04:52 /var/lib/mysql/undo_001
-rw-r-----. 1 mysql mysql 10M Oct  8 04:54 /var/lib/mysql/undo_002
```

7. 一旦缩小了 UNDO 表空间，就将 innodb_purge_rseg_truncate_frequency 设置为默认值 128：

```
mysql> SELECT @@GLOBAL.innodb_purge_rseg_truncate_frequency;
+-----------------------------------------------+
| @@GLOBAL.innodb_purge_rseg_truncate_frequency |
+-----------------------------------------------+
|                                             1 |
+-----------------------------------------------+
1 row in set (0.00 sec)

mysql> SET @@GLOBAL.innodb_purge_rseg_truncate_frequency=128;
Query OK, 0 rows affected (0.00 sec)

mysql> SELECT @@GLOBAL.innodb_purge_rseg_truncate_frequency;
+-----------------------------------------------+
| @@GLOBAL.innodb_purge_rseg_truncate_frequency |
+-----------------------------------------------+
|                                           128 |
+-----------------------------------------------+
1 row in set (0.01 sec)
```

11.7　管理通用表空间

在 MySQL 8 之前，有两种类型的表空间：系统表空间和单个表空间。这两种类型各有优缺点。为了克服它们的缺点，MySQL 8 引入了通用表空间。与系统表空间类似，通用表空间是共享的表空间，可以存储多个表的数据。但是，通用表空间更好控制。为了存储表的元数据信息，多个表使用更小的通用表空间所消耗的内存比相同数量的表使用单独的独立表空间所耗费的内存要少。

其局限性如下：

- 与系统表空间类似，截断（truncate）或删除（drop）存储在通用表空间中的表会在通用表空间的.ibd 数据文件内部创建空闲空间（free space），但只能用于新的 InnoDB 数据。这些空间不会被释放给操作系统，就像用于独立表空间的空间一样。
- 对于属于通用表空间的表，不支持对表空间传输。

在本节中，你将学习如何创建一个通用表空间，并从其中添加和删除表。

实际的用法

最初，InnoDB 维护着一个.frm 文件，此文件中包含表结构。MySQL 需要打开和关闭.frm 文件，而这会降低性能。在 MySQL 8 中，.frm 文件被删除，所有元数据都使用事务性数据字典处理。这就使通用表空间有了用武之地。

假设你用的是 MySQL 5.7，或者是用于 SaaS 或多租户的更早的 MySQL 版本，这些版本的 MySQL 中每个客户都有一个单独的模式，每个客户都有数百个表。如果你的客户不断增长，你将会碰到不少性能问题。但是 MySQL 8 删除了.frm 文件，性能得到了极大的提高。此外，你可以为每个模式（客户）创建一个单独的表空间。

11.7.1 如何操作

让我们首先从创建通用表空间开始。

创建通用表空间

你可以在 MySQL 数据目录中或在它的外部创建一个通用表空间。

在 MySQL 数据目录中创建一个通用表空间：

```
mysql> CREATE TABLESPACE `ts1` ADD DATAFILE 'ts1.ibd' Engine=InnoDB;
Query OK, 0 rows affected (0.02 sec)
```

要在 MySQL 数据目录之外创建表空间，请在/var/lib/mysql_general_ts 上挂载新磁盘并将其所有权改为 mysql：

```
shell> sudo chown mysql:mysql /var/lib/mysql_general_ts

mysql> CREATE TABLESPACE `ts2` ADD DATAFILE
'/var/lib/mysql_general_ts/ts2.ibd' Engine=InnoDB;Query OK, 0 rows affected
(0.02 sec)
```

向通用表空间中添加表

你可以在创建表空间时向其中添加一个表，或者运行 ALTER 命令将一个表从一个表空间移到另一个表空间：

```
mysql> CREATE TABLE employees.table_gen_ts1 (id INT PRIMARY KEY) TABLESPACE
ts1;
Query OK, 0 rows affected (0.01 sec)
```

假设你想将 employees 表移到 TABLESPACE ts2:

```
mysql> USE employees;
Database changed

mysql> ALTER TABLE employees TABLESPACE ts2;
Query OK, 0 rows affected (3.93 sec)
Records: 0  Duplicates: 0  Warnings: 0
```

你可能注意到 ts2.ibd 文件变大了:

```
shell> sudo ls -lhtr /var/lib/mysql_general_ts/ts2.ibd
-rw-r-----. 1 mysql mysql 32M Oct  8 17:07
/var/lib/mysql_general_ts/ts2.ibd
```

在表空间之间移动非分区表

你可以按照如下步骤来移动表。

1. 以下是将表从一个通用表空间移动到另一个通用表空间的方法。假设你希望将 employees 表从 ts2 移到 ts1:

   ```
   mysql> ALTER TABLE employees TABLESPACE ts1;
   Query OK, 0 rows affected (3.83 sec)
   Records: 0  Duplicates: 0  Warnings: 0

   shell> sudo ls -lhtr /var/lib/mysql/ts1.ibd
   -rw-r-----. 1 mysql mysql 32M Oct  8 17:16 /var/lib/mysql/ts1.ibd
   ```

2. 以下是将表移动到独立表空间的方法。假设你想要将 employee 表从 ts1 移到独立表空间:

   ```
   mysql> ALTER TABLE employees TABLESPACE innodb_file_per_table;
   Query OK, 0 rows affected (4.05 sec)
   Records: 0  Duplicates: 0  Warnings: 0
   ```

```
shell> sudo ls -lhtr /var/lib/mysql/employees/employees.ibd
-rw-r-----. 1 mysql mysql 32M Oct  8 17:18
/var/lib/mysql/employees/employees.ibd
```

3. 以下是将表移到系统表空间的方法。假设你希望将 employees 表从独立表空间移到系统表空间：

```
mysql> ALTER TABLE employees TABLESPACE innodb_system;
Query OK, 0 rows affected (5.28 sec)
    Records: 0  Duplicates: 0  Warnings: 0
```

管理通用表空间中的分区表

你可以在多个表空间中创建带有分区的表：

```
mysql> CREATE TABLE table_gen_part_ts1 (id INT, value varchar(100)) ENGINE
= InnoDB
       PARTITION BY RANGE(id) (
        PARTITION p1 VALUES LESS THAN (1000000) TABLESPACE ts1,
        PARTITION p2 VALUES LESS THAN (2000000) TABLESPACE ts2,
        PARTITION p3 VALUES LESS THAN (3000000) TABLESPACE
innodb_file_per_table,
        PARTITION pmax VALUES LESS THAN (MAXVALUE) TABLESPACE
innodb_system);
Query OK, 0 rows affected (0.19 sec)
```

可以在另一个表空间中添加一个新分区，不然如果没有明确指定，将在表的默认表空间中创建新分区。在分区表上的 ALTER TABLE tbl_name TABLESPACE tablespace_name 操作只会修改表的默认表空间，不移动表分区。但是，在更改了默认的表空间之后，如果没有使用 TABLESPACE 子句明确定义另一个表空间，那些重新构建表的操作（如使用 ALGORITHM=COPY 的 ALTER TABLE 操作）就会将分区移到默认的表空间。

如果你希望跨表空间移动分区，则需要对分区进行 REORGANIZE 操作。例如，你希望将分区 p3 移动到 ts2：

```
mysql> ALTER TABLE table_gen_part_ts1 REORGANIZE PARTITION p3 INTO
(PARTITION p3 VALUES LESS THAN (3000000) TABLESPACE ts2);
```

删除通用表空间

你可以使用 DROP TABLESPACE 命令来删除表空间。但是，表空间内的所有表都将被删除或移动：

```
mysql> DROP TABLESPACE ts2;
ERROR 3120 (HY000): Tablespace `ts2` is not empty.
```

你必须在删除表空间之前将 ts2 表空间中 table_gen_part_ts1 表的 p2 和 p3 分区移到其他表空间：

```
mysql> ALTER TABLE table_gen_part_ts1 REORGANIZE PARTITION p2 INTO
(PARTITION p2 VALUES LESS THAN (3000000) TABLESPACE ts1);

mysql> ALTER TABLE table_gen_part_ts1 REORGANIZE PARTITION p3 INTO
(PARTITION p3 VALUES LESS THAN (3000000) TABLESPACE ts1);
```

现在可以删除表空间了：

```
mysql> DROP TABLESPACE ts2;
Query OK, 0 rows affected (0.01 sec)
```

11.8　压缩 InnoDB 表

你可以创建以压缩格式存储数据的表。压缩有助于提高性能和可伸缩性。压缩意味着在磁盘和内存之间传输的数据更少，并且占用的磁盘和内存空间更少。

根据 MySQL 文档：

由于处理器和缓存的速度比磁盘存储设备快，因此许多工作负载都是跟磁盘相关的。数据压缩可以减小数据库的大小、减少 I/O 和提高吞吐量，因为 CPU 的利用率提高了。压缩对于读密集型应用程序来说特别有价值，因为在内存中有足够的 RAM 来保存需要频繁使用的数据。而因为索引数据也被压缩了，所以使用二级索引表的好处就更多了。

要启用压缩，你需要使用 ROW_FORMAT=COMPRESSED KEY_BLOCK_SIZE 选项来创建或更改表。你可以更改 KEY_BLOCK_SIZE 参数，该参数使用磁盘上较小的页大小，而不是配置的 innodb_page_size 值。如果表位于系统表空间中，那么压缩就不起作用。

要在通用表空间中创建压缩表，必须为通用表空间定义 FILE_BLOCK_SIZE，这个值是在创建表空间时指定的。FILE_BLOCK_SIZE 必须是与 innodb_page_size 值相关的一个有效的压缩页面大小值。压缩表的页面大小必须等于 FILE_BLOCK_SIZE/1024，且由 CREATE TABLE 或 ALTER TABLE KEY_BLOCK_SIZE 子句定义。

在缓冲池中，压缩的数据保存在小页面中，页面大小由 KEY_BLOCK_SIZE 值指定。为了提取或更新列值，MySQL 还使用未压缩的数据在缓冲池中创建一个未压缩的页面。在缓冲池中，对未压缩页的任何更新也将被重写回等价的压缩页。你可能需要调整缓冲池的大小，以容纳压缩过的和未压缩的页面的额外数据，尽管在需要空间时，未压缩的页面会从缓冲池中被逐出，但是在下次需要访问它们时，会再次解压。

什么时候使用压缩

一般情况下，对于包含的字符串列不是特别多且数据读取的次数要比写入的次数多得多的表，压缩最有效。因为没有可靠的方法来预测压缩是否有利于某一种情况，所以总是需要利用特定的工作负载和运行在具有代表性配置上的数据集来进行测试。

11.8.1 如何操作

你需要选择参数 KEY_BLOCK_SIZE。innodb_page_size 的值是 16,000；理想情况下，其一半是 8000，以这个值作为起点不错。要优化压缩，请参考文档 https://dev.mysql.com/doc/refman/8.0/en/innodb-compression-tuning.html。

为 file_per_table 表启用压缩

1. 确保已启用 file_per_table：

```
mysql> SET GLOBAL innodb_file_per_table=1;
```

2. 在创建语句中指定 ROW_FORMAT=COMPRESSED KEY_BLOCK_SIZE=8：

```
mysql> CREATE TABLE compressed_table (id INT PRIMARY KEY)
ROW_FORMAT=COMPRESSED KEY_BLOCK_SIZE=8;
Query OK, 0 rows affected (0.07 sec)
```

如果这个表已经存在，则执行 ALTER：

```
mysql> ALTER TABLE event_history ROW_FORMAT=COMPRESSED KEY_BLOCK_SIZE=8;
```

```
Query OK, 0 rows affected (0.67 sec)
Records: 0  Duplicates: 0  Warnings: 0
```

如果你试图压缩系统表空间中的表，将会收到一个报错信息：

```
mysql> ALTER TABLE employees ROW_FORMAT=COMPRESSED KEY_BLOCK_SIZE=8;
ERROR 1478 (HY000): InnoDB: Tablespace `innodb_system` cannot contain a
COMPRESSED table
```

为 file_per_table 表禁用压缩

要禁用压缩，请执行 ALTER 表，并指定 ROW_FORMAT=DYNAMIC 或 ROW_FORMAT=COMPACT，然后设定 KEY_BLOCK_SIZE = 0。

例如，你不想在 event_history 表上进行压缩：

```
mysql> ALTER TABLE event_history ROW_FORMAT=DYNAMIC KEY_BLOCK_SIZE=0;
Query OK, 0 rows affected (0.53 sec)
Records: 0  Duplicates: 0  Warnings: 0
```

为通用表空间启用压缩

首先，你需要通过 FILE_BLOCK_SIZE 来创建压缩表空间，注意，不能更改表空间的 FILE_BLOCK_SIZE。

如果希望创建一个压缩表，则需要在启用压缩的通用表空间中创建这个表；而且 KEY_BLOCK_SIZE 必须等于 FILE_BLOCK_SIZE/1024。如果你没有提到 KEY_BLOCK_SIZE，那么将自动从 FILE_BLOCK_SIZE 中获取该值。

你可以用不同的 FILE_BLOCK_SIZE 值创建多个压缩的通用表空间，并将表添加到所需的表空间。

1. 创建一个通用压缩表空间。你可以以 FILE_BLOCK_SIZE 为 8 KB 创建一个表空间，用 FILE_BLOCK_SIZE 为 4 KB 再创建一个，并将所有 KEY_BLOCK_SIZE 为 8 的表移动到第一个表空间（FILE_BLOCK_SIZE 为 8 KB），将 KEY_BLOCK_SIZE 为 4 的表移动到第二个表空间（FILE_BLOCK_SIZE 为 4 KB）：

   ```
   mysql> CREATE TABLESPACE `ts_8k` ADD DATAFILE 'ts_8k.ibd'
   FILE_BLOCK_SIZE = 8192 Engine=InnoDB;
   ```

```
Query OK, 0 rows affected (0.01 sec)

mysql> CREATE TABLESPACE `ts_4k` ADD DATAFILE 'ts_4k.ibd'
FILE_BLOCK_SIZE = 4096 Engine=InnoDB;
Query OK, 0 rows affected (0.04 sec)
```

2. 通过 ROW_FORMAT=COMPRESSED 创建这些表空间的压缩表：

```
mysql> CREATE TABLE compress_table_1_8k (id INT PRIMARY KEY)
TABLESPACE ts_8k ROW_FORMAT=COMPRESSED;
Query OK, 0 rows affected (0.01 sec)
```

如果没有使用 ROW_FORMAT=COMPRESSED，将会出现错误：

```
mysql> CREATE TABLE compress_table_2_8k (id INT PRIMARY KEY)
TABLESPACE ts_8k;
ERROR 1478 (HY000): InnoDB: Tablespace `ts_8k` uses block size 8192
and cannot contain a table with physical page size 16384
```

或者，你可以指定 KEY_BLOCK_SIZE=FILE_BLOCK_SIZE/1024：

```
mysql> CREATE TABLE compress_table_8k (id INT PRIMARY KEY)
TABLESPACE ts_8k ROW_FORMAT=COMPRESSED KEY_BLOCK_SIZE=8;
Query OK, 0 rows affected (0.01 sec)
```

如果你指定的是除 FILE_BLOCK_SIZE/1024 之外的其他值，将会收到报错信息：

```
mysql> CREATE TABLE compress_table_2_8k (id INT PRIMARY KEY)
TABLESPACE ts_8k ROW_FORMAT=COMPRESSED KEY_BLOCK_SIZE=4;
ERROR 1478 (HY000): InnoDB: Tablespace `ts_8k` uses block size 8192
and cannot contain a table with physical page size 4096
```

3. 只有当 KEY_BLOCK_SIZE 取了合适的值时，才可以将表从 file_per_table 表空间移到压缩的通用表空间。否则，将会出现错误：

```
mysql> CREATE TABLE compress_tables_4k (id INT PRIMARY KEY)
TABLESPACE innodb_file_per_table ROW_FORMAT=COMPRESSED
KEY_BLOCK_SIZE=4;
Query OK, 0 rows affected (0.02 sec)
```

```
mysql> ALTER TABLE compress_tables_4k TABLESPACE ts_4k;
Query OK, 0 rows affected (0.02 sec)
Records: 0  Duplicates: 0  Warnings: 0

mysql> ALTER TABLE compress_tables_4k TABLESPACE ts_8k;
ERROR 1478 (HY000): InnoDB: Tablespace `ts_8k` uses block size 8192
and cannot contain a table with physical page size 4096
```

第12章 日志管理

本章涵盖以下几个方面的内容：

- 错误日志的管理
- 通用查询日志和慢查询日志的管理
- 二进制日志的管理

12.1 引言

本章我们将学习如何管理不同类型的日志：错误日志、通用查询日志、慢查询日志、二进制日志、中继日志和 DDL 日志。

12.2 管理错误日志

根据 MySQL 文档：

错误日志包含了 `mysqld` 的启动和宕机次数的记录，还包含一些诊断信息，例如：错误、警告，以及服务器在启动、运行及关闭期间发出的提示信息。

错误日志子系统由两个组件组成，它们筛选和写入日志事件，还有一个名为 `log_error_services` 的系统变量，这个系统变量对组件进行配置以实现所需的日志记录结果。`global.log_error_services` 的默认值为 `log_filter_internal; log_sink_internal`。

```
mysql> SELECT @@global.log_error_services;
+----------------------------------------+
| @@global.log_error_services            |
+----------------------------------------+
| log_filter_internal; log_sink_internal |
+----------------------------------------+
```

该值表示日志事件首先穿过内置的筛选器组件 log_filter_internal，然后穿过内置的日志写入器组件 log_sink_internal。组件顺序是非常重要的，因为服务器是按照组件被列出的顺序执行的。在 log_error_services 的值中指定的任何可加载（非内置）组件都必须首先通过 INSTALL COMPONENT 安装。本节将对此进行介绍。

想要了解所有类型的错误日志，请参考文档：https://dev.mysql.com/doc/refman/8.0/en/error-log.html。

12.2.1　如何操作

在某种程度上而言，错误日志很简单。首先让我们来看如何配置一个错误日志。

配置错误日志

错误日志记录由 log_error 变量（在启动脚本里为--log-error）控制。

如果没有给出--log-error，默认的目标文件是控制台。

如果在没有命名文件的情况下给出了--log-error，则默认的目标文件是一个在数据目录中名为 host_name.err 的文件。

如果--log-error 被指定来命名一个文件，默认的目标文件就是该文件（如果文件名没有后缀，则添加一个.err 后缀），如果没有用一个绝对路径名来指定别的位置，那么这个文件就位于数据目录下。

系统变量 log_error_verbosity 控制着服务器将错误、警告和注释信息记录到错误日志的冗余情况。可以使用的 log_error_verbosity 值有 1（只输出错误）、2（输出错误和警告）和 3（输出错误、警告和注释），其默认值为 3。

要改变错误日志的位置，请修改配置文件并重新启动 MySQL：

```
shell> sudo mkdir /var/log/mysql
shell> sudo chown -R mysql:mysql /var/log/mysql

shell> sudo vi /etc/my.cnf
[mysqld]
log-error=/var/log/mysql/mysqld.log

shell> sudo systemctl restart mysql
```

验证错误日志：

```
mysql> SHOW VARIABLES LIKE 'log_error';
+---------------+---------------------------+
| Variable_name | Value                     |
+---------------+---------------------------+
| log_error     | /var/log/mysql/mysqld.log |
+---------------+---------------------------+
1 row in set (0.00 sec)
```

要调整冗余信息，可以动态更改 `log_error_verbosity` 变量。建议采用默认值 3，这样，错误、警告和注释消息就都可以记录下来了：

```
mysql> SET @@GLOBAL.log_error_verbosity=2;
Query OK, 0 rows affected (0.00 sec)

mysql> SELECT @@GLOBAL.log_error_verbosity;
+------------------------------+
| @@GLOBAL.log_error_verbosity |
+------------------------------+
|                            2 |
+------------------------------+
1 row in set (0.00 sec)
```

轮转错误日志

假设错误日志文件变大了，你想要轮转（rotate）它。可以简单地移动文件并执行 `FLUSH LOGS` 命令：

```
shell> sudo mv /var/log/mysql/mysqld.log /var/log/mysql/mysqld.log.0;
```

```
shell> mysqladmin -u root -p<password> flush-logs
mysqladmin: [Warning] Using a password on the command line interface can be
insecure.

shell> ls -lhtr /var/log/mysql/mysqld.log
-rw-r-----. 1 mysql mysql 0 Oct 10 14:03 /var/log/mysql/mysqld.log

shell> ls -lhtr /var/log/mysql/mysqld.log.0
-rw-r-----. 1 mysql mysql 3.4K Oct 10 14:03 /var/log/mysql/mysqld.log.0
```

你可以使用一些脚本自动执行前面的步骤，并且把它们放入 cron 中。如果错误日志文件的位置无法由服务器写入，则日志刷新操作将无法创建新的日志文件：

```
shell> sudo mv /var/log/mysqld.log /var/log/mysqld.log.0 && mysqladmin
flush-logs -u root -p<password>
mysqladmin: [Warning] Using a password on the command line interface can be
insecure.
mysqladmin: refresh failed; error: 'Unknown error'
```

使用系统日志来记录日志

想要使用系统日志来记录日志，需要加载名为 log_sink_syseventlog 的系统日志写入器。你可以使用内置筛选器 log_filter_internal 进行筛选。

1. 加载系统日志写入器：

```
mysql> INSTALL COMPONENT 'file://component_log_sink_syseventlog';
Query OK, 0 rows affected (0.43 sec)
```

2. 使它在重新启动后一直生效：

```
mysql> SET PERSIST log_error_services = 'log_filter_internal;
log_sink_syseventlog';
Query OK, 0 rows affected (0.00 sec)

mysql> SHOW VARIABLES LIKE 'log_error_services';
+--------------------+---------------------------------------------+
| Variable_name      | Value                                       |
+--------------------+---------------------------------------------+
```

```
| log_error_services | log_filter_internal; log_sink_syseventlog |
+--------------------+-------------------------------------------+
1 row in set (0.00 sec)
```

3. 你可以验证，日志将被定向到系统日志。在 CentOS 和 Red Hat 系统中，可以用 `/var/log/ messages` 查看；在 Ubuntu 中，可以用 `/var/log/syslog` 查看。

   这里为了进行演示，会重新启动服务器。你可以在系统日志中看到这些日志：

```
shell> sudo grep mysqld /var/log/messages | tail
Oct 10 14:50:31 centos7 mysqld[20953]: InnoDB: Buffer pool(s) dump
completed at 171010 14:50:31
Oct 10 14:50:32 centos7 mysqld[20953]: InnoDB: Shutdown completed;
log sequence number 350327631
Oct 10 14:50:32 centos7 mysqld[20953]: InnoDB: Removed temporary
tablespace data file: "ibtmp1"
Oct 10 14:50:32 centos7 mysqld[20953]: Shutting down plugin
'MEMORY'
Oct 10 14:50:32 centos7 mysqld[20953]: Shutting down plugin 'CSV'
Oct 10 14:50:32 centos7 mysqld[20953]: Shutting down plugin
'sha256_password'
Oct 10 14:50:32 centos7 mysqld[20953]: Shutting down plugin
'mysql_native_password'
Oct 10 14:50:32 centos7 mysqld[20953]: Shutting down plugin
'binlog'
Oct 10 14:50:32 centos7 mysqld[20953]: /usr/sbin/mysqld: Shutdown
complete
Oct 10 14:50:33 centos7 mysqld[21220]: /usr/sbin/mysqld: ready for
connections. Version: '8.0.3-rc-log' socket:
'/var/lib/mysql/mysql.sock' port: 3306 MySQL Community Server
(GPL)
```

如果你运行了多个 `mysqld` 进程，可以使用 `[]` 中指定的 PID 来区分它们。或者，设置 `log_syslog_tag` 变量，这个变量将服务器标识符附加到一个主要的字符后面，产生一个 `mysqld-tag_val` 标识符。

例如，你可以使用类似 `instance1` 的东西标记实例：

```
mysql> SELECT @@GLOBAL.log_syslog_tag;
+-------------------------+
| @@GLOBAL.log_syslog_tag |
+-------------------------+
|                         |
+-------------------------+
1 row in set (0.00 sec)

mysql> SET @@GLOBAL.log_syslog_tag='instance1';
Query OK, 0 rows affected (0.00 sec)

mysql> SELECT @@GLOBAL.log_syslog_tag;
+-------------------------+
| @@GLOBAL.log_syslog_tag |
+-------------------------+
| instance1               |
+-------------------------+
1 row in set (0.01 sec)

shell> sudo systemctl restart mysqld

shell> sudo grep mysqld /var/log/messages | tail
Oct 10 14:59:20 centos7 mysqld-instance1[21220]: InnoDB: Buffer pool(s)
dump completed at 171010 14:59:20
Oct 10 14:59:21 centos7 mysqld-instance1[21220]: InnoDB: Shutdown
completed; log sequence number 350355306
Oct 10 14:59:21 centos7 mysqld-instance1[21220]: InnoDB: Removed temporary
tablespace data file: "ibtmp1"
Oct 10 14:59:21 centos7 mysqld-instance1[21220]: Shutting down plugin
'MEMORY'
Oct 10 14:59:21 centos7 mysqld-instance1[21220]: Shutting down plugin 'CSV'
Oct 10 14:59:21 centos7 mysqld-instance1[21220]: Shutting down plugin
'sha256_password'
Oct 10 14:59:21 centos7 mysqld-instance1[21220]: Shutting down plugin
'mysql_native_password'
Oct 10 14:59:21 centos7 mysqld-instance1[21220]: Shutting down plugin
'binlog'
```

```
Oct 10 14:59:21 centos7 mysqld-instance1[21220]: /usr/sbin/mysqld: Shutdown
complete
Oct 10 14:59:22 centos7 mysqld[21309]: /usr/sbin/mysqld: ready for
connections. Version: '8.0.3-rc-log'  socket: '/var/lib/mysql/mysql.sock'
port: 3306  MySQL Community Server (GPL)
```

你会注意到，instance1 标记被附加到日志中，因此可以从多个实例中很容易地识别出来。

如果你希望切换回原始日志，可以把 log_error_services 设置为'log_filter_internal; log_sink_internal'：

```
mysql> SET @@global.log_error_services='log_filter_internal;
log_sink_internal';
Query OK, 0 rows affected (0.00 sec)
```

用 JSON 格式记录错误日志

想要使用JSON格式记录日志，需要加载名为 log_sink_json 的 JSON 日志记录器。你可以使用内置的筛选器 log_filter_internal 来进行筛选。

1. 安装 JSON 日志写入器：

```
mysql> INSTALL COMPONENT 'file://component_log_sink_json';
Query OK, 0 rows affected (0.05 sec)
```

2. 使它在重新启动后一直生效：

```
mysql> SET PERSIST log_error_services = 'log_filter_internal;
log_sink_json';
Query OK, 0 rows affected (0.00 sec)
```

3. JSON 日志记录器根据默认的错误日志目标文件决定其输出目标文件，该目标文件由 log_error 系统变量给出：

```
mysql> SHOW VARIABLES LIKE 'log_error';
+---------------+--------------------------+
| Variable_name | Value                    |
+---------------+--------------------------+
| log_error     | /var/log/mysql/mysqld.log |
```

```
+--------------+--------------------------+
1 row in set (0.00 sec)
```

4. 这种日志文件类似于 mysqld.log.00.json。重启以后，JSON 日志文件是这样的：

```
shell> sudo less /var/log/mysql/mysqld.log.00.json
{ "prio" : 2, "err_code" : 4356, "subsystem" : "", "SQL_state" :
"HY000", "source_file" : "sql_plugin.cc", "function" :
"reap_plugins", "msg" : "Shutting down plugin 'sha256_password'",
"time" : "2017-10-15T12:29:08.862969Z", "err_symbol" :
"ER_PLUGIN_SHUTTING_DOWN_PLUGIN", "label" : "Note" }
{ "prio" : 2, "err_code" : 4356, "subsystem" : "", "SQL_state" :
"HY000", "source_file" : "sql_plugin.cc", "function" :
"reap_plugins", "msg" : "Shutting down plugin
'mysql_native_password'", "time" : "2017-10-15T12:29:08.862975Z",
"err_symbol" : "ER_PLUGIN_SHUTTING_DOWN_PLUGIN", "label" : "Note" }
{ "prio" : 2, "err_code" : 4356, "subsystem" : "", "SQL_state" :
"HY000", "source_file" : "sql_plugin.cc", "function" :
"reap_plugins", "msg" : "Shutting down plugin 'binlog'", "time" :
"2017-10-15T12:29:08.863758Z", "err_symbol" :
"ER_PLUGIN_SHUTTING_DOWN_PLUGIN", "label" : "Note" }
{  "prio" : 2, "err_code" : 1079, "subsystem" : "", "SQL_state" :
"HY000",  "source_file" : "mysqld.cc", "function" : "clean_up",
"msg" : "/usr/sbin/mysqld: Shutdown complete\u000a", "time" :
"2017-10-15T12:29:08.867077Z", "err_symbol" :
"ER_SHUTDOWN_COMPLETE", "label" : "Note" }
{ "log_type" : 1, "prio" : 0, "err_code" : 1408, "msg" :
"/usr/sbin/mysqld: ready for connections. Version: '8.0.3-rc-log'
socket: '/var/lib/mysql/mysql.sock'  port: 3306  MySQL Community
Server (GPL)", "time" : "2017-10-15T12:29:10.952502Z", "err_symbol"
: "ER_STARTUP", "SQL_state" : "HY000", "label" : "Note" }
```

如果你希望切换回原始日志，可以把 log_error_services 设置为'log_filter_internal; log_sink_internal'：

```
mysql> SET @@global.log_error_services='log_filter_internal;
log_sink_internal';
```

```
Query OK, 0 rows affected (0.00 sec)
```

如果想要了解关于错误日志配置的更多信息，请参考文档：https://dev.mysql.com/doc/refman/8.0/en/error-log-component-configuration.html。

12.3 管理通用查询日志和慢查询日志

有两种方法可以记录查询。一种方法是通过通用查询日志记录，而另一种方法是通过慢查询日志记录。在本节中，你将了解如何配置它们。

12.3.1 如何操作

我们会在下面的几节详细讨论操作步骤。

通用查询日志

根据 MySQL 文档：

通用查询日志一般都记录了 mysqld 在做的事情。当客户端连接或断开连接时，服务器会将这个信息写入日志中，并记录从客户端收到的每一条 SQL 语句。如果怀疑客户端存在错误，并且想知道客户端发送给 mysqld 的具体内容时，通用查询日志是非常有用的。

1. 指定记录日志的文件。如果你不指定的话，将会在数据目录中创建名为 hostname.log 的文件。如果没有给出一个绝对路径名来指定别的目录，服务器就会在数据目录中创建这个文件。

```
mysql> SET
@@GLOBAL.general_log_file='/var/log/mysql/general_query_log';
Query OK, 0 rows affected (0.04 sec)
```

2. 启用通用查询日志：

```
mysql> SET GLOBAL general_log = 'ON';
Query OK, 0 rows affected (0.00 sec)
```

3. 你可以看到被记录的一些查询内容：

```
shell> sudo cat /var/log/mysql/general_query_log
```

```
/usr/sbin/mysqld, Version: 8.0.3-rc-log (MySQL Community Server
(GPL)). started with:
Tcp port: 3306 Unix socket: /var/lib/mysql/mysql.sock
Time Id Command Argument
2017-10-11T04:21:00.118944Z        220 Connect root@localhost on
using Socket
2017-10-11T04:21:00.119212Z        220 Query select
@@version_comment limit 1
2017-10-11T04:21:03.217603Z        220 Query SELECT DATABASE()
2017-10-11T04:21:03.218275Z        220 Init DB employees
2017-10-11T04:21:03.219339Z        220 Query show databases
2017-10-11T04:21:03.220189Z        220 Query show tables
2017-10-11T04:21:03.227635Z        220 Field List current_dept_emp
2017-10-11T04:21:03.233820Z        220 Field List departments
2017-10-11T04:21:03.235937Z        220 Field List dept_emp
2017-10-11T04:21:03.236089Z        220 Field List
dept_emp_latest_date
2017-10-11T04:21:03.236337Z        220 Field List dept_manager
2017-10-11T04:21:03.237291Z        220 Field List employees
2017-10-11T04:21:03.247921Z        220 Field List titles
2017-10-11T04:21:03.248217Z        220 Field List titles_only
~
~
2017-10-11T04:21:09.483117Z 220 Query select count(*) from
employees
2017-10-11T04:21:10.523421Z 220 Quit
```

通用查询日志会产生一个非常大的日志文件。在生产服务器上启用它时要非常小心。它会极大地影响服务器的性能。

慢查询日志

根据 MySQL 文档：

> 慢查询日志包含了执行时间超过 `long_query_time` 秒，以及至少扫描了 `min_examined_row_limit` 行的 SQL 语句。

要记录所有查询，可以将 `long_query_time` 的值设置为 0。`long_query_time`

的默认值是 10 秒，min_examined_row_limit 的默认值为 0。

默认情况下，不使用索引的查询和管理语句（例如 ALTER TABLE、ANALYZE TABLE、CHECK TABLE、CREATE INDEX、DROP INDEX、OPTIMIZE TABLE 和 REPAIR TABLE）是不会被记录的。不过，可以通过 log_slow_admin_statements 和 log_queries_not_using_indexes 来改变这种设置。

要启用慢查询日志，可以动态设置 slow_query_log=1，你可以使用 slow_query_log_file 来设置文件名。要指定日志目标文件，可以使用--log-output。

1. 验证 long_query_time 并按照你的需求调整它：

```
mysql> SELECT @@GLOBAL.LONG_QUERY_TIME;
+--------------------------+
| @@GLOBAL.LONG_QUERY_TIME |
+--------------------------+
|                10.000000 |
+--------------------------+
1 row in set (0.00 sec)

mysql> SET @@GLOBAL.LONG_QUERY_TIME=1;
Query OK, 0 rows affected (0.00 sec)

mysql> SELECT @@GLOBAL.LONG_QUERY_TIME;
+--------------------------+
| @@GLOBAL.LONG_QUERY_TIME |
+--------------------------+
|                 1.000000 |
+--------------------------+
1 row in set (0.00 sec)
```

2. 验证慢查询文件。默认情况下，它应该是在数据目录中使用 hostname-slow 的日志：

```
mysql> SELECT @@GLOBAL.slow_query_log_file;
+-----------------------------------+
| @@GLOBAL.slow_query_log_file      |
+-----------------------------------+
```

```
| /var/lib/mysql/server1-slow.log |
+---------------------------------+
1 row in set (0.00 sec)

mysql> SET
@@GLOBAL.slow_query_log_file='/var/log/mysql/mysql_slow.log';
Query OK, 0 rows affected (0.00 sec)

mysql> SELECT @@GLOBAL.slow_query_log_file;
+-------------------------------+
| @@GLOBAL.slow_query_log_file  |
+-------------------------------+
| /var/log/mysql/mysql_slow.log |
+-------------------------------+
1 row in set (0.00 sec)

mysql> FLUSH LOGS;
Query OK, 0 rows affected (0.03 sec)
```

3. 启用慢查询日志：

```
mysql> SELECT @@GLOBAL.slow_query_log;
+-------------------------+
| @@GLOBAL.slow_query_log |
+-------------------------+
|                       0 |
+-------------------------+
1 row in set (0.00 sec)

mysql> SET @@GLOBAL.slow_query_log=1;
Query OK, 0 rows affected (0.01 sec)

mysql> SELECT @@GLOBAL.slow_query_log;
+-------------------------+
| @@GLOBAL.slow_query_log |
+-------------------------+
|                       1 |
+-------------------------+
```

```
1 row in set (0.00 sec)
```

4. 验证这些查询确实已被记录（必须执行几个长时间运行的查询，以便在慢查询日志中看到它们）：

```
mysql> SELECT SLEEP(2);
+----------+
| SLEEP(2) |
+----------+
|        0 |
+----------+
1 row in set (2.00 sec)

shell> sudo less /var/log/mysql/mysql_slow.log
/usr/sbin/mysqld, Version: 8.0.3-rc-log (MySQL Community Server
(GPL)). started with:
Tcp port: 3306  Unix socket: /var/lib/mysql/mysql.sock
Time                 Id Command    Argument
# Time: 2017-10-15T12:43:55.038601Z
# User@Host: root[root] @ localhost []  Id:     7
# Query_time: 2.000845  Lock_time: 0.000000 Rows_sent: 1
Rows_examined: 0
SET timestamp=1508071435;
SELECT SLEEP(2);
```

选择查询日志的输出目标文件

你可以通过指定变量 `log_output` 的值，将查询记录到 MySQL 的文件或者表中，`log_output` 变量可以是 `FILE` 或者 `TABLE`，也可以同时是 `FILE` 和 `TABLE`。

如果将 `log_output` 指定为 `FILE`，则通用查询日志和慢查询日志将分别被写入由 `general_log_file` 和 `slow_query_log_file` 指定的文件。

如果将 `log_output` 指定为 `TABLE`，则通用查询日志和慢查询日志将分别被写入 `mysql.general_log` 和 `mysql.slow_log` 表中。日志内容可以通过 SQL 语句访问。

例如：

```
mysql> SET @@GLOBAL.log_output='TABLE';
Query OK, 0 rows affected (0.00 sec)

mysql> SET @@GLOBAL.general_log='ON';
Query OK, 0 rows affected (0.02 sec)
```

执行一些查询，然后查询 mysql.general_log 表：

```
mysql> SELECT * FROM mysql.general_log WHERE command_type='Query' \G
~
~
*************************** 3. row ***************************
  event_time: 2017-10-25 10:56:56.416746
   user_host: root[root] @ localhost []
   thread_id: 2421
   server_id: 32
command_type: Query
    argument: show databases
*************************** 4. row ***************************
  event_time: 2017-10-25 10:56:56.418896
   user_host: root[root] @ localhost []
   thread_id: 2421
   server_id: 32
command_type: Query
    argument: show tables
*************************** 5. row ***************************
  event_time: 2017-10-25 10:57:08.207964
   user_host: root[root] @ localhost []
   thread_id: 2421
   server_id: 32
command_type: Query
    argument: select * from salaries limit 1
*************************** 6. row ***************************
  event_time: 2017-10-25 10:57:47.041475
   user_host: root[root] @ localhost []
   thread_id: 2421
   server_id: 32
command_type: Query
```

```
    argument: SELECT * FROM mysql.general_log WHERE command_type='Query'
```

你可以用类似的方法使用 slow_log 表：

```
mysql> SET @@GLOBAL.slow_query_log=1;
Query OK, 0 rows affected (0.00 sec)

mysql> SET @@GLOBAL.long_query_time=1;
Query OK, 0 rows affected (0.00 sec)

mysql> SELECT SLEEP(2);
+----------+
| SLEEP(2) |
+----------+
|        0 |
+----------+
1 row in set (2.00 sec)

mysql> SELECT * FROM mysql.slow_log \G
*************************** 1. row ***************************
    start_time: 2017-10-25 11:01:44.817421
     user_host: root[root] @ localhost []
    query_time: 00:00:02.000530
     lock_time: 00:00:00.000000
     rows_sent: 1
 rows_examined: 0
            db: employees
last_insert_id: 0
     insert_id: 0
     server_id: 32
      sql_text: SELECT SLEEP(2)
     thread_id: 2421
1 row in set (0.00 sec)
```

如果慢查询日志表变得非常大，你可以创建一个新表来替换它：

1. 创建一个新表，mysql.general_log_new：

```
mysql> DROP TABLE IF EXISTS mysql.general_log_new;
```

```
Query OK, 0 rows affected, 1 warning (0.19 sec)

mysql> CREATE TABLE mysql.general_log_new LIKE mysql.general_log;
Query OK, 0 rows affected (0.10 sec)
```

2. 使用 RENAME TABLE 命令交换新表和旧表：

```
mysql> RENAME TABLE mysql.general_log TO mysql.general_log_1,
mysql.general_log_new TO mysql.general_log;
Query OK, 0 rows affected (0.00 sec)
```

12.4 管理二进制日志

本节将介绍在复制环境中如何管理二进制日志。在第 6 章已经介绍了使用 PURGE BINARY LOGS 命令和 expire_logs_days 变量简单地处理二进制日志。

在复制环境中使用这些方法是不安全的，因为如果有任何一个从库没有消费完二进制日志，而你删除了这些日志，那么这个从库将会失去同步，你就要重新构建它。

删除二进制日志的安全方法是，检查每一个从库读取的二进制日志的情况，然后删除它们。可以使用小工具 mysqlbinlogpurge 来实现。

12.4.1 如何操作

在一台服务器上执行 mysqlbinlogpurge 脚本，并指定主库和从库。该脚本连接所有的从库，并找出从库中最新的二进制日志被应用到什么位置，然后在主库上清除直到那个位置的二进制日志。你需要一个超级用户来连接所有的从库。

1. 随便连接到一台服务器，并执行 mysqlbinlogpurge 脚本：

```
shell> mysqlbinlogpurge --master=dbadmin:<pass>@master:3306 --
slaves=dbadmin:<pass>@slave1:3306,dbadmin:<pass>@slave2:3306
mysql> SHOW BINARY LOGS;
+--------------------+-----------+
| Log_name           | File_size |
+--------------------+-----------+
| master-bin.000001  |       177 |
```

```
~
| master-bin.000018   |     47785 |
| master-bin.000019   |       203 |
| master-bin.000020   |       203 |
| master-bin.000021   |       177 |
| master-bin.000022   |       203 |
| master-bin.000023   |  57739432 |
+--------------------+-----------+
23 rows in set (0.00 sec)

shell> mysqlbinlogpurge --master=dbadmin:<pass>@master:3306 --
slaves=dbadmin:<pass>@slave1:3306,dbadmin:<pass>@slave2:3306

# Latest binlog file replicated by all slaves: master-bin.000022
# Purging binary logs prior to 'master-bin.000023'
```

2. 如果你不想在命令行里指定所有从库而是希望自动发现它们，则应该在所有的从库上设置 report_host 和 report_port，并重新启动 MySQL 服务器。在每一个从库上：

```
shell> sudo vi /etc/my.cnf
[mysqld]
report-host     = slave1
report-port     = 3306

shell> sudo systemctl restart mysql

mysql> SHOW VARIABLES LIKE 'report%';
+-----------------+---------------+
| Variable_name   | Value         |
+-----------------+---------------+
| report_host     | slave1        |
| report_password |               |
| report_port     | 3306          |
| report_user     |               |
+-----------------+---------------+
4 rows in set (0.00 sec)
```

3. 执行带 discover-slaves-login 选项的 mysqlbinlogpurge：

```
mysql> SHOW BINARY LOGS;
+--------------------+-----------+
| Log_name           | File_size |
+--------------------+-----------+
| centos7-bin.000025 |       203 |
| centos7-bin.000026 |       203 |
| centos7-bin.000027 |       203 |
| centos7-bin.000028 |       154 |
+--------------------+-----------+
4 rows in set (0.00 sec)

shell> mysqlbinlogpurge --master=dbadmin:<pass>@master -discover
slaves-login=dbadmin:<pass>
# Discovering slaves for master at master:3306
# Discovering slave at slave1:3306
# Found slave: slave1:3306
# Discovering slave at slave2:3306
# Found slave: slave2:3306
# Latest binlog file replicated by all slaves: master-bin.000027
# Purging binary logs prior to 'master-bin.000028'
```

第 13 章 性能调优

在本章中，我们将介绍以下内容：

- explain 计划
- 基准查询和服务器
- 添加索引
- 不可见索引
- 降序索引
- 使用 `pt-query-digest` 分析慢查询
- 优化数据类型
- 删除重复和冗余索引
- 检查索引使用情况
- 控制查询优化器
- 使用索引提示（hint）
- 使用生成的列为 JSON 建立索引
- 使用资源组
- 使用 performance_schema
- 使用 sys schema

13.1 引言

本章将带你完成查询和 schema 的调优。数据库就是用于执行查询的，提高查询速度是调优的最终目标。数据库的性能取决于许多因素，主要是查询、schema、配置项和硬件。

在本章中，我们将使用 `employees` 数据库来解释所有示例。在前面的章节中，你可能已经以多种方式转换过 `employees` 数据库。因此，建议在动手操作本章的示例之前再次加载 `employees` 中的示例数据。可以参阅第 2 章的 2.6 节，以了解如何加载示例数据。

13.2 explain 计划

MySQL 执行查询的方式是影响数据库性能的主要因素之一。你可以使用 `EXPLAIN` 命令来验证 MySQL 的执行计划。从 MySQL 5.7.2 开始，可以使用 `EXPLAIN` 命令来检查当前在其他会话中执行的查询。执行 `EXPLAIN FORMAT = JSON` 命令，将得到详细信息。

13.2.1 如何操作

让我们来看细节。

13.2.2 使用 EXPLAIN

explain 计划提供了关于查询优化器如何执行查询的信息。只需要将 `EXPLAIN` 关键字前缀加到查询中即可：

```
mysql> EXPLAIN SELECT dept_name FROM dept_emp JOIN employees ON
dept_emp.emp_no=employees.emp_no JOIN departments ON
departments.dept_no=dept_emp.dept_no WHERE employees.first_name='Aamer'\G
*************************** 1. row ***************************
           id: 1
  select_type: SIMPLE
        table: employees
   partitions: NULL
         type: ref
possible_keys: PRIMARY,name
          key: name
      key_len: 58
          ref: const
         rows: 228
     filtered: 100.00
```

```
         Extra: Using index
*************************** 2. row ***************************
            id: 1
   select_type: SIMPLE
         table: dept_emp
    partitions: NULL
          type: ref
possible_keys: PRIMARY,dept_no
           key: PRIMARY
       key_len: 4
           ref: employees.employees.emp_no
          rows: 1
      filtered: 100.00
         Extra: Using index
*************************** 3. row ***************************
            id: 1
   select_type: SIMPLE
         table: departments
    partitions: NULL
          type: eq_ref
possible_keys: PRIMARY
           key: PRIMARY
       key_len: 16
           ref: employees.dept_emp.dept_no
          rows: 1
      filtered: 100.00
         Extra: NULL
3 rows in set, 1 warning (0.00 sec)
```

使用 EXPLAIN JSON

以 JSON 格式使用 explain 计划，能提供有关查询执行情况的完整信息：

```
mysql> EXPLAIN FORMAT=JSON SELECT dept_name FROM dept_emp JOIN employees ON
dept_emp.emp_no=employees.emp_no JOIN departments ON
departments.dept_no=dept_emp.dept_no WHERE employees.first_name='Aamer'\G
*************************** 1. row ***************************
EXPLAIN: {
```

```
    "query_block": {
      "select_id": 1,
      "cost_info": {
        "query_cost": "286.13"
    },
    "nested_loop": [
      {
        "table": {
          "table_name": "employees",
          "access_type": "ref",
          "possible_keys": [
            "PRIMARY",
            "name"
          ],
          "key": "name",
          "used_key_parts": [
            "first_name"
          ],
          "key_length": "58",
          "ref": [
            "const"
          ],
          "rows_examined_per_scan": 228,
          "rows_produced_per_join": 228,
          "filtered": "100.00",
          "using_index": true,
          "cost_info": {
            "read_cost": "1.12",
            "eval_cost": "22.80",
            "prefix_cost": "23.92",
            "data_read_per_join": "30K"
          },
          "used_columns": [
            "emp_no",
            "first_name"
          ]
        }
```

```
    },
    {
    "table": {
      "table_name": "dept_emp",
      "access_type": "ref",
      "possible_keys": [
       "PRIMARY",
       "dept_no"
      ],
      "key": "PRIMARY",
      "used_key_parts": [
       "emp_no"
      ],
      "key_length": "4",
      "ref": [
        "employees.employees.emp_no"
      ],
      "rows_examined_per_scan": 1,
      "rows_produced_per_join": 252,
      "filtered": "100.00",
      "using_index": true,
      "cost_info": {
        "read_cost": "148.78",
        "eval_cost": "25.21",
        "prefix_cost": "197.91",
        "data_read_per_join": "7K"
      },
      "used_columns": [
        "emp_no",
        "dept_no"
      ]
    }
  },
  {
    "table": {
      "table_name": "departments",
      "access_type": "eq_ref",
```

```
      "possible_keys": [
        "PRIMARY"
      ],
      "key": "PRIMARY",
      "used_key_parts": [
        "dept_no"
      ],
      "key_length": "16",
      "ref": [
        "employees.dept_emp.dept_no"
      ],
      "rows_examined_per_scan": 1,
      "rows_produced_per_join": 252,
      "filtered": "100.00",
      "cost_info": {
        "read_cost": "63.02",
        "eval_cost": "25.21",
        "prefix_cost": "286.13",
        "data_read_per_join": "45K"
      },
      "used_columns": [
        "dept_no",
        "dept_name"
      ]
    }
  }
```

使用 EXPLAIN 进行连接

可以为正在运行的会话执行 explain 计划。不过，你需要指定 connection ID。

要获得 connection ID，请执行：

```
mysql> SELECT CONNECTION_ID();
+-----------------+
| CONNECTION_ID() |
+-----------------+
|             778 |
```

```
+-----------------+
1 row in set (0.00 sec)

mysql> EXPLAIN FORMAT=JSON FOR CONNECTION 778\G ***************************
1. row ***************************
EXPLAIN: {
  "query_block": {
    "select_id": 1,
    "cost_info": {
      "query_cost": "881.04"
    },
    "nested_loop": [
      {
        "table": {
        "table_name": "employees",
        "access_type": "index",
        "possible_keys": [
          "PRIMARY"
        ],
        "key": "name",
        "used_key_parts": [
          "first_name",
          "last_name"
        ],
        "key_length": "124",
        "rows_examined_per_scan": 1,
        "rows_produced_per_join": 1,
        "filtered": "100.00",
        "using_index": true,
        "cost_info": {
          "read_cost": "880.24",
          "eval_cost": "0.10",
          "prefix_cost": "880.34",
          "data_read_per_join": "136"
        },
 ~
 ~
```

```
1 row in set (0.00 sec)
```

如果此连接没有运行任何 SELECT/UPDATE/INSERT/DELETE/REPLACE 查询，则会引发错误：

```
mysql> EXPLAIN FOR CONNECTION 779;
ERROR 3012 (HY000): EXPLAIN FOR CONNECTION command is supported only for
SELECT/UPDATE/INSERT/DELETE/REPLACE
```

请参阅文档 https://dev.mysql.com/doc/refman/8.0/en/explain-output.html 以了解关于 explain 计划格式的更多信息。文档 https://www.percona.com/blog/category/explain-2/explain-formatjson-is-cool/ 非常清楚地解释了 JSON 格式的 explain 计划。

13.3 基准查询和服务器

假设你想知道哪个查询执行起来更快。explain 计划给了你一个结果，但有时你并不能依据它来做决定。如果查询时间为几十秒左右，你可以在服务器上执行这些查询以判断哪一个更快。但是，如果查询时间大约为几毫秒，则很难基于单次执行来做判断。

你可以使用 mysqlslap 工具（就在 MySQL 客户端的安装包中），它模拟 MySQL 服务器的客户端负载，并报告每个阶段所耗费的时间，就像多个客户端正在访该问服务器一样。本节你将了解 mysqlslap 的用法；在后面几节，你将了解 mysqlslap 的强大功能。

13.3.1 如何操作

假设你想评估一条查询的执行时间。如果在 MySQL 客户端中执行该操作，则能够知道大致的执行时间（精确到 100 ms）：

```
mysql> pager grep rows
PAGER set to 'grep rows'
mysql> SELECT e.emp_no, salary FROM salaries s JOIN employees e ON
s.emp_no=e.emp_no WHERE (first_name='Adam');
2384 rows in set (0.00 sec)
```

你可以使用 mysqlslap 模拟客户端负载，并在多个迭代中并行运行上述 SQL 语句：

```
shell> mysqlslap -u <user> -p<pass> --create-schema=employees --
query="SELECT e.emp_no, salary FROM salaries s JOIN employees e ON
s.emp_no=e.emp_no WHERE (first_name='Adam');" -c 1000 i 100
mysqlslap: [Warning] Using a password on the command line interface can be
insecure.
Benchmark
    Average number of seconds to run all queries: 3.216 seconds
    Minimum number of seconds to run all queries: 3.216 seconds
    Maximum number of seconds to run all queries: 3.216 seconds
    Number of clients running queries: 1000
    Average number of queries per client: 1
```

以上查询是用 1,000 个并发和 100 个迭代执行的，平均花费了 3.216 秒。

你可以在文件中指定多个 SQL 并指定分隔符。`Mysqlslap` 会运行文件中的所有查询：

```
shell> cat queries.sql
SELECT e.emp_no, salary FROM salaries s JOIN employees e ON
s.emp_no=e.emp_no WHERE (first_name='Adam');
SELECT * FROM employees WHERE first_name='Adam' OR last_name='Adam';
SELECT * FROM employees WHERE first_name='Adam';

shell> mysqlslap -u <user> -p<pass> --create-schema=employees --
concurrency=10 --iterations=10 --query=query.sql --query=queries.sql --
delimiter=";"
mysqlslap: [Warning] Using a password on the command line interface can be
insecure.
Benchmark
    Average number of seconds to run all queries: 5.173 seconds
    Minimum number of seconds to run all queries: 5.010 seconds
    Maximum number of seconds to run all queries: 5.257 seconds
    Number of clients running queries: 10
    Average number of queries per client: 3
```

你甚至可以自动生成表和 SQL 语句。通过这种方式，你可以将结果与之前的服务器设置进行比较：

```
shell> mysqlslap -u <user> -p<pass> --concurrency=100 --iterations=10 --
number-int-cols=4 --number-char-cols=10 --auto-generate-sql
mysqlslap: [Warning] Using a password on the command line interface can be
insecure.
Benchmark
Average number of seconds to run all queries: 1.640 seconds
Minimum number of seconds to run all queries: 1.511 seconds
Maximum number of seconds to run all queries: 1.791 seconds
Number of clients running queries: 100
Average number of queries per client: 0
```

你还可以使用 performance_schema 来查看所有与查询相关的指标。我们将在 13.15 节讲解 performance_schema。

13.4 添加索引

如果没有索引，MySQL 在查找相关行时就必须逐行扫描整个表。如果这个表在你要筛选的列上有索引，MySQL 就可以快速找到大数据文件中的行而无须扫描整个文件。

MySQL 可以使用索引来筛选 WHERE、ORDER BY 和 GROUP BY 子句中的行，也可以用索引来连接表。如果一个列上有多个索引，MySQL 会选择给出了最多筛选行的索引。

你可以执行 ALTER TABLE 命令来添加或删除索引。索引的添加和删除都是在线操作，不会妨碍表上的 DML，但在大表上这么做会花费大量时间。

13.4.1 主键（聚簇索引）和二级索引

在继续往下学习之前，了解什么是主键（或聚簇索引）和二级索引是非常重要的。

为了提升对涉及主键列的查询和排序的速度，InnoDB 基于主键来存储行。按照 Oracle 的说法，这也被称为 **index-orgnized** 表。其他所有的索引都被称为辅助键，它们存储主键的值（不直接引用行）。

假设有这样的表：

```
mysql> CREATE TABLE index_example (
col1 int PRIMARY KEY,
```

```
col2 char(10),
KEY `col2`(`col2`)
);
```

这个表的行是根据 col1 的值进行排序和存储的。如果搜索 col1 的任何值，它可以直接指向物理行，这就是聚簇索引非常快的原因。col2 上的索引也包含了 col1 的值，如果搜索 col2，则会返回 col1 的值，反过来在聚簇索引中搜索 col1 就可以返回实际行的值。

关于主键的选择，有如下一些小技巧：

- 它应该是 UNIQUE（唯一）和 NOT NULL（非空）的。
- 选择最小的可能键，因为所有的二级索引都会存储主键。所以如果主键很大，整个索引也会占用更多的空间。
- 选择一个单调递增的值。物理行是根据主键进行排序的。所以，如果你选择一个随机键，需要做多次行重排，这会导致性能下降。AUTO_INCREMENT 非常适合主键。
- 最好选择一个主键。如果找不到任何主键，请添加一个 AUTO_INCREMENT 列。如果你不选择任何内容，InnoDB 会在内部生成一个带有 6 字节行 ID 的隐藏聚簇索引。

13.4.2 如何操作

你可以通过查看表定义来查看表的索引。你会注意到在 first_name 和 last_name 上有一个索引。如果通过指定 first_name 或指定 first_name 和 last_name 来筛选行，MySQL 就可以使用这个索引来提升查询的速度。

但是，如果仅指定 last_name，则不能使用索引。这是因为优化器只能使用索引最左边的前缀。请参阅文档：https://dev.mysql.com/doc/refman/8.0/en/multiple-column-indexes.html 以了解更详细的例子。

```
mysql> ALTER TABLE employees ADD INDEX name(first_name, last_name);
Query OK, 0 rows affected (2.23 sec)
Records: 0 Duplicates: 0 Warnings: 0
mysql> SHOW CREATE TABLE employees\G
*************************** 1. row ***************************
```

```
       Table: employees
Create Table: CREATE TABLE `employees` (
  `emp_no` int(11) NOT NULL,
  `birth_date` date NOT NULL,
  `first_name` varchar(14) NOT NULL,
  `last_name` varchar(16) NOT NULL,
  `gender` enum('M','F') NOT NULL,
  `hire_date` date NOT NULL,
  PRIMARY KEY (`emp_no`),
  KEY `name` (`first_name`,`last_name`)
ENGINE=InnoDB DEFAULT CHARSET=utf8mb4
1 row in set (0.00 sec)
```

添加索引

你可以通过执行 ALTER TABLE ADD INDEX 命令来添加索引。例如，如果要在 last_name 上添加索引，请参阅以下代码：

```
mysql> ALTER TABLE employees ADD INDEX (last_name);
Query OK, 0 rows affected (1.28 sec)
Records: 0  Duplicates: 0  Warnings: 0

mysql> SHOW CREATE TABLE employees\G
*************************** 1. row ***************************
       Table: employees
Create Table: CREATE TABLE `employees` (
  `emp_no` int(11) NOT NULL,
  `birth_date` date NOT NULL,
  `first_name` varchar(14) NOT NULL,
  `last_name` varchar(16) NOT NULL,
  `gender` enum('M','F') NOT NULL,
  `hire_date` date NOT NULL,
  PRIMARY KEY (`emp_no`),
  KEY `name` (`first_name`,`last_name`),
  KEY `last_name` (`last_name`)
ENGINE=InnoDB DEFAULT CHARSET=utf8mb4
1 row in set (0.01 sec)
```

你可以指定索引的名字；如果不指定，则最左边的前缀将被用作索引名。如果发生重名的情况，则会在名字后面附加上_2、_3 等，依此类推。

例如：

```
mysql> ALTER TABLE employees ADD INDEX index_last_name (last_name);
```

唯一索引

如果你希望索引是唯一的，可以指定关键字 UNIQUE。例如：

```
mysql> ALTER TABLE employees ADD UNIQUE INDEX unique_name (last_name,
first_name);
# There are few duplicate entries in employees database, the above
statement is shown for illustration purpose only.
```

前缀索引

对于字符串列，可以创建仅使用列值的前导部分而非整个列的索引。你需要指定前导部分的长度：

```
## `last_name` varchar(16) NOT NULL
mysql> ALTER TABLE employees ADD INDEX (last_name(10));
Query OK, 0 rows affected (1.78 sec)
Records: 0 Duplicates: 0  Warnings: 0
```

last_name 的最大长度是 16 个字符，但索引仅基于其前 10 个字符创建。

删除索引

可以使用 ALTER TABLE 命令来删除索引：

```
mysql> ALTER TABLE employees DROP INDEX last_name;
Query OK, 0 rows affected (0.02 sec)
Records: 0 Duplicates: 0  Warnings: 0
```

生成列的索引

对于封装在函数中的列不能使用索引。假设你在 hire_date 上添加一个索引：

```
mysql> ALTER TABLE employees ADD INDEX(hire_date);
Query OK, 0 rows affected (0.93 sec)
Records: 0 Duplicates: 0 Warnings: 0
```

hire_date 上的索引可用于在 WHERE 子句中带有 hire_date 的查询：

```
mysql> EXPLAIN SELECT COUNT(*) FROM employees WHERE
hire_date>'2000-01-01'\G
*************************** 1. row ***************************
           id: 1
  select_type: SIMPLE
        table: employees
   partitions: NULL
         type: range
possible_keys: hire_date
          key: hire_date
      key_len: 3
          ref: NULL
         rows: 14
     filtered: 100.00
        Extra: Using where; Using index
1 row in set, 1 warning (0.00 sec)
```

相反，如果将 hire_date 放入函数中，MySQL 就必须扫描整个表：

```
mysql> EXPLAIN SELECT COUNT(*) FROM employees WHERE YEAR(hire_date)>=2000\G
*************************** 1. row ***************************
           id: 1
  select_type: SIMPLE
        table: employees
   partitions: NULL
         type: index
possible_keys: NULL
          key: hire_date
      key_len: 3
          ref: NULL
         rows: 291892
     filtered: 100.00
```

```
        Extra: Using where; Using index
1 row in set, 1 warning (0.00 sec)
```

所以，尽量避免将已被索引的列放入函数中。如果无法避免使用函数，请创建一个虚拟列并在虚拟列上添加一个索引：

```
mysql> ALTER TABLE employees ADD hire_date_year YEAR AS (YEAR(hire_date))
VIRTUAL, ADD INDEX (hire_date_year);
Query OK, 0 rows affected (1.16 sec)
Records: 0 Duplicates: 0   Warnings: 0

mysql> SHOW CREATE TABLE employees\G
*************************** 1. row ***************************
       Table: employees
Create Table: CREATE TABLE  `employees` (
  `emp_no` int(11) NOT NULL,
  `birth_date` date NOT NULL,
  `first_name` varchar(14) NOT NULL,
  `last_name` varchar(16) NOT NULL,
  `gender` enum('M','F') NOT NULL,
  `hire_date` date NOT NULL,
  `hire_date_year` year(4) GENERATED ALWAYS AS (year(`hire_date`)) VIRTUAL,
  PRIMARY KEY (`emp_no`),
  KEY `name` (`first_name`,`last_name`),
  KEY `hire_date` (`hire_date`),
  KEY `hire_date_year` (`hire_date_year`)
) ENGINE=InnoDB DEFAULT CHARSET=utf8mb4
1 row in set (0.00 sec)
```

现在，你无须在查询中使用 YEAR() 函数，可以直接在 WHERE 子句中使用 hire_date_year：

```
mysql> EXPLAIN SELECT COUNT(*) FROM employees WHERE hire_date_year>=2000\G
*************************** 1. row ***************************
           id: 1
  select_type: SIMPLE
        table: employees
   partitions: NULL
```

```
         type: range
possible_keys: hire_date_year
          key: hire_date_year
      key_len: 2
          ref: NULL
         rows: 15
     filtered: 100.00
        Extra: Using where; Using index
1 row in set, 1 warning (0.00 sec)
```

请注意，即使你使用了 YEAR(hire_date)，优化器也会认识到表达式 YEAR()与 hire_date_year 的定义匹配，并且 hire_date_year 已被索引，所以它在构建执行计划时会考虑该索引：

```
mysql> EXPLAIN SELECT COUNT(*) FROM employees WHERE YEAR(hire_date)>=2000\G
*************************** 1. row ***************************
           id: 1
  select_type: SIMPLE
        table: employees
   partitions: NULL
         type: range
possible_keys: hire_date_year
          key: hire_date_year
      key_len: 2
          ref: NULL
         rows: 15
     filtered: 100.00
        Extra: Using where
1 row in set, 1 warning (0.00 sec)
```

13.5　不可见索引

如果你想删除未使用的索引，可以不立即删除，而是先将其标记为不可见，然后监控应用程序的行为，稍后再删除它。之后，如果你还需要该索引，则可以将其标记为可见，这与先删除索引再重新添加相比会快很多。

要解释不可见索引，你需要先添加一个正常索引（如果还没有的话）。例如：

```
mysql> ALTER TABLE employees ADD INDEX (last_name);
Query OK, 0 rows affected (1.81 sec)
Records: 0 Duplicates: 0 Warnings: 0
```

13.5.1 如何操作

如果你希望删除 last_name 上的索引，但又不是直接删除它，可以使用 ALTER TABLE 命令将其标记为不可见：

```
mysql> EXPLAIN SELECT * FROM employees WHERE last_name='Aamodt'\G
*************************** 1. row ***************************
           id: 1
  select_type: SIMPLE
        table: employees
   partitions: NULL
         type: ref
possible_keys: last_name
          key: last_name
      key_len: 66
          ref: const
         rows: 205
     filtered: 100.00
        Extra: NULL
1 row in set, 1 warning (0.00 sec)

mysql> ALTER TABLE employees ALTER INDEX last_name INVISIBLE;
Query OK, 0 rows affected (0.01 sec)
Records: 0 Duplicates: 0  Warnings: 0

mysql> EXPLAIN SELECT * FROM employees WHERE last_name='Aamodt'\G
*************************** 1. row ***************************
           id: 1
  select_type: SIMPLE
        table: employees
   partitions: NULL
         type: ALL
```

```
possible_keys: NULL
          key: NULL
      key_len: NULL
          ref: NULL
         rows: 299733
     filtered: 10.00
        Extra: Using where
1 row in set, 1 warning (0.00 sec)

mysql> SHOW CREATE TABLE employees\G
*************************** 1. row ***************************
       Table: employees
Create Table: CREATE TABLE
  `employees` ( `emp_no` int(11) NOT NULL,
  `birth_date` date NOT NULL,
  `first_name` varchar(14) NOT NULL,
  `last_name` varchar(16) NOT NULL,
  `gender` enum('M','F') NOT NULL,
  `hire_date` date NOT NULL,
  PRIMARY KEY (`emp_no`),
  KEY `name` (`first_name`,`last_name`),
  KEY `last_name` (`last_name`) /*!80000 INVISIBLE */
) ENGINE=InnoDB DEFAULT CHARSET=utf8mb4
1 row in set (0.00 sec)
```

你会注意到，在筛选 `last_name` 的查询中使用了 `last_name` 索引；将其标记为不可见后，它就无法使用了。你可以再次将其标记为可见：

```
mysql> ALTER TABLE employees ALTER INDEX last_name VISIBLE;
Query OK, 0 rows affected (0.01 sec)
Records: 0 Duplicates: 0 Warnings: 0
```

13.6　降序索引

在 MySQL 8 之前，索引的定义中可以包含顺序（升序或降序），但它只是被解析并没有被实现。索引值始终以升序存储。MySQL 8.0 引入了对降序索引的支持。因此，索引定义中指定的顺序不会被忽略。降序索引实际上按降序存储关键值。请记住，对于降序查询，反向扫描升序索引效率不高。

在多列索引中，可以指定某些列降序。这样做对同时具有升序和降序的 ORDER BY 子句的查询很有用。

假设你想要按照 first_name 升序和 last_name 降序对 employees 表进行排序。MySQL 不能使用 first_name 和 last_name 上的索引。

如果没有降序索引：

```
mysql> SHOW CREATE TABLE employees\G
*************************** 1. row ***************************
       Table: employees
Create Table: CREATE TABLE `employees` (
  `emp_no` int(11) NOT NULL,
  `birth_date` date NOT NULL,
  `first_name` varchar(14) NOT NULL,
  `last_name` varchar(16) NOT NULL,
  `gender` enum('M','F') NOT NULL,
  `hire_date` date NOT NULL,
  PRIMARY KEY (`emp_no`),
  KEY `name` (`first_name`,`last_name`),
  KEY `last_name` (`last_name`) /*!80000 INVISIBLE */
) ENGINE=InnoDB DEFAULT CHARSET=utf8mb4
```

在 explain 计划中，你会注意到索引名（first_name 和 last_name）没有被使用：

```
mysql> EXPLAIN SELECT * FROM employees ORDER BY first_name ASC, last_name
DESC LIMIT 10\G
*************************** 1. row ***************************
           id: 1
  select_type: SIMPLE
        table: employees
   partitions: NULL
         type: ALL
possible_keys: NULL
          key: NULL
```

```
    key_len: NULL
        ref: NULL
       rows: 299733
   filtered: 100.00
      Extra: Using filesort
```

13.6.1　如何操作

1. 添加降序索引：

```
mysql> ALTER TABLE employees ADD INDEX name_desc(first_name ASC,
last_name DESC);
Query OK, 0 rows affected (1.61 sec)
Records: 0 Duplicates: 0 Warnings: 0
```

2. 添加降序索引后，查询就可以使用该索引了：

```
mysql> EXPLAIN SELECT * FROM employees ORDER BY first_name ASC, last_name
DESC LIMIT 10\G
*************************** 1. row ***************************
           id: 1
  select_type: SIMPLE
        table: employees
   partitions: NULL
         type: index
possible_keys: NULL
          key: name_desc
      key_len: 124
          ref: NULL
         rows: 10
     filtered: 100.00
        Extra: NULL
```

3. 同一个索引可以用于其他排序方式，即通过向后索引扫描以 first_name 降序和 last_name 升序进行排序：

```
mysql> EXPLAIN SELECT * FROM employees ORDER BY first_name DESC,
last_name ASC LIMIT 10\G
*************************** 1. row ***************************
```

```
           id: 1
  select_type: SIMPLE
        table: employees
   partitions: NULL
         type: index
possible_keys: NULL
          key: name_desc
      key_len: 124
          ref: NULL
         rows: 10
     filtered: 100.00
        Extra: Backward index scan
```

13.7 使用 pt-query-digest 分析慢查询

pt-query-digest 是 Percona 工具包的一部分，用于对查询进行分析。可以通过以下任何方式收集查询：

- 慢查询日志
- 通用查询日志
- 进程列表
- 二进制日志
- TCP 转储

Percona 工具包的安装在第 10 章的 10.2 节中介绍过。在本节中，你将学习如何使用 pt-query-digest。每种方法都有缺点。慢查询日志不会包括所有查询，除非你将 long_query_time 指定为 0，但是这会显著降低系统运行速度。通用查询日志不包括查询时间。从处理列表中无法获得完整的查询。使用二进制日志只能分析写入操作，而使用 TCP 转储会导致服务器性能下降。通常，该工具用于 long_query_time 为 1 秒或更长的慢查询日志。

13.7.1 如何操作

我们来看使用 pt-query-digest 分析慢查询的细节。

慢查询日志

第 12 章的 12.3 节介绍过如何启用和配置慢查询日志。一旦启用慢查询日志并收集查询，就可以通过传递慢查询日志来运行 pt-query-digest。

假设慢查询文件位于/var/lib/mysql/mysql-slow.log：

```
shell> sudo pt-query-digest /var/lib/mysql/ubuntu-slow.log > query_digest
```

摘要报告（digest report）中的查询按照查询执行的次数与查询时间的乘积排列。所有查询的详细信息，例如查询校验和（每一种查询类型有一个唯一的值）、平均时间、百分比时间和执行次数等都会显示出来。你可以通过搜索查询校验和来深入研究特定查询。

摘要报告的内容如下所示：

```
# 286.8s user time, 850ms system time, 232.75M rss, 315.73M vsz
# Current date: Sat Nov 18 05:16:55 2017
# Hostname: db1
# Files: /var/lib/mysql/db1-slow.log
# Rate limits apply
# Overall: 638.54k total, 2.06k unique, 0.49 QPS, 0.14x concurrency ______
# Time range: 2017-11-03 01:02:40 to 2017-11-18 05:16:47
# Attribute          total     min     max     avg     95%  stddev  median
# ============     ======= ======= ======= ======= ======= ======= =======
# Exec time        179486s     3us   2713s   281ms    21ms     15s   176us
# Lock time          1157s       0     36s     2ms   194us   124ms    49us
# Rows sent         18.25M       0 753.66k   29.96  212.52   1.63k    0.99
# Rows examine     157.39G       0   3.30G 258.45k   3.35k  24.78M    0.99
# Rows affecte       3.66M       0 294.77k    6.01    0.99   1.16k       0
# Bytes sent         3.08G       0  95.15M   5.05k  13.78k 206.42k  174.84
# Merge passes       2.84k       0      97    0.00       0    0.16       0
# Tmp tables       129.02k       0    1009    0.21    0.99    1.43
# Tmp disk tbl      25.20k       0     850    0.04       0    1.09       0
# Tmp tbl size      26.21G       0 218.27M  43.04k       0   2.06M       0
# Query size       178.92M       6 452.25k  293.81  592.07   5.26k   72.65
# InnoDB:
# IO r bytes        79.06G       0   2.09G 200.37k       0 12.94M       0
# IO r ops           7.26M       0 233.16k   18.39       0  1.36k       0
```

```
# IO r wait      96525s          0    3452s   233ms         0     18s         0
# pages distin 526.99M          0 608.33k   1.30k  964.41   9.15k      1.96
# queue wait          0          0        0       0         0       0         0
# rec lock wai      46s          0       9s   111us         0    28ms         0
# Boolean:
# Filesort        5%      yes,      94%      no
# Filesort on     0%      yes,      99%      no
# Full join       3%      yes,      96%      no
# Full scan      40%      yes,      59%      no
# Tmp table      13%      yes,      86%      no
# Tmp table on    2%      yes,      97%      no
```

对此查询的分析如下所示：

```
# Rank Query ID           Response time    Calls  R/Call    V/M   Item
# ==== ================== ================ ====== ========= ===== ======
#    1 0x55F499860A034BCB 76560.4220 42.7%     47 1628.9451 18.06 SELECT
orders
#    2 0x3A2F0B98DA39BCB9 10490.4155  5.8%   2680    3.9143 33... SELECT
orders order_status
#    3 0x25119C7C31A24011 7378.8763   4.1%   1534    4.8102 30.11 SELECT
orders users
#    4 0x41106CE92AD9DFED 5412.7326   3.0%  15589    0.3472  2.98 SELECT
sessions
#    5 0x860DCDE7AE0AD554 5187.5257   2.9%    500   10.3751 54.99 SELECT
orders sessions
#    6 0x5DF64920B008AD63 4517.5041   2.5%     58   77.8880 22.23 UPDATE
SELECT
#    7 0xC9F9A31DE77B93A1 4473.0208   2.5%     58   77.1210 96... INSERT
SELECT tmpMove
#    8 0x8BF88451DA989BFF 4036.4413   2.2%     13  310.4955 16... UPDATE
SELECT orders tmpDel
```

从前面的输出中可以推断出，对于 #1 查询（0x55F499860A034BCB），其所有的执行累计响应时间为 76,560 秒，占全部查询的累计响应时间的 42.7%。执行次数为 47 次，平均查询时间为 1628 秒。

你可以通过搜索校验和来查看任何查询详情，显示完整的查询、explain 计划的命令和表状态。例如：

```
Query 1: 0.00 QPS, 0.06x concurrency, ID 0x55F499860A034BCB at byte
249542900
# This item is included in the report because it matches --limit.
# Scores: V/M = 18.06
# Time range: 2017-11-03 01:39:19 to 2017-11-18 01:46:50
# Attribute    pct   total     min     max     avg     95% stddev median
# ============ == ======= ======= ======= ======= ======= ======= ======
# Count          0      47
# Exec time     42  76560s   1182s   1854s   1629s   1819s    172s  1649s
# Lock time      0      3s   102us   994ms    70ms   293ms   174ms  467us
# Rows sent      0  78.78k     212   5.66k   1.68k   4.95k   1.71k 652.75
# Rows examine 85 135.34G   2.11G   3.30G   2.88G   3.17G 303.82M  2.87G
# Rows affecte  0       0       0       0       0       0       0      0
# Bytes sent     0   3.22M  10.20k 226.13k  70.14k 201.74k  66.71k 31.59k
# Merge passes  0       0       0       0       0       0       0      0
# Tmp tables    0       0       0       0       0       0       0      0
# Tmp disk tbl  0       0       0       0       0       0       0      0
# Tmp tbl size  0       0       0       0       0       0       0      0
# Query size    0  11.66k     254     254     254     254       0    254
# InnoDB:
# IO r bytes    1   1.11G       0  53.79M  24.20M  51.29M  21.04M 20.30M
# IO r ops      1 142.14k       0   6.72k   3.02k   6.63k   2.67k  2.50k
# IO r wait     0     92s       0     14s      2s      5s      3s     1s
# pages distin  0 325.46k   6.10k   7.30k   6.92k   6.96k  350.84  6.96k
# queue wait    0       0       0       0       0       0       0      0
# rec lock wai  0       0       0       0       0       0       0      0
# Boolean:
# Full scan    100% yes,   0% no
# String:
# Databases    lashrenew_... (32/68%), betsy_db (15/31%)
# Hosts        10.37.69.197
# InnoDB trxID CF22C985 (1/2%), CF23455A (1/2%)... 45 more
# Last errno   0
# rate limit   query:100
# Users        db1_... (32/68%), dba (15/31%)
```

```
# Query_time distribution
#   1us
#  10us
# 100us
#   1ms
#  10ms
# 100ms
#    1s
#   10s+  ################################################################
# Tables
#    SHOW TABLE STATUS FROM `db1` LIKE 'orders'\G
#    SHOW CREATE TABLE `db1`.`orders`\G
#    SHOW TABLE STATUS FROM `db1` LIKE 'shipping_tracking_history'\G
#    SHOW CREATE TABLE `db1`.`shipping_tracking_history`\G
# EXPLAIN /*!50100 PARTITIONS*/
SELECT tracking_num, carrier, order_id, userID FROM orders o WHERE
tracking_num!=""
and NOT EXISTS (SELECT 1 FROM shipping_tracking_history sth WHERE
sth.order_id=o.order_id AND sth.is_final=1)
AND o.date_finalized>date_add(curdate(),interval -1 month)\G
```

通用查询日志

你可以使用 pt-query-digest 通过传递参数--type genlog 来分析通用查询日志。由于通用日志不报告查询的次数，因此只显示累计的数字：

```
shell> sudo pt-query-digest --type genlog /var/lib/mysql/db1.log >
general_query_digest
```

输出结果如下所示：

```
# 400ms user time, 0 system time, 28.84M rss, 99.35M vsz
# Current date: Sat Nov 18 09:02:08 2017
# Hostname: db1
# Files: /var/lib/mysql/db1.log
# Overall: 511 total, 39 unique, 30.06 QPS, 0x concurrency ________________
# Time range: 2017-11-18 09:01:09 to 09:01:26
# Attribute          total     min     max     avg     95%  stddev  median
```

```
# ============    ======= ======= ======= ======= ======= ======= =======
# Exec time             0       0       0       0       0       0       0
# Query size       92.18k      10   3.22k  184.71  363.48  348.86  102.22
```

对此查询的分析如下所示：

```
# Profile
# Rank  Query ID    Response time Calls R/Call  V/M Item
# ==== ================== ============= ===== ====== ===== ==============
#    1 0x625BF8F82D174492   0.0000 0.0%   130 0.0000  0.00         SELECT
facebook_like_details
#    2 0xAA353644DE4C4CB4   0.0000 0.0%    44 0.0000  0.00    ADMIN QUIT
#    3 0x5D51E5F01B88B79E   0.0000 0.0%    44 0.0000  0.00 ADMIN CONNECT
```

进程列表

你可以使用 pt-query-digest 而非日志文件从进程列表（process list）中读取查询：

```
shell> pt-query-digest --processlist h=localhost --iterations 10 --run-time
1m -u <user> -p<pass>
```

run-time 指定每次迭代应该运行多长时间。在前面的示例中，该工具会每分钟会生成一份报告，并持续 10 分钟。

二进制日志

要使用 pt-query-digest 分析二进制日志，应该先用 mysqlbinlog 工具将其转换为文本格式：

```
shell> sudo mysqlbinlog /var/lib/mysql/binlog.000639 > binlog.00063
shell> pt-query-digest --type binlog binlog.000639 > binlog_digest
```

TCP 转储

你可以使用 tcpdump 命令捕获 TCP 流量，并将其发送给 pt-query-digest 进行分析：

```
shell> sudo tcpdump -s 65535 -x -nn -q -tttt -i any -c 1000 port 3306 >
mysql.tcp.txt
```

```
shell> pt-query-digest --type tcpdump mysql.tcp.txt > tcpdump_digest
```

在 `pt-query-digest` 中有很多选项可供选择，例如用特定时间窗口筛选查询，筛选某个特定的查询以及生成报告。更多有关的详细信息，请参阅 Percona 文档：`https://www.percona.com/doc/percona-toolkit/LATEST/pt-query-digest.html`。

13.7.2 延伸阅读

请参阅 `https://engineering.linkedin.com/blog/2017/09/query-analyzer--a-tool-for-analyzing-mysql-queries-without-overh`，以了解更多关于无须任何开销而分析所有查询的新方法。

13.8 优化数据类型

你应该这样定义表，它既能保存所有可能值，同时在磁盘上占用的空间又最小。

如果表占用的存储空间越小，则：

- 向磁盘写入或读取的数据就越少，查询起来就越快。
- 在处理查询时，磁盘上的内容会被加载到主内存中。所以，表越小，占用的主存空间就越小。
- 被索引占用的空间就越小。

13.8.1 如何操作

1. 如果要存储员工编号，而其可能的最大值为 500,000，则最佳数据类型为 `MEDIUMINT UNSIGNED`（3 个字节）。如果将它存储为 4 个字节的 `INT` 类型，则每一行都浪费了一个字节。

2. 如果要存储员工名字（first name），由于其长度不等，可能的最大长度为 20 个字符，则最好将其声明为 `varchar(20)` 类型。如果将员工名字存储为 `char(20)` 类型，但是只有几个人的名字长为 20 个字符，其余的长度不到 10 个字符，就会浪费 10 个字符的空间。

3. 在声明类型为 varchar 的列时，应该考虑其长度。尽管类型 varchar 在磁盘上进行了优化，但这个类型的数据被加载到内存时却会占用全部的长度空间。例如，如果将 first_name 存储在类型 varchar(255) 中，并且其实际长度为 10，则在磁盘上它占用 10 + 1（用于存储长度的一个附加字节）个字节；但在内存中，它会占用全部的 255 个字节。

4. 如果类型为 varchar 列的长度超过 255 个字符，则需要用 2 个字节来存储长度。

5. 如果不允许存储空值，则应将列声明为 NOT NULL。这样做就避免了测试每个值是否为空的开销，并且还节省了一些存储空间——每列能节省 1 位。

6. 如果字符串的长度是固定的，请存储为 char 而非 varchar 类型，因为类型 varchar 需要一个或两个字节来存储字符串的长度。

7. 如果这些值是固定的，则使用 ENUM 而非 varchar 类型。例如，如果要存储可能处于等待状态，或者已批准、已拒绝、已部署、尚未部署，以及已失效或被删除的值，则可以使用 ENUM 类型。它需要 1 或 2 个字节即可，不像类型 char(10) 那样占用 10 个字节。

8. 优先选择使用整数类型而非字符串类型。

9. 尝试利用前缀索引。

10. 尝试利用 InnoDB 压缩。

请参阅 https://dev.mysql.com/doc/refman/8.0/en/storage-requirements.html，以了解更多关于每一种数据类型的存储需求的内容，还可参阅 https://dev.mysql.com/doc/refman/8.0/en/integer-types.html 来了解每一种整数类型的范围。

如果你想了解优化的数据类型，可以使用 PROCEDURE ANALYZE 功能。虽然它不够准确，但它能给出一个差不多的建议。遗憾的是，在 MySQL 8 中已弃用这个功能：

```
mysql> SELECT user_id, first_name FROM user PROCEDURE ANALYSE(1,100)\G
*************************** 1. row ***************************
             Field_name: db1.user.user_id
              Min_value: 100000@nat.test123.net
```

```
                Max_value: test1234@nat.test123.net
               Min_length: 22
               Max_length: 33
          Empties_or_zeros: 0
                    Nulls: 0
  Avg_value_or_avg_length: 25.8003
                      Std: NULL
        Optimal_fieldtype: VARCHAR(33) NOT NULL
*************************** 2. row ***************************
               Field_name: db1.user.first_name
                Min_value: *Alan
                Max_value: Zuniga 102031
               Min_length: 3
               Max_length: 33
          Empties_or_zeros: 0
                    Nulls: 0
  Avg_value_or_avg_length: 10.1588
                      Std: NULL
        Optimal_fieldtype: VARCHAR(33) NOT NULL
2 rows in set (0.02 sec)
```

13.9　删除重复和冗余索引

你可以在一列上定义多个索引。如果不小心的话，你可能会重复定义相同的索引（相同的列、相同的列顺序或相同的键顺序），这被称为**重复索引**。如果只有部分索引（最左边的列）是重复的，则称为**冗余索引**。重复索引没有什么用处。在某些情况下，冗余索引可能很有用（本节末尾的提示中提到了一个用例），但是这两者都会减慢插入的速度。因此，找出并移除这两种索引非常重要。

有三种工具可以帮助找出重复索引：

- `pt-duplicate-key-checker`，它是 Percona 工具包的一部分。第 10 章的 10.2 节介绍了如何安装 Percona 工具包。
- `mysqlindexcheck`，它是 MySQL 工具集的一部分。本书第 1 章介绍了如何安装 MySQL 工具集。

- 使用 sys schema（将在 13.10 节中介绍）。

考虑下面的 employees 表：

```
mysql> SHOW CREATE TABLE employees\G
*************************** 1. row ***************************
       Table: employees
Create Table: CREATE TABLE
  `employees` (
   `emp_no` int(11) NOT NULL,
  `birth_date` date NOT NULL, `
  first_name` varchar(14) NOT NULL,
  `last_name` varchar(16) NOT NULL,
  `gender` enum('M','F') NOT NULL,
  `hire_date` date NOT NULL,
  PRIMARY KEY (`emp_no`),
  KEY `last_name` (`last_name`) /*!80000 INVISIBLE */,
  KEY `full_name` (`first_name`,`last_name`),
  KEY `full_name_desc` (`first_name` DESC,`last_name`),
  KEY `first_name` (`first_name`),
  KEY `full_name_1` (`first_name`,`last_name`),
  KEY `first_name_emp_no` (`first_name`,`emp_no`)
) ENGINE=InnoDB DEFAULT CHARSET=utf8mb4
```

索引 full_name_1 是 full_name 的重复项，因为它们都位于相同的列上，列的顺序相同，键的顺序相同（升序或降序）。

索引 first_name 是一个冗余索引，因为 first_name 索引最左边的后缀已经涵盖了列 first_name。

索引 first_name_emp_no 是一个冗余索引，因为它最右侧的后缀中包含了主键。InnoDB 二级索引已包含主键，因此将主键声明为辅助索引的一部分是多余的。但是，它在通过 first_name 进行筛选并按 emp_no 进行排序的查询中非常有用：

```
SELECT * FROM employees WHERE first_name='Adam' ORDER BY emp_no;
```

full_name_desc 索引不是 full_name 的重复项，因为键的排序不同。

13.9.1 如何操作

让我们来看删除重复索引和冗余索引的详细步骤。

pt-duplicate-key-checker

pt-duplicate-key-checker 给出了精确的 ALTER 语句来删除重复的键：

```
shell> pt-duplicate-key-checker -u <user> -p<pass>

# A software update is available:
# ########################################################################
# employees.employees
# ########################################################################

# full_name_1 is a duplicate of full_name
# Key definitions:
#   KEY `full_name_1` (`first_name`,`last_name`),
#   KEY `full_name` (`first_name`,`last_name`),
# Column types:
#       `first_name` varchar(14) not null
#       `last_name` varchar(16) not null
# To remove this duplicate index, execute:
ALTER TABLE `employees`.`employees` DROP INDEX `full_name_1`;

# first_name is a left-prefix of full_name
# Key definitions:
#   KEY `first_name` (`first_name`),
#   KEY `full_name` (`first_name`,`last_name`),
# Column types:
#       `first_name` varchar(14) not null
#       `last_name` varchar(16) not null
# To remove this duplicate index, execute:
ALTER TABLE `employees`.`employees` DROP INDEX `first_name`;

# Key first_name_emp_no ends with a prefix of the clustered index
# Key definitions:
#   KEY `first_name_emp_no` (`first_name`,`emp_no`)
```

```
#   PRIMARY KEY (`emp_no`),
# Column types:
#       `first_name` varchar(14) not null
#       `emp_no` int(11) not null
# To shorten this duplicate clustered index, execute:
ALTER TABLE `employees`.`employees` DROP INDEX `first_name_emp_no`, ADD
INDEX `first_name_emp_no` (`first_name`);
```

该工具建议你从最右侧的后缀中删除主键来缩短重复的聚簇索引。请注意，这样做可能会产生另一个重复索引。如果你希望忽略重复的聚簇索引，则可以传递--noclustered 选项。

要检查特定数据库的重复索引，可以传递--databases <database name>选项：

```
shell> pt-duplicate-key-checker -u <user> -p<pass> --database employees
```

要删除键，你甚至可以将 pt-duplicate-key-checker 的输出传递给 mysql：

```
shell> pt-duplicate-key-checker -u <user> -p<pass> | mysql -u <user> -
p<pass>
```

mysqlindexcheck

请注意，mysqlindexcheck 会忽略降序索引。例如，full_name_desc（first_name 降序和 last_name）被视为 full_name（first_name 和 last_name）的重复索引：

```
shell> mysqlindexcheck --server=<user>:<pass>@localhost:3306 employees --
show-drops
WARNING: Using a password on the command line interface can be insecure.
# Source on localhost: ... connected.
#  The following indexes are duplicates or redundant for table
employees.employees:
#
CREATE INDEX `full_name_desc` ON `employees`.`employees` (`first_name`,
`last_name`) USING BTREE
#     may be redundant or duplicate of:
CREATE INDEX `full_name` ON `employees`.`employees` (`first_name`,
`last_name`) USING BTREE
```

```
#
CREATE INDEX `first_name` ON `employees`.`employees` (`first_name`) USING
BTREE
#     may be redundant or duplicate of:
CREATE INDEX `full_name` ON `employees`.`employees` (`first_name`,
`last_name`) USING BTREE
#
CREATE INDEX `full_name_1` ON `employees`.`employees` (`first_name`,
`last_name`) USING BTREE
#     may be redundant or duplicate of:
CREATE INDEX `full_name` ON `employees`.`employees` (`first_name`,
`last_name`) USING BTREE
#
# DROP statements:
#
ALTER TABLE `employees`.`employees` DROP INDEX `full_name_desc`;
ALTER TABLE `employees`.`employees` DROP INDEX `first_name`;
ALTER TABLE `employees`.`employees` DROP INDEX `full_name_1`;
#
# The following index for table employees.employees contains the clustered
index and might be redundant:
#
CREATE INDEX `first_name_emp_no` ON `employees`.`employees` (`first_name`,
`emp_no`) USING BTREE
#
# DROP/ADD statement:
#
ALTER TABLE `employees`.`employees` DROP INDEX `first_name_emp_no`, ADD
INDEX `first_name_emp_no` (first_name);
#
```

> 如之前所述，在某些情况下，冗余索引可能很有用。你需要考虑你的应用程序是否属于这种情况。

创建索引以了解以下示例：

```
mysql> ALTER TABLE employees DROP PRIMARY KEY, ADD PRIMARY KEY(emp_no,
hire_date), ADD INDEX `name` (`first_name`,`last_name`);

mysql> ALTER TABLE salaries ADD INDEX from_date(from_date), ADD INDEX
from_date_2(from_date,emp_no);
```

考虑以下 employees 和 salaries 表：

```
mysql> SHOW CREATE TABLE employees\G
*************************** 1. row ***************************
       Table: employees
Create Table: CREATE TABLE `employees` (
  `emp_no` int(11) NOT NULL,
  `birth_date` date NOT NULL,
  `first_name` varchar(14) NOT NULL,
  `last_name` varchar(16) NOT NULL,
  `gender` enum('M','F') NOT NULL,
  `hire_date` date NOT NULL,
  PRIMARY KEY (`emp_no`,`hire_date`),
  KEY `name` (`first_name`,`last_name`)
) /*!50100 TABLESPACE `innodb_system` */ ENGINE=InnoDB DEFAULT
CHARSET=utf8mb4

mysql> SHOW CREATE TABLE salaries\G
*************************** 1. row ***************************
Create Table: CREATE TABLE `salaries` (
  `emp_no` int(11) NOT NULL,
  `salary` int(11) NOT NULL,
  `from_date` date NOT NULL,
  `to_date` date NOT NULL,
  PRIMARY KEY (`emp_no`,`from_date`),
  KEY `from_date` (`from_date`),
  KEY `from_date_2` (`from_date`,`emp_no`)
) ENGINE=InnoDB DEFAULT CHARSET=utf8mb4
```

看起来 from_date 是 from_date_2 的冗余索引，但请检查后面的查询的 eplain 计划！它正在使用这两个索引的交集。from_date 索引用于筛选，from_date_2 用于连接 employees 表。对每一个表，优化器仅扫描一行：

```
mysql> EXPLAIN SELECT e.emp_no, salary FROM salaries s JOIN employees e ON
s.emp_no=e.emp_no WHERE from_date='2001-05-23'\G
*************************** 1. row ***************************
           id: 1
  select_type: SIMPLE
        table: s
   partitions: NULL
         type: index_merge
possible_keys: PRIMARY,from_date_2,from_date
          key: from_date_2,from_date
      key_len: 3,3
          ref: NULL
         rows: 1
     filtered: 100.00
        Extra: Using intersect(from_date_2,from_date); Using where
*************************** 2. row ***************************
           id: 1
  select_type: SIMPLE
        table: e
   partitions: NULL
         type: ref
possible_keys: PRIMARY
          key: PRIMARY
      key_len: 4
          ref: employees.s.emp_no
         rows: 1
     filtered: 100.00
        Extra: Using index
2 rows in set, 1 warning (0.00 sec)
```

现在删除冗余索引 `from_date` 并检查 explain 计划。你可以看到，优化器正在扫描 `salaries` 表中的 90 行和 `employees` 表中的一行。但看看 `ref` 列，它显示优化器会将常量值与 `key` 列（`from_date_2`）中指定的索引进行比较，以便从表中选择相应的行。你可以通过传递优化器提示（hint）或索引提示来测试这种行为，而非删除索引，这些内容将在下一节中介绍。

```
mysql> EXPLAIN SELECT e.emp_no, salary FROM salaries s JOIN employees e ON
s.emp_no=e.emp_no WHERE from_date='2001-05-23'\G
*************************** 1. row ***************************
           id: 1
  select_type: SIMPLE
        table: s
   partitions: NULL
         type: ref

possible_keys: PRIMARY,from_date_2
          key: from_date_2
      key_len: 3
          ref: const
         rows: 90
     filtered: 100.00
        Extra: NULL
*************************** 2. row ***************************
           id: 1
  select_type: SIMPLE
        table: e
   partitions: NULL
         type: ref
possible_keys: PRIMARY
          key: PRIMARY
      key_len: 4
          ref: employees.s.emp_no
         rows: 1
     filtered: 100.00
        Extra: Using index
2 rows in set, 1 warning (0.00 sec)
```

现在你需要确定哪些查询执行起来更快。

- **计划 1**：使用 `from_date` 与 `from_date_2` 的交集，ref 为空则扫描 1 行。
- **计划 2**：使用 `from_date_2`，ref 为 `constant` 则扫描 90 行。

你可以使用 `mysqlslap` 工具来查找（不要直接在生产主机上运行），并确保并发数小于max_connections。

计划1的基准如下：

```
shell> mysqlslap -u <user> -p<pass> --create-schema='employees' -c 500 -i
100 --query="SELECT e.emp_no, salary FROM salaries s JOIN employees e ON
s.emp_no=e.emp_no WHERE from_date='2001-05-23'"
mysqlslap: [Warning] Using a password on the command line interface can be
insecure.
Benchmark
    Average number of seconds to run all queries: 0.466 seconds
    Minimum number of seconds to run all queries: 0.424 seconds
    Maximum number of seconds to run all queries: 0.568 seconds
    Number of clients running queries: 500
    Average number of queries per client: 1
```

计划2的基准如下：

```
shell> mysqlslap -u <user> -p<pass> --create-schema='employees' -c 500 -i
100 --query="SELECT e.emp_no, salary FROM salaries s JOIN employees e ON
s.emp_no=e.emp_no WHERE from_date='2001-05-23'"
mysqlslap: [Warning] Using a password on the command line interface can be
insecure.
Benchmark
    Average number of seconds to run all queries: 0.435 seconds
    Minimum number of seconds to run all queries: 0.376 seconds
    Maximum number of seconds to run all queries: 0.504 seconds
    Number of clients running queries: 500
    Average number of queries per client: 1
```

事实证明，计划1和计划2的查询平均执行时间分别为0.466秒和0.435秒。由于结果非常接近，因此可以删除冗余索引。使用计划2。

这只是一个例子，帮助你在你的应用场景中学习和应用这个概念。

13.10 检查索引的使用情况

在前面的章节中，你了解了如何移除冗余索引和重复索引。在设计应用程序时，你可能想过基于列来筛选查询和添加索引。但过了一段时间，由于应用程序发生变化，你可能不再需要该索引。在本节，你将了解如何找出那些未使用的索引。

有两种方法可以找到未使用的索引：

- 使用 pt-index-usage（在本节介绍）
- 使用 sys schema（在 13.11 节介绍）

13.10.1 如何操作

我们可以使用 Percona 工具包中的 pt-index-usage 工具来获取对索引的分析。它从慢查询日志中提取查询，对每个查询运行 exlpain 计划，并标识未使用的索引。如果你有查询列表，可以将它们保存为慢查询格式并将其传递给该工具。请注意，这只是一种近似操作，因为慢查询日志并不包括所有的查询语句：

```
shell> sudo pt-index-usage
slow -u <user> -p<password> /var/lib/mysql/db1-slow.log > unused_indexes
```

13.11 控制查询优化器

查询优化器的任务是找出执行一条 SQL 查询的最佳计划。一条查询可以有多种执行计划，特别是在连接表时，需要考量的计划数量呈指数级增长。在本节中，你将学习如何根据需要调整优化器。

以 employees 表为例，添加必要的索引：

```
mysql> CREATE TABLE `employees_index_example` (
`emp_no` int(11) NOT NULL,
  `birth_date` date NOT NULL,
  `first_name` varchar(14) NOT NULL,
  `last_name` varchar(16) NOT NULL,
  `gender` enum('M','F') NOT NULL,
  `hire_date` date NOT NULL,
  PRIMARY KEY (`emp_no`),
  KEY `last_name` (`last_name`) /*!80000 INVISIBLE */,
  KEY `full_name` (`first_name`,`last_name`),
```

```
  KEY `full_name_desc` (`first_name` DESC,`last_name`),
  KEY `first_name` (`first_name`),
  KEY `full_name_1` (`first_name`,`last_name`),
  KEY `first_name_emp_no` (`first_name`,`emp_no`),
  KEY `last_name_2` (`last_name`(10))
) ENGINE=InnoDB DEFAULT CHARSET=utf8mb4;
Query OK, 0 rows affected, 1 warning (0.08 sec)

mysql> SHOW WARNINGS;
+---------+------+------------------------------------------------------------
----------------------------------------------------------------------------
--------------------+
| Level | Code | Message
|
+---------+------+------------------------------------------------------------
----------------------------------------------------------------------------
--------------------+
Warning | 1831 | Duplicate index 'full_name_1' defined on the table
'employees.employees_index_example'. This is deprecated and will be
disallowed in a future release. |
+---------+------+------------------------------------------------------------
----------------------------------------------------------------------------
--------------------+
1 row in set (0.00 sec)

mysql> INSERT INTO employees_index_example SELECT
emp_no,birth_date,first_name,last_name,gender,hire_date FROM employees;

mysql> RENAME TABLE employees TO employees_old;
mysql> RENAME TABLE employees_index_example TO employees;
```

假设你想查看是否有 first_name 或 last_name 为 Adam 的员工。

explain 计划如下：

```
mysql> EXPLAIN SELECT emp_no FROM employees WHERE first_name='Adam' OR
last_name='Adam'\G
*************************** 1. row ***************************
```

```
           id: 1
  select_type: SIMPLE
        table: employees
   partitions: NULL
         type: index_merge
possible_keys:
full_name,full_name_desc,first_name,full_name_1,first_name_emp_no,last_n
ame _2
          key: first_name,last_name_2
      key_len: 58,42
          ref: NULL
         rows: 252
     filtered: 100.00
        Extra: Using sort_union(first_name,last_name_2); Using where
1 row in set, 1 warning (0.00 sec)
```

你会注意到优化器有很多选项可用于完成这个查询。它可以使用 possible_keys 中列出的任何索引：(full_name, full_name_desc, first_name, full_name_1, first_name_emp_no, last_name_2)。优化器会验证所有的计划，并确定哪个计划的成本最低。

一条查询的成本包括：从磁盘访问数据，从内存访问数据，创建临时表，在内存中对结果进行排序，等等。MySQL 为每个操作分配一个相对值，并将每个计划的成本相加后汇总。最终它会执行那个成本最低的执行计划。

13.11.1 如何操作

你可以通过向查询传递提示或通过在全局或会话级别调整变量来控制优化器。你甚至可以改变操作的成本。建议将这些值保留为默认值，除非你有自己的主意。

optimizer_search_depth

根据 Jørgen 的观点[1]：

1 参见 http://jorgenloland.blogspot.in/2012/04/improvements-for-many-table-join-in.html。

“MySQL 使用贪婪搜索算法来查找连接表的最佳顺序，如果你只加入几个表，计算所有连接顺序组合的成本，然后选择最佳的计划，是没有问题的。但是，因为可能的组合有(#tables)!种，这个计算成本很快就会变得非常高。例如，对于 5 个表，有 120 种组合（5!），这个级别的计算量还可以接受；对于 10 个表，有 360 万种组合（10!）；对于 15 个表，有 13,070 亿种组合（15!）。出于这个原因，MySQL 进行了折中：使用启发式算法来寻找可行的计划，这样做应该会显著减少 MySQL 需要计算的计划数量，但是你也有可能找不到最好的执行计划。”

MySQL 文档如是说：

“optimizer_search_depth 变量告诉优化器应该看到每个不完整计划多远的‘未来’，以便评估其是否应该进一步扩展。optimizer_search_depth 的值越小可能导致查询编译时间缩短几个数量级。如果 optimizer_search_depth 的值与查询中表的数量接近，那么 13 个或更多的表的查询可能需要数小时甚至数天的时间来编译，同时，如果使用 optimizer_search_depth 的值等于 3 或 4 进行编译，对于同样的查询，优化器可以在 1 分钟内完成编译。如果你不知道 optimizer_search_depth 的值取多少合适，可以将它设置为 0，以便让优化器自己来确定这个值。”

optimizer_search_depth 的默认值是 62，这是一个非常贪婪的设置，但是由于启发式算法的逻辑，MySQL 很快就选定了执行计划。MySQL 文档中并没有解释清楚为什么要将默认值设置为 62 而不是 0。

如果要连接 7 个以上的表，可以将 optimizer_search_depth 设置为 0，或者传递优化器提示（将在下一节介绍）。自动选择选择最小值（表的数量，7），将搜索深度限定在合理的值：

```
mysql> SHOW VARIABLES LIKE 'optimizer_search_depth';
+------------------------+-------+
| Variable_name          | Value |
+------------------------+-------+
| optimizer_search_depth | 62    |
+------------------------+-------+
1 row in set (0.00 sec)
mysql> SET @@SESSION.optimizer_search_depth=0;
```

```
Query OK, 0 rows affected (0.00 sec)
```

如何知道查询的时间是花在评估计划里了

假设你要连接 10 个表（大部分由 ORM 自动生成），并运行 explain 计划。如果花了很长的时间，意味着这个查询在评估计划上耗时太长。调整 optimizer_search_depth 的值（可能设置为 0），并检查 explain 计划需要花多长时间。在调整 optimizer_search_depth 的值时，还要记下执行计划的变化。

optimizer_switch

系统变量 optimizer_switch 是一组标志（flag）。你可以将每个标志设置为 ON 或 OFF，以启用或禁用相应的优化器行为。你可以在会话级别或全局级别动态设置这个变量。如果在会话级别调整 optimizer_switch，该会话中的所有查询都会受到影响；如果在全局级别调整的话，所有查询都会受到影响。

例如，你已注意到前面的查询 SELECT emp_no FROM employees WHERE first_name ='Adam' OR last_name ='Adam' 正在使用 sort_union(first_name, last_name_2)。如果你认为该查询的优化方案不正确，可以调整 optimizer_switch 以切换到另一个优化方案：

```
mysql> SHOW VARIABLES LIKE 'optimizer_switch'\G
*************************** 1. row ***************************
Variable_name: optimizer_switch
        Value:
index_merge=on,index_merge_union=on,index_merge_sort_union=on,index_merge_i
ntersection=on,engine_condition_pushdown=on,index_condition_pushdown=on,mrr
=on,mrr_cost_based=on,block_nested_loop=on,batched_key_access=off,materiali
zation=on,semijoin=on,loosescan=on,firstmatch=on,duplicateweedout=on,subque
ry_materialization_cost_based=on,use_index_extensions=on,condition_fanou
t_f ilter=on,derived_merge=on
1 row in set (0.00 sec)
```

最开始时，index_merge_union=on：

```
mysql> EXPLAIN SELECT emp_no FROM employees WHERE first_name='Adam' OR
last_name='Adam'\G
*************************** 1. row ***************************
```

```
             id: 1
    select_type: SIMPLE
          table: employees
     partitions: NULL
           type: index_merge
possible_keys:
full_name,full_name_desc,first_name,full_name_1,first_name_emp_no,last_n
ame _2
            key: first_name,last_name_2
        key_len: 58,42
            ref: NULL
           rows: 252
       filtered: 100.00
          Extra: Using sort_union(first_name,last_name_2); Using where
1 row in set, 1 warning (0.00 sec)
```

优化器可以使用 sort_union 来优化：

```
mysql> SET @@SESSION.optimizer_switch="index_merge_sort_union=off";
Query OK, 0 rows affected (0.00 sec)
```

可以在会话级别关闭 index_merge_sort_union 优化，以保证只有此会话中的查询才会受影响：

```
mysql> SHOW VARIABLES LIKE 'optimizer_switch'\G ***************************
1. row ***************************
Variable_name: optimizer_switch
        Value:
index_merge=on,index_merge_union=on,index_merge_sort_union=off,index_merge_
intersection=on,engine_condition_pushdown=on,index_condition_pushdown=on,mr
r=on,mrr_cost_based=on,block_nested_loop=on,batched_key_access=off,material
ization=on,semijoin=on,loosescan=on,firstmatch=on,duplicateweedout=on,subqu
ery_materialization_cost_based=on,use_index_extensions=on,condition_fanout_
filter=on,derived_merge=on
1 row in set (0.00 sec)
```

关闭 index_merge_sort_union 后，你会注意到查询计划的变化。它不再使用 sort_union 优化：

```
mysql> EXPLAIN SELECT emp_no FROM employees WHERE first_name='Adam' OR
last_name='Adam'\G
*************************** 1. row ***************************
           id: 1
  select_type: SIMPLE
        table: employees
   partitions: NULL
         type: index
possible_keys:
full_name,full_name_desc,first_name,full_name_1,first_name_emp_no,last_n
ame _2
          key: full_name
      key_len: 124
          ref: NULL
rows: 299379
filtered: 19.00
Extra: Using where; Using index
1 row in set, 1 warning (0.00 sec)
```

你会进一步发现，在这种情况下，使用 sort_union 是最佳选择。关于所有类型的优化器开关的更多详细信息，请参阅 https://dev.mysql.com/doc/refman/8.0/en/switchable-optimizations.html。

优化器提示（hint）

你可以提示优化器使用或不使用某些优化，而不是在会话级别调整优化器开关或 optimizer_search_depth 变量。优化器提示的作用范围仅限于那些能让你更精细地控制查询的语句，而优化器开关的作用范围则是在会话内或全局。

再次以前面的查询为例，如果你觉得使用 sort_union 不是最佳方法，可以在这个查询中将其作为一个提示来传递而关闭它：

```
mysql> EXPLAIN SELECT /*+ NO_INDEX_MERGE(employees first_name,last_name_2)
*/ * FROM employees WHERE first_name='Adam' OR last_name='Adam'\G
*************************** 1. row ***************************
           id: 1
  select_type: SIMPLE
```

```
         table: employees
    partitions: NULL
          type: ALL
possible_keys:
full_name,full_name_desc,first_name,full_name_1,first_name_emp_no,last_n
ame _2
           key: NULL
       key_len: NULL
           ref: NULL
          rows: 299379
      filtered: 19.00
         Extra: Using where
1 row in set, 1 warning (0.00 sec)
```

在 13.9 节，我们删除了冗余索引，以找出哪一个查询计划更好。在这里，你可以使用优化器提示忽略 from_date 和 from_date_2 的交集：

```
mysql> EXPLAIN SELECT /*+ NO_INDEX_MERGE(s from_date,from_date_2) */
e.emp_no, salary FROM salaries s JOIN employees e ON s.emp_no=e.emp_no WHERE
from_date='2001-05-23'\G
*************************** 1. row ***************************
           id: 1
  select_type: SIMPLE
        table: s
   partitions: NULL
         type: ref
possible_keys: PRIMARY,from_date,from_date_2
          key: from_date
      key_len: 3
          ref: const
         rows: 90
     filtered: 100.00
        Extra: NULL
*************************** 2. row ***************************
           id: 1
  select_type: SIMPLE
        table: e
```

```
   partitions: NULL
         type: ref
possible_keys: PRIMARY
          key: PRIMARY
      key_len: 4
          ref: employees.s.emp_no
         rows: 1
     filtered: 100.00
        Extra: Using index
2 rows in set, 1 warning (0.00 sec)
```

还有一个使用优化器提示的好例子，是设置 JOIN 顺序：

```
mysql> EXPLAIN SELECT e.emp_no, salary FROM salaries s JOIN employees e ON
s.emp_no=e.emp_no WHERE (first_name='Adam' OR last_name='Adam') ORDER BY
from_date DESC\G
*************************** 1. row ***************************
           id: 1
  select_type: SIMPLE
        table: e
   partitions: NULL
         type: index_merge
possible_keys:
PRIMARY,full_name,full_name_desc,first_name,full_name_1,first_name_emp_no
,l ast_name_2
          key: first_name,last_name_2
      key_len: 58,42
          ref: NULL
         rows: 252
     filtered: 100.00
Extra: Using sort_union(first_name,last_name_2); Using where; Using
temporary; Using filesort
*************************** 2. row ***************************
           id: 1
  select_type: SIMPLE
        table: s
   partitions: NULL
         type: ref
```

```
 possible_keys: PRIMARY
           key: PRIMARY
       key_len: 4
           ref: employees.e.emp_no
          rows: 9
      filtered: 100.00
         Extra: NULL
2 rows in set, 1 warning (0.00 sec)
```

在前面的查询中，优化器首先使用 employees 表，然后连接 salaries 表。你可以通过传递提示/ *+ JOIN _ ORDER (s, e) * /来改变连接顺序：

```
mysql> EXPLAIN SELECT /*+ JOIN_ORDER(s, e) */ e.emp_no, salary FROM salaries
s JOIN employees e ON s.emp_no=e.emp_no WHERE (first_name='Adam' OR
last_name='Adam') ORDER BY from_date DESC\G
*************************** 1. row ***************************
           id: 1
  select_type: SIMPLE
        table: s
   partitions: NULL
         type: ALL
possible_keys: PRIMARY
          key: NULL
      key_len: NULL
          ref: NULL
         rows: 2838426
     filtered: 100.00
        Extra: Using filesort
*************************** 2. row ***************************
           id: 1
  select_type: SIMPLE
        table: e
   partitions: NULL
         type: eq_ref
possible_keys:
PRIMARY,full_name,full_name_desc,first_name,full_name_1,first_name_emp_no,l
ast_name_2
```

```
         key: PRIMARY
     key_len: 4
         ref: employees.s.emp_no
        rows: 1
    filtered: 19.00
       Extra: Using where
2 rows in set, 1 warning (0.00 sec)
```

现在你会注意到，salaries 表被首先使用，这就避免了创建临时表，但是优化器会对 salaries 表进行全表扫描。

优化器提示的另一个用例如下：不为每条语句或会话设置会话变量，而只为某条语句设置会话变量。假设你正在使用 ORDER BY 子句对查询结果进行排序，但 ORDER BY 子句上没有索引。优化器利用 sort_buffer_size 来提升排序速度。默认情况下，sort_buffer_size 的值为 256K（即 256 KB）。如果 sort_buffer_size 的值不够大，则排序算法必须执行的合并传递次数就会增加。你可以通过会话变量 sort_merge_passes 来统计合并传递次数：

```
mysql> SHOW SESSION status LIKE 'sort_merge_passes';
+-------------------+-------+
| Variable_name     | Value |
+-------------------+-------+
| Sort_merge_passes | 0     |
+-------------------+-------+
1 row in set (0.00 sec)

mysql> pager grep "rows in set"; SELECT * FROM employees ORDER BY hire_date
DESC;nopager;
PAGER set to 'grep "rows in set"'
300025 rows in set (0.45 sec)

PAGER set to stdout
mysql> SHOW SESSION status LIKE 'sort_merge_passes';
+-------------------+-------+
| Variable_name     | Value |
+-------------------+-------+
| Sort_merge_passes | 8     |
```

```
+-------------------+-------+
1 row in set (0.00 sec)
```

你会注意到 MySQL 没有足够大的 sort_buffer_size，它必须执行 8 次 sort_merge_passes。可以通过优化器提示将 sort_buffer_size 设置为较大的值，如 16M（即 16 MB），并检查 sort_merge_passes：

```
mysql> SHOW SESSION status LIKE 'sort_merge_passes';
+-------------------+-------+
| Variable_name     | Value |
+-------------------+-------+
| Sort_merge_passes | 0     |
+-------------------+-------+
1 row in set (0.00 sec)

mysql> pager grep "rows in set"; SELECT /*+ SET_VAR(sort_buffer_size = 16M)
*/ * FROM employees ORDER BY hire_date DESC;nopager;
PAGER set to 'grep "rows in set"'
300025 rows in set (0.45 sec)

PAGER set to stdout
mysql> SHOW SESSION status LIKE 'sort_merge_passes';
+-------------------+-------+
| Variable_name     | Value |
+-------------------+-------+
| Sort_merge_passes | 0     |
+-------------------+-------+
1 row in set (0.00 sec)
```

你会注意到，当 sort_buffer_size 被设置为 16M 时，sort_merge_pass 为 0。

强烈建议使用索引而不是依赖 sort_buffer_size 来优化查询。你可以考虑增加 sort_buffer_size 的值，以加快用查询优化或改进索引无法提升的 ORDER BY 或 GROUP BY 操作的速度。

使用 SET_VAR，可以在语句级别设置 optimizer_switch：

```
mysql> EXPLAIN SELECT /*+ SET_VAR(optimizer_switch =
```

```
'index_merge_sort_union=off') */ e.emp_no, salary FROM salaries s JOIN
employees e ON s.emp_no=e.emp_no WHERE from_date='2001-05-23'\G
*************************** 1. row ***************************
           id: 1
  select_type: SIMPLE
        table: e
   partitions: NULL
         type: index
possible_keys: PRIMARY
          key: name
      key_len: 124
          ref: NULL
         rows: 299379
     filtered: 100.00
        Extra: Using index
*************************** 2. row ***************************
           id: 1
  select_type: SIMPLE
        table: s
   partitions: NULL
         type: eq_ref
possible_keys: PRIMARY
          key: PRIMARY
      key_len: 7
          ref: employees.e.emp_no,const
         rows: 1
     filtered: 100.00
        Extra: NULL
2 rows in set, 1 warning (0.00 sec)
```

你还可以设置查询的最长执行时间，这意味着查询将在指定的时间（MAX_EXECUTION_TIME(milliseconds)）后自动终止：

```
mysql> SELECT /*+ MAX_EXECUTION_TIME(100) */ * FROM employees ORDER BY
hire_date DESC;
ERROR 1028 (HY000): Sort aborted: Query execution was interrupted, maximum
statement execution time exceeded
```

你可以向优化器提示许多其他事情，请参阅 https://dev.mysql.com/refman/8.0/en/ optimizer-hints.html 以获取完整列表和更多示例。

调整优化器成本模型

为了生成执行计划，优化器使用了一个成本模型，基于执行查询期间发生的各种操作来估算成本。优化器有一组预编译的默认成本常数，可用于做关于执行计划的决策。你可以通过更新或插入 mysql.engine_cost 表并执行 FLUSH OPTIMIZER_COSTS 命令来调整这些常数：

```
mysql> SELECT * FROM mysql.engine_cost\G
*************************** 1. row ***************************
  engine_name: InnoDB
  device_type: 0
    cost_name: io_block_read_cost
   cost_value: 1
  last_update: 2017-11-20 16:24:56
      comment: NULL
default_value: 1
*************************** 2. row ***************************
  engine_name: InnoDB
  device_type: 0
    cost_name: memory_block_read_cost
   cost_value: 0.25
  last_update: 2017-11-19 13:58:32
      comment: NULL
default_value: 0.25
2 rows in set (0.00 sec)
```

假设你有一个超高速磁盘，你可以减少 io_block_read_cost 的 cost_value 值：

```
mysql> UPDATE mysql.engine_cost SET cost_value=0.5 WHERE
cost_name='io_block_read_cost';
Query OK, 1 row affected (0.08 sec)
Rows matched: 1 Changed: 1 Warnings: 0

mysql> FLUSH OPTIMIZER_COSTS;
```

```
Query OK, 0 rows affected (0.01 sec)

mysql> SELECT * FROM mysql.engine_cost\G
*************************** 1. row ***************************
  engine_name: InnoDB
  device_type: 0
    cost_name: io_block_read_cost
   cost_value: 0.5
  last_update: 2017-11-20 17:02:43
      comment: NULL
default_value: 1
*************************** 2. row ***************************
  engine_name: InnoDB
  device_type: 0
    cost_name: memory_block_read_cost
   cost_value: 0.25
  last_update: 2017-11-19 13:58:32
      comment: NULL
default_value: 0.25
2 rows in set (0.00 sec)
```

要了解有关优化器成本模型的更多信息，请参阅 https://dev.mysql.com/doc/refman/8.0/en/cost-model.html。

13.12 使用索引提示（hint）

使用索引提示，可以提示优化器使用或忽略索引，它不同于优化器提示，在优化器提示中，你提示优化器使用或忽略某些优化方法。索引提示和优化器提示可以单独使用或一起使用，以实现所需的计划。索引提示是紧接着表名指定的。

当你执行涉及多个表连接（join）的复杂查询时，如果优化器在评估计划时花费的时间太长，你可以确定最佳计划，并且给这个查询发出提示。但一定要确保你建议的计划是最优的，而且它在所有情况下都有效。

13.12.1 如何操作

以评估冗余索引使用情况时的那个查询为例，它使用的是 from_date 与 from_date_2 的交集。通过优化器提示（/*+NO_INDEX_MERGE(s from_date,from_date_2) */），你可以避免使用交集。你也可以提示优化器忽略 from_date_2 索引来实现相同的结果：

```
mysql> EXPLAIN SELECT e.emp_no, salary FROM salaries s IGNORE
INDEX(from_date_2) JOIN employees e ON s.emp_no=e.emp_no WHERE
from_date='2001-05-23'\G
*************************** 1. row ***************************
           id: 1
  select_type: SIMPLE
        table: s
   partitions: NULL
         type: ref
possible_keys: PRIMARY,from_date
          key: from_date
      key_len: 3
          ref: const
         rows: 90
     filtered: 100.00
        Extra: NULL
*************************** 2. row ***************************
           id: 1
  select_type: SIMPLE
        table: e
   partitions: NULL
         type: ref
possible_keys: PRIMARY
          key: PRIMARY
      key_len: 4
          ref: employees.s.emp_no
         rows: 1
     filtered: 100.00
        Extra: Using index
2 rows in set, 1 warning (0.00 sec)
```

另一个用例是提示优化器并节省评估多个计划的成本。考虑以下 employees 表和查询（与 13.11 节开头讨论的相同）：

```
mysql> SHOW CREATE TABLE employees\G
*************************** 1. row ***************************
       Table: employees
Create Table: CREATE TABLE `employees` (
  `emp_no` int(11) NOT NULL,
  `birth_date` date NOT NULL,
  `first_name` varchar(14) NOT NULL,
  `last_name` varchar(16) NOT NULL,
  `gender` enum('M','F') NOT NULL,
  `hire_date` date NOT NULL,
  PRIMARY KEY (`emp_no`),
  KEY `last_name` (`last_name`) /*!80000 INVISIBLE */,
  KEY `full_name` (`first_name`,`last_name`),
  KEY `full_name_desc` (`first_name` DESC,`last_name`),
  KEY `first_name` (`first_name`),
  KEY `full_name_1` (`first_name`,`last_name`),
  KEY `first_name_emp_no` (`first_name`,`emp_no`),
  KEY `last_name_2` (`last_name`(10))
  ENGINE=InnoDB DEFAULT CHARSET=utf8mb4
1 row in set (0.00 sec)

mysql> EXPLAIN SELECT emp_no FROM employees WHERE first_name='Adam' OR
last_name='Adam'\G
*************************** 1. row ***************************
           id: 1
  select_type: SIMPLE
        table: employees
   partitions: NULL
         type: index_merge
possible_keys:
full_name,full_name_desc,first_name,full_name_1,first_name_emp_no,last_name
_2
          key: first_name,last_name_2
      key_len: 58,42
```

```
           ref: NULL
          rows: 252
      filtered: 100.00
         Extra: Using sort_union(first_name,last_name_2); Using where
1 row in set, 1 warning (0.00 sec)
```

你可以看到，优化器必须评估索引 full_name、full_name_desc、first_name、full_name_1、first_name_emp_no、last_name_2 才能得出最佳计划。你可以通过传递 USE INDEX(first_name,last_name_2)来提示优化器，这就能避免扫描其他索引：

```
mysql> EXPLAIN SELECT emp_no FROM employees USE
INDEX(first_name,last_name_2) WHERE first_name='Adam' OR last_name='Adam'\G
*************************** 1. row ***************************
            id: 1
   select_type: SIMPLE
         table: employees
    partitions: NULL
          type: index_merge
possible_keys: first_name,last_name_2
           key: first_name,last_name_2
       key_len: 58,42
           ref: NULL
          rows: 252
      filtered: 100.00
         Extra: Using sort_union(first_name,last_name_2); Using where
1 row in set, 1 warning (0.00 sec)
```

由于这是一条简单的查询，而且表非常小，性能增益可以忽略不计。如果这条查询很复杂并且每小时执行数百万次时，这种方法对性能的提高会非常显著。

13.13 使用生成列为 JSON 建立索引

在 JSON 列上不能直接建立索引。因此，如果要在 JSON 列上使用索引，可以使用虚拟列和在虚拟列上创建的索引来提取信息。

13.13.1　如何操作

1. 参考在第 3 章 3.2 节创建的 emp_details 表：

```
mysql> SHOW CREATE TABLE emp_details\G
*************************** 1. row ***************************
       Table: emp_details
Create Table: CREATE TABLE `emp_details` (
  `emp_no` int(11) NOT NULL,
  `details` json DEFAULT NULL,
  PRIMARY KEY (`emp_no`)
) ENGINE=InnoDB DEFAULT CHARSET=utf8mb4
1 row in set (0.00 sec)
```

2. 插入一些虚拟记录：

```
mysql> INSERT IGNORE INTO emp_details(emp_no, details) VALUES
     ('1', '{ "location": "IN", "phone": "+11800000000", "email":
"abc@example.com", "address": { "line1": "abc", "line2": "xyz
street", "city": "Bangalore", "pin": "560103"}}'),
     ('2', '{ "location": "IN", "phone": "+11800000000", "email":
"def@example.com", "address": { "line1": "abc", "line2": "xyz
street", "city": "Delhi", "pin": "560103"}}'),
     ('3', '{ "location": "IN", "phone": "+11800000000", "email":
"ghi@example.com", "address": { "line1": "abc", "line2": "xyz
street", "city": "Mumbai", "pin": "560103"}}'),
     ('4', '{ "location": "IN", "phone": "+11800000000", "email":
"jkl@example.com", "address": { "line1": "abc", "line2": "xyz
street", "city": "Delhi", "pin": "560103"}}'),
     ('5', '{ "location": "US", "phone": "+11800000000", "email":
"mno@example.com", "address": { "line1": "abc", "line2": "xyz
street", "city": "Sunnyvale", "pin": "560103"}}');
Query OK, 5 rows affected (0.00 sec)
Records: 5 Duplicates: 0  Warnings: 0
```

3. 假设你要检索在城市 Bangalore 的员工编号 emp_no：

```
mysql> EXPLAIN SELECT emp_no FROM emp_details WHERE
details->>'$.address.city'="Bangalore"\G
```

```
*************************** 1. row ***************************
           id: 1
  select_type: SIMPLE
        table: emp_details
   partitions: NULL
         type: ALL
possible_keys: NULL
          key: NULL
      key_len: NULL
          ref: NULL
         rows: 5
     filtered: 100.00
        Extra: Using where
1 row in set, 1 warning (0.00 sec)
```

你会注意到，该查询无法使用索引也无法扫描所有行。

4. 可以将城市作为虚拟列进行检索，并在这个虚拟列上添加索引：

```
mysql> ALTER TABLE emp_details ADD COLUMN city varchar(20) AS
(details->>'$.address.city'), ADD INDEX (city);
Query OK, 0 rows affected (0.22 sec)
Records: 0 Duplicates: 0  Warnings: 0

mysql> SHOW CREATE TABLE emp_details\G
*************************** 1. row ***************************
       Table: emp_details
Create Table: CREATE TABLE `emp_details` (
  `emp_no` int(11) NOT NULL,
  `details` json DEFAULT NULL,
  `city` varchar(20) GENERATED ALWAYS AS
(json_unquote(json_extract(`details`,_utf8'$.address.city')))
VIRTUAL,
  PRIMARY KEY (`emp_no`),
  KEY `city` (`city`)
ENGINE=InnoDB DEFAULT CHARSET=utf8mb4
1 row in set (0.01 sec)
```

5. 如果现在检查 explain 计划，你会注意到该查询能够使用 city 上的索引，并且只扫描了一行：

```
mysql> EXPLAIN SELECT emp_no FROM emp_details WHERE
details->>'$.address.city'="Bangalore"\G
*************************** 1. row ***************************
           id: 1
  select_type: SIMPLE
        table: emp_details
   partitions: NULL
         type: ref
possible_keys: city
          key: city
      key_len: 83
          ref: const
         rows: 1
     filtered: 100.00
        Extra: NULL
1 row in set, 1 warning (0.00 sec)
```

要了解有关生成列上的辅助索引的详细信息，请参阅 https://dev.mysql.com/doc/refman/8.0/en/create-table-secondary-indexes.html。

13.14　使用资源组

你可以使用资源组来限制查询仅使用一定数量的系统资源。目前，只有 CPU 时间是由**虚拟 CPU**（**VCPU**）代表的可管理资源，它包括 CPU 核、超线程、硬件线程等。你可以创建一个资源组并为其分配 VCPU。除了 CPU 外，资源组的属性是线程优先的。

你可以将资源组分配给线程，在会话级别设置默认资源组，或者将资源组作为优化器提示来传递。例如，你希望运行一些优先级最低的查询（例如，报告式的查询），可以将它们分配给资源最少的资源组。

13.14.1　如何操作

1. 为 mysqld 设置 CAP_SYS_NICE 功能：

```
shell> ps aux | grep mysqld | grep -v grep
mysql 5238  0.0 28.1 1253368 488472 ? Sl   Nov19   4:04
/usr/sbin/mysqld --daemonize --pid-file=/var/run/mysqld/mysqld.pid

shell> sudo setcap cap_sys_nice+ep /usr/sbin/mysqld

shell> getcap /usr/sbin/mysqld
/usr/sbin/mysqld = cap_sys_nice+ep
```

2. 使用 `CREATE RESOURCE GROUP` 语句创建一个资源组。必须指定资源组的名称、VCPU 的数量、线程优先级和类型（可以是 `USER` 或 `SYSTEM`）。如果没有指定 VCPU，所有的 CPU 都会被用到：

```
mysql> CREATE RESOURCE GROUP report_group
TYPE = USER
VCPU = 2-3
THREAD_PRIORITY = 15
ENABLE;
# You should have at least 4 CPUs for the above resource group to
create. If you have less CPUs, you can use VCPU = 0-1 for testing
the example.
```

VCPU 表示 CPU 编号为 0～5，包括编号为 0、1、2、3、4 和 5 的 CPU；而 0～3、8～9 和 11 则表示编号为 0、1、2、3、8、9 和 11 的 CPU。

`THREAD_PRIORITY` 设定 CPU 的优先级。对于系统资源组，这个值的范围为–20～0，对于用户组，范围为 0～19。–20 表示优先级最高，19 表示优先级最低。

你还可以启用或禁用资源组。默认情况下，资源组在创建时被启用。被禁用的组不能分配线程。

3. 创建资源组后，可以对其进行验证：

```
mysql> SELECT * FROM INFORMATION_SCHEMA.RESOURCE_GROUPS\G
*************************** 1. row ***************************
   RESOURCE_GROUP_NAME: USR_default
   RESOURCE_GROUP_TYPE: USER
RESOURCE_GROUP_ENABLED: 1
```

```
              VCPU_IDS: 0-0
        THREAD_PRIORITY: 0
*************************** 2. row ***************************
    RESOURCE_GROUP_NAME: SYS_default
    RESOURCE_GROUP_TYPE: SYSTEM
RESOURCE_GROUP_ENABLED: 1
              VCPU_IDS: 0-0
THREAD_PRIORITY: 0
*************************** 3. row ***************************
    RESOURCE_GROUP_NAME: report_group
    RESOURCE_GROUP_TYPE: USER
RESOURCE_GROUP_ENABLED: 1
              VCPU_IDS: 2-3
        THREAD_PRIORITY: 15
```

USR_default 和 SYS_default 是默认资源组，不能删除或修改。

4. 将资源组分配给一个线程：

```
mysql> SET RESOURCE GROUP report_group FOR <thread_id>;
```

5. 设置会话资源组。该会话中的所有查询都将在 report_group 下执行：

```
mysql> SET RESOURCE GROUP report_group;
```

6. 利用 RESOURCE_GROUP 优化器提示来执行一条使用了 report_group 的语句：

```
mysql> SELECT /*+ RESOURCE_GROUP(report_group) */ * FROM employees;
```

更改和删除资源组

你可以动态调整资源组的 CPU 数或线程优先级。如果系统负载很重，可以降低线程优先级：

```
mysql> ALTER RESOURCE GROUP report_group VCPU = 3 THREAD_PRIORITY = 19;
Query OK, 0 rows affected (0.12 sec)
```

同样，当系统负载较轻时，可以提高优先级：

```
mysql> ALTER RESOURCE GROUP report_group VCPU = 0-12 THREAD_PRIORITY = 0;
Query OK, 0 rows affected (0.12 sec)
```

可以禁用资源组：

```
mysql> ALTER RESOURCE GROUP report_group DISABLE FORCE;
Query OK, 0 rows affected (0.00 sec)
```

也可以使用 DROP RESOURCE GROUP 语句删除资源组：

```
mysql> DROP RESOURCE GROUP report_group FORCE;
```

如果使用了 FORCE 命令，则运行的线程将被移到默认资源组（系统线程将被移到 SYS_default，用户线程将被移到 USR_default）。

如果没有使用 FORCE 命令，资源组中的现有线程将继续运行，直到它们终止，但不能为这个组分配新线程。

资源组仅限在本地服务器使用，服务器之间不复制任何与资源组相关的语句。要了解有关资源组的更多信息，请参阅 https://dev.mysql.com/doc/refman/8.0/en/resource-groups.html。

13.15 使用 performance_schema

你可以使用 performance_schema 在运行时检查服务器的内部执行情况。不应该把它与用于检查元数据的 information schema 混为一谈。

performance_schema 中有许多影响服务器计时的事件消费者，例如函数调用、对操作系统的等待、SQL 语句执行中的某个阶段（例如解析或排序）、一条语句或一组语句。所有收集的信息都存储在 performance_schema 中，不会被复制。

默认情况下，performance_schema 是启用的；如果要禁用它，可以在 my.cnf 文件中设置 performance_schema = OFF。默认情况下，并非所有的消费者和计数器都处于启用状态；你可以通过更新 performance_schema.setup_instruments 和 performance_schema.setup_consumers 表来关闭/打开它们。

13.15.1　如何操作

我们来看一下如何使用 performance_schema。

启用/禁用 performance_schema

要禁用 performance_schema，请将其设置为 0：

```
shell> sudo vi /etc/my.cnf
[mysqld]
performance_schema = 0
```

启用/禁用消费者和计数器

你可以在 setup_consumers 表中看到可用的消费者列表，如下所示：

```
mysql> SELECT * FROM performance_schema.setup_consumers;
+----------------------------------+---------+
| NAME   | ENABLED |
+----------------------------------+---------+
| events_stages_current            | NO      |
| events_stages_history            | NO      |
| events_stages_history_long       | NO      |
| events_statements_current        | YES     |
| events_statements_history        | YES     |
| events_statements_history_long   | NO      |
| events_transactions_current      | YES     |
| events_transactions_history      | YES     |
| events_transactions_history_long | NO      |
| events_waits_current             | NO      |
| events_waits_history             | NO      |
| events_waits_history_long        | NO      |
| global_instrumentation           | YES     |
| thread_instrumentation           | YES     |
| statements_digest                | YES     |
+----------------------------------+---------+
15 rows in set (0.00 sec)
```

假设要启用 events_waits_current:

```
mysql> UPDATE performance_schema.setup_consumers SET ENABLED='YES' WHERE
NAME='events_waits_current';
```

同样，你可以从 setup_instruments 表禁用或启用计数器，大约有 1182 种计数器（视 MySQL 版本而定）:

```
mysql> SELECT NAME, ENABLED, TIMED FROM setup_instruments LIMIT 10;
+---------------------------------------------------------+---------+------
-+
| NAME                                                    | ENABLED | TIMED
|
+---------------------------------------------------------+---------+------
-+
| wait/synch/mutex/pfs/LOCK_pfs_share_list                | NO      | NO
|
| wait/synch/mutex/sql/TC_LOG_MMAP::LOCK_tc               | NO      | NO
|
| wait/synch/mutex/sql/MYSQL_BIN_LOG::LOCK_commit         | NO      | NO
|
| wait/synch/mutex/sql/MYSQL_BIN_LOG::LOCK_commit_queue   | NO      | NO
|
| wait/synch/mutex/sql/MYSQL_BIN_LOG::LOCK_done           | NO      | NO
|
| wait/synch/mutex/sql/MYSQL_BIN_LOG::LOCK_flush_queue    | NO      | NO
|
| wait/synch/mutex/sql/MYSQL_BIN_LOG::LOCK_index          | NO      | NO
|
| wait/synch/mutex/sql/MYSQL_BIN_LOG::LOCK_log            | NO      | NO
|
| wait/synch/mutex/sql/MYSQL_BIN_LOG::LOCK_binlog_end_pos | NO      | NO
|
| wait/synch/mutex/sql/MYSQL_BIN_LOG::LOCK_sync           | NO      | NO
|
+---------------------------------------------------------+---------+------
-+
10 rows in set (0.00 sec)
```

performance_schema 表

performance_schema 中有 5 种主要的表类型。它们是当前事件表、事件历史表、事件摘要表、对象实例表和设置（配置）表：

```
mysql> SHOW TABLES LIKE '%current%';
+------------------------------------------+
| Tables_in_performance_schema (%current%) |
+------------------------------------------+
| events_stages_current                    |
| events_statements_current                |
| events_transactions_current              |
| events_waits_current                     |
+------------------------------------------+
4 rows in set (0.00 sec)

mysql> SHOW TABLES LIKE '%history%';
+------------------------------------------+
| Tables_in_performance_schema (%history%) |
+------------------------------------------+
| events_stages_history                    |
| events_stages_history_long               |
| events_statements_history                |
| events_statements_history_long           |
| events_transactions_history              |
| events_transactions_history_long         |
| events_waits_history                     |
| events_waits_history_long                |
+------------------------------------------+
8 rows in set (0.00 sec)

mysql> SHOW TABLES LIKE '%summary%';
+------------------------------------------------------+
| Tables_in_performance_schema (%summary%)             |
+------------------------------------------------------+
| events_errors_summary_by_account_by_error            |
| events_errors_summary_by_host_by_error               |
```

```
~
~
| table_io_waits_summary_by_table                         |
| table_lock_waits_summary_by_table                       |
+---------------------------------------------------------+
41 rows in set (0.00 sec)

mysql> SHOW TABLES LIKE '%setup%';
+------------------------------------------+
| Tables_in_performance_schema (%setup%)  |
+------------------------------------------+
| setup_actors                             |
| setup_consumers                          |
| setup_instruments                        |
| setup_objects                            |
| setup_threads                            |
| setup_timers                             |
+------------------------------------------+
6 rows in set (0.00 sec)
```

假设你要找出被访问得最多的文件：

```
mysql> SELECT EVENT_NAME, COUNT_STAR from file_summary_by_event_name ORDER
BY count_star DESC LIMIT 10;
+-----------------------------------------------------+------------+
| EVENT_NAME                                          | COUNT_STAR |
+-----------------------------------------------------+------------+
| wait/io/file/innodb/innodb_data_file                | 35014      |
| wait/io/file/sql/io_cache                           | 13454      |
| wait/io/file/sql/binlog                             | 8785       |
| wait/io/file/innodb/innodb_log_file                 | 2070       |
| wait/io/file/sql/query_log                          | 1257       |
| wait/io/file/innodb/innodb_temp_file                | 96         |
| wait/io/file/innodb/innodb_tablespace_open_file     | 88         |
| wait/io/file/sql/casetest                           | 15         |
| wait/io/file/sql/binlog_index                       | 14         |
| wait/io/file/mysys/cnf                              | 5          |
+-----------------------------------------------------+------------+
```

```
10 rows in set (0.00 sec)
```

或者你想知道哪一个文件的写入时间最长：

```
mysql> SELECT EVENT_NAME, SUM_TIMER_WRITE FROM file_summary_by_event_name
ORDER BY SUM_TIMER_WRITE DESC LIMIT 10;
+-----------------------------------------------+-----------------+
| EVENT_NAME                                    | SUM_TIMER_WRITE |
+-----------------------------------------------+-----------------+
| wait/io/file/innodb/innodb_data_file          | 410909759715    |
| wait/io/file/innodb/innodb_log_file           | 366157166830    |
| wait/io/file/sql/io_cache                     | 341899621700    |
| wait/io/file/sql/query_log                    | 203975010330    |
| wait/io/file/sql/binlog                       | 85261691515     |
| wait/io/file/innodb/innodb_temp_file          | 25291378385     |
| wait/io/file/innodb/innodb_tablespace_open_file | 674778195     |
| wait/io/file/sql/SDI                          | 18981690        |
| wait/io/file/sql/pid                          | 10233405        |
| wait/io/file/archive/FRM                      | 0               |
+-----------------------------------------------+-----------------+
```

你可以使用 events_statements_summary_by_digest 表来获取查询报告，就像你对 pt-query-digest 所做的那样。按所花费的时间列出排名靠前的查询：

```
mysql> SELECT SCHEMA_NAME, digest, digest_text, round(sum_timer_wait/
1000000000000, 6) as avg_time, count_star FROM
performance_schema.events_statements_summary_by_digest ORDER BY
sum_timer_wait DESC LIMIT 1\G
*************************** 1. row ***************************
SCHEMA_NAME: NULL
     digest: 719f469393f90c27d84681a1d0ab3c19
digest_text: SELECT `sleep` (?)
   avg_time: 60.000442
 count_star: 1
1 row in set (0.00 sec)
```

按执行次数列出排名靠前的查询：

```
mysql> SELECT SCHEMA_NAME, digest, digest_text, round(sum_timer_wait/
1000000000000, 6) as avg_time, count_star FROM
performance_schema.events_statements_summary_by_digest ORDER BY count_star
DESC LIMIT 1\G
*************************** 1. row ***************************
SCHEMA_NAME: employees
     digest: f5296ec6642c0fb977b448b350a2ba9b
digest_text: INSERT INTO `salaries` VALUES (...) /* , ... */
   avg_time: 32.736742
 count_star: 114
1 row in set (0.01 sec)
```

假设你要查找特定查询的统计信息。你可以使用 performance_schema 检查所有统计信息，而不是依赖于 mysqlslap 基准数据：

```
mysql> SELECT * FROM events_statements_summary_by_digest WHERE DIGEST_TEXT
LIKE '%SELECT%employee%ORDER%' LIMIT 1\G
*************************** 1. row ***************************
                SCHEMA_NAME: employees
                     DIGEST: d3b56f71f362f1bf6b067bfa358c04ab
                DIGEST_TEXT: EXPLAIN SELECT /*+ SET_VAR (
`sort_buffer_size` = ? ) */ `e` . `emp_no` , `salary` FROM `salaries` `s`
JOIN `employees` `e` ON `s` . `emp_no` = `e` . `emp_no` WHERE (
`first_name` = ? OR `last_name` = ? ) ORDER BY `from_date` DESC
                 COUNT_STAR: 1
             SUM_TIMER_WAIT: 643710000
             MIN_TIMER_WAIT: 643710000
             AVG_TIMER_WAIT: 643710000
             MAX_TIMER_WAIT: 643710000
              SUM_LOCK_TIME: 288000000
                 SUM_ERRORS: 0
               SUM_WARNINGS: 1
          SUM_ROWS_AFFECTED: 0
              SUM_ROWS_SENT: 2
          SUM_ROWS_EXAMINED: 0
SUM_CREATED_TMP_DISK_TABLES: 0
     SUM_CREATED_TMP_TABLES: 0
```

```
        SUM_SELECT_FULL_JOIN: 0
~
                 FIRST_SEEN: 2017-11-23 08:40:28.565406
                  LAST_SEEN: 2017-11-23 08:40:28.565406
                QUANTILE_95: 301995172
                QUANTILE_99: 301995172
               QUANTILE_999: 301995172
          QUERY_SAMPLE_TEXT: EXPLAIN SELECT /*+ SET_VAR(sort_buffer_size =
16M) */ e.emp_no, salary FROM salaries s JOIN employees e ON
s.emp_no=e.emp_no WHERE (first_name='Adam' OR last_name='Adam') ORDER BY
from_date DESC
          QUERY_SAMPLE_SEEN: 2017-11-23 08:40:28.565406
    QUERY_SAMPLE_TIMER_WAIT: 643710000
```

13.16　使用 sys schema

sys schema 帮助你以一种更简单和更易理解的形式解释从 performance_schema 收集来的数据。为了使 sys schema 能工作，应该启用 performance_schema。如果想最大限度地使用 sys schema，你需要启用 performance_schema 上的所有消费者和计时器，但这会影响服务器的性能。所以，仅启动你在寻找的消费者。

带有 x$ 前缀的视图以皮秒为单位显示数据，供其他工具做进一步的处理；其他表是人类可阅读的。

13.16.1　如何操作

从 sys schema 中启用一个计数器：

```
mysql> CALL sys.ps_setup_enable_instrument('statement');
+------------------------+
| summary                |
+------------------------+
| Enabled 22 instruments |
+------------------------+
1 row in set (0.08 sec)
```

```
Query OK, 0 rows affected (0.08 sec)
```

如果要重置为默认值，请执行以下操作：

```
mysql> CALL sys.ps_setup_reset_to_default(TRUE)\G
*************************** 1. row ***************************
status: Resetting: setup_actors
DELETE FROM performance_schema.setup_actors WHERE NOT (HOST = '%' AND USER
= '%' AND `ROLE` = '%')
1 row in set (0.01 sec)
~
*************************** 1. row ***************************
status: Resetting: threads
UPDATE performance_schema.threads SET INSTRUMENTED = 'YES'
1 row in set (0.03 sec)

Query OK, 0 rows affected (0.03 sec)
```

sys schema 中有许多表，本节展示了其中一些最常用的表。

按类型列出每个主机的语句（INSERT 和 SELECT）

```
mysql> SELECT statement, total, total_latency, rows_sent, rows_examined,
rows_affected, full_scans FROM sys.host_summary_by_statement_type WHERE
host='localhost' ORDER BY total DESC LIMIT 5;
+------------+--------+--------------+-----------+--------------+-----
-------+------------+
| statement  | total  | total_latency | rows_sent | rows_examined |
rows_affected | full_scans |
+------------+--------+--------------+-----------+--------------+-----
-------+------------+
| select     | 208526 | 1.14 d       | 27484761  | 799220003     |
0 |       9265 |
| Quit       | 199551 | 4.76 s       | 0         | 0             |
0 |          0 |
| insert     |   9848 | 12.75 m      | 0         | 0             |
5075058 |          0 |
| Ping       | 4674   | 278.76 ms    | 0         | 0             |
```

```
0 |          0 |
| set_option | 2552   | 634.76 ms     | 0          | 0              |
0 |          0 |
+------------+--------+---------------+-----------+--------------+-----
-------+------------+
6 rows in set (0.00 sec)
```

按类型列出每个用户的语句

```
mysql> SELECT statement, total, total_latency, rows_sent, rows_examined,
rows_affected, full_scans FROM sys.user_summary_by_statement_type ORDER BY
total DESC LIMIT 5;
+------------+--------+---------------+-----------+--------------+-----
-------+------------+
| statement  | total  | total_latency | rows_sent | rows_examined |
rows_affected | full_scans |
+------------+--------+---------------+-----------+--------------+-----
-------+------------+
| select     | 208535 | 1.14 d        | 27485256  | 799246972     |
0 |       9273 |
| Quit       | 199551 | 4.76 s        |         0 |             0 |
0 |          0 |
| insert     |   9848 | 12.75 m       |         0 |             0 |
5075058 |             0 |
| Ping       | 4674   | 278.76 ms     |         0 |             0 |
0 |         0 |
| set_option |   2552 | 634.76 ms     |         0 |             0 |
0 |         0 |
+------------+--------+---------------+-----------+--------------+-----
-------+------------+
5 rows in set (0.01 sec)
```

冗余索引

```
mysql> SELECT * FROM sys.schema_redundant_indexes WHERE
table_name='employees'\G
*************************** 1. row ***************************
              table_schema: employees
```

```
                table_name: employees
       redundant_index_name: first_name
    redundant_index_columns: first_name
  redundant_index_non_unique: 1
         dominant_index_name: first_name_emp_no
      dominant_index_columns: first_name,emp_no
   dominant_index_non_unique: 1
              subpart_exists: 0
              sql_drop_index: ALTER TABLE `employees`.`employees` DROP INDEX
`first_name`
~
*************************** 8. row ***************************
                table_schema: employees
                  table_name: employees
        redundant_index_name: last_name_2
     redundant_index_columns: last_name
  redundant_index_non_unique: 1
         dominant_index_name: last_name
      dominant_index_columns: last_name
   dominant_index_non_unique: 1
              subpart_exists: 1
              sql_drop_index: ALTER TABLE `employees`.`employees` DROP INDEX
`last_name_2`
8 rows in set (0.00 sec)
```

未使用的索引

```
mysql> SELECT * FROM sys.schema_unused_indexes WHERE
object_schema='employees';
+---------------+----------------+-------------------+
| object_schema | object_name    | index_name        |
+---------------+----------------+-------------------+
| employees     | departments    | dept_name         |
| employees     | dept_emp       | dept_no           |
| employees     | dept_manager   | dept_no           |
| employees     | employees      | name              |
| employees     | employees1     | last_name         |
```

```
| employees      | employees1      | full_name          |
| employees      | employees1      | full_name_desc     |
| employees      | employees1      | first_name         |
| employees      | employees1      | full_name_1        |
| employees      | employees1      | first_name_emp_no  |
| employees      | employees1      | last_name_2        |
| employees      | employees_mgr   | manager_id         |
| employees      | employees_test  | name               |
| employees      | emp_details     | city               |
+----------------+-----------------+--------------------+
14 rows in set (0.00 sec)
```

每个主机执行的语句

```
mysql> SELECT * FROM sys.host_summary ORDER BY statements DESC LIMIT 1\G
*************************** 1. row ***************************
                  host: localhost
            statements: 431214
     statement_latency: 1.15 d
 statement_avg_latency: 231.14 ms
           table_scans: 9424
              file_ios: 671972
       file_io_latency: 4.13 m
   current_connections: 3
     total_connections: 200193
          unique_users: 1
        current_memory: 0 bytes
total_memory_allocated: 0 bytes
1 row in set (0.02 sec)
```

对表的统计

```
mysql> SELECT * FROM sys.schema_table_statistics LIMIT 1\G
*************************** 1. row ***************************
     table_schema: employees
       table_name: employees
    total_latency: 14.03 h
     rows_fetched: 731760045
```

```
      fetch_latency: 14.03 h
      rows_inserted: 300025
     insert_latency: 2.81 s
       rows_updated: 0
     update_latency: 0 ps
       rows_deleted: 0
     delete_latency: 0 ps
   io_read_requests: NULL
            io_read: NULL
    io_read_latency: NULL
  io_write_requests: NULL
           io_write: NULL
   io_write_latency: NULL
   io_misc_requests: NULL
    io_misc_latency: NULL
1 row in set (0.01 sec)
```

对带缓冲区（buffer）的表的统计

```
mysql> SELECT * FROM sys.schema_table_statistics_with_buffer LIMIT 1\G
*************************** 1. row ***************************
              table_schema: employees
                table_name: employees
              rows_fetched: 731760045
             fetch_latency: 14.03 h
             rows_inserted: 300025
            insert_latency: 2.81 s
              rows_updated: 0
            update_latency: 0 ps
              rows_deleted: 0
            delete_latency: 0 ps
~
   innodb_buffer_allocated: 6.80 MiB
        innodb_buffer_data: 6.23 MiB
        innodb_buffer_free: 582.77 KiB
       innodb_buffer_pages: 435
innodb_buffer_pages_hashed: 0
```

```
   innodb_buffer_pages_old: 435
  innodb_buffer_rows_cached: 147734
1 row in set (0.13 sec)
```

语句分析

此输出类似于 performance_schema.events_statements_summary_by_digest 和 pt-query- digest 的输出。

根据查询的执行次数，排在前几位的查询如下：

```
mysql> SELECT * FROM sys.statement_analysis ORDER BY exec_count DESC LIMIT
1\G
*************************** 1. row ***************************
            query: SELECT `e` . `emp_no` , `salar ... emp_no` WHERE
`from_date` = ?
               db: employees
        full_scan:
       exec_count: 159997
        err_count: 0
       warn_count: 0
    total_latency: 1.98 h
      max_latency: 661.58 ms
      avg_latency: 44.54 ms
     lock_latency: 1.28 m
        rows_sent: 14400270
    rows_sent_avg: 90
    rows_examined: 28800540
rows_examined_avg: 180
    rows_affected: 0
rows_affected_avg: 0
       tmp_tables: 0
  tmp_disk_tables: 0
      rows_sorted: 0
sort_merge_passes: 0
           digest: 94c925c0f00e06566d0447822066b1fe
       first_seen: 2017-11-23 05:39:09
        last_seen: 2017-11-23 05:45:45
```

```
1 row in set (0.01 sec)
```

消耗了最大的 tmp_disk_tables 的语句为：

```
mysql> SELECT * FROM sys.statement_analysis ORDER BY tmp_disk_tables DESC
LIMIT 1\G
*************************** 1. row ***************************
            query: SELECT `cat` . `name` AS `TABL ... SE `col` . `type`
WHEN ? THEN
               db: employees
        full_scan:
       exec_count: 195
        err_count: 0
       warn_count: 0
    total_latency: 249.55 ms
      max_latency: 2.84 ms
      avg_latency: 1.28 ms
     lock_latency: 97.95 ms
        rows_sent: 732
    rows_sent_avg: 4
    rows_examined: 4245
rows_examined_avg: 22
    rows_affected: 0
rows_affected_avg: 0
       tmp_tables: 195
  tmp_disk_tables: 195
      rows_sorted: 732
sort_merge_passes: 0
digest: 8e8c46a210908a2efc2f1e96dd998130
first_seen: 2017-11-19 05:27:24
last_seen: 2017-11-20 17:24:34
1 row in set (0.01 sec)
```

要了解有关 sys schema 对象的更多信息，请参阅 https://dev.mysql.com/doc/refman/8.0/en/sys-schema-object-index.html。

第 14 章 安全

在本章，我们将介绍以下几个方面的内容：

- 安全安装
- 限定网络和用户
- 使用 mysql_config_editor 无密码认证
- 重置 root 密码
- 使用 X509 设置加密连接
- 设置 SSL 复制

14.1 引言

本章将介绍 MySQL 安全方面的内容，其中包括限定网络、使用强密码、使用 SSL、数据库的访问控制，安全安装以及安全插件。

14.2 安全安装

一旦 MySQL 安装完成后，建议使用 `mysql_secure_installation` 工具进行安全性安装设置。

14.2.1 如何操作

```
shell> mysql_secure_installation
```

```
Securing the MySQL server deployment.

Enter password for user root:
The 'validate_password' plugin is installed on the server.
The subsequent steps will run with the existing configuration
of the plugin.
Using existing password for root.

Estimated strength of the password: 100
Change the password for root ? ((Press y|Y for Yes, any other key for No) :

 ... skipping.
By default, a MySQL installation has an anonymous user,
allowing anyone to log into MySQL without having to have
a user account created for them. This is intended only for
testing, and to make the installation go a bit smoother.
You should remove them before moving into a production
environment.

Remove anonymous users? (Press y|Y for Yes, any other key for No) : y
Success.

Normally, root should only be allowed to connect from
'localhost'. This ensures that someone cannot guess at
the root password from the network.

Disallow root login remotely? (Press y|Y for Yes, any other key for No) : y
Success.

By default, MySQL comes with a database named 'test' that
anyone can access. This is also intended only for testing,
and should be removed before moving into a production
environment.

Remove test database and access to it? (Press y|Y for Yes, any other key
for No) : y
```

```
 - Dropping test database...
Success.

 - Removing privileges on test database...
Success.

Reloading the privilege tables will ensure that all changes
made so far will take effect immediately.

Reload privilege tables now? (Press y|Y for Yes, any other key for No) : y
Success.

All done!
```

默认情况下，mysqld 进程在 mysql 用户下运行。你还可以更改 mysqld 所使用的所有目录（例如 datadir、binlog 目录<如果有的话>，其他磁盘中的表空间等）的所有权，在另一个用户下运行 mysqld，并在 my.cnf 中添加 user = <user>。请参阅 https://dev.mysql.com/doc/refman/8.0/en/changing-mysql-user.html 以了解有关更改 MySQL 用户的更多信息。

强烈建议不要以 UNIX root 用户的身份运行 mysqld。原因之一是任何具有 FILE 权限的用户都可以使服务器以 root 用户身份创建文件。

FILE 权限

在向任何用户授予 FILE 权限时都需谨慎，因为用户可以利用 mysqld 守护进程的权限在文件系统中的任何位置（包括服务器的数据目录）写入文件。但是，它们不能覆盖现有的文件。另外，用户可以将 MySQL（或运行 mysqld 的用户）可访问的任何文件读取到数据库表中。FILE 是一个全局权限，这意味着你不能将其作用范围限制为某一个数据库：

```
mysql> SHOW GRANTS;
+---------------------------------------------------------------------+
| Grants for company_admin@%                                          |
+---------------------------------------------------------------------+
| GRANT FILE ON *.* TO `company_admin`@`%`                            |
| GRANT SELECT, INSERT, CREATE ON `company`.* TO `company_admin`@`%`  |
+---------------------------------------------------------------------+
```

```
2 rows in set (0.00 sec)

mysql> USE company;
Database changed
mysql> CREATE TABLE hack (ibdfile longblob);
Query OK, 0 rows affected (0.05 sec)

mysql> LOAD DATA INFILE '/var/lib/mysql/employees/salaries.ibd' INTO TABLE
hack CHARACTER SET latin1 FIELDS TERMINATED BY '@@@@@';
Query OK, 366830 rows affected (18.98 sec)
Records: 366830  Deleted: 0  Skipped: 0  Warnings: 0

mysql> SELECT * FROM hack;
```

请注意，具有 FILE 权限的 company 用户可以从 employees 表中读取数据。

你没必要担心前面的黑客行为，因为在默认情况下，使用 secure_file_priv 变量可以将文件的读/写位置限制为/var/lib/mysql-files。如果将 secure_file_priv 变量设置为 NULL、一个空字符串、MySQL 数据目录或 MySQL 有权访问的任何敏感目录（例如，MySQL 数据目录之外的表空间），就会出现问题。如果将 secure_file_priv 设置为不存在的目录，则会导致错误。

建议将 secure_file_priv 设置为其默认值：

```
mysql> SHOW VARIABLES LIKE 'secure_file_priv';
+------------------+-----------------------+
| Variable_name | Value |
+------------------+-----------------------+
| secure_file_priv | /var/lib/mysql-files/ |
+------------------+-----------------------+
1 row in set (0.00 sec)
```

永远不要允许任何人访问 mysql.user 表。要详细了解安全准则，请参阅 https://dev.mysql.com/doc/refman/8.0/en/security-guidelines.html 和 https://dev.mysql.com/doc/refman/8.0/EN/security-against-attack.html。

14.3 限定网络和用户

不要将数据库开放给整个网络，这意味着从其他网络访问应该无法访问运行 MySQL 的端口（3306）。它应该只对应用程序服务器开放。你可以使用 iptables 或 host.access 文件设置防火墙来限制对端口 3306 的访问。如果是在云上使用 MySQL，云服务商也会提供防火墙。

14.3.1 如何操作

若要测试对端口 3306 的访问是否是受限的，可以使用 telnet：

```
shell> telnet <mysql ip> 3306
# if telnet is not installed you can install it or use nc (netcat)
```

如果 telnet 挂起或连接被拒绝，则表示该端口已关闭。请注意，如果你看到类似下面这样的输出，则表示该端口未被屏蔽：

```
shell> telnet 35.186.158.188 3306
Trying 35.186.158.188...
Connected to 188.158.186.35.bc.googleusercontent.com.
Escape character is '^]'.
FHost '183.82.17.137' is not allowed to connect to this MySQL
serverConnection closed by foreign host.
```

这意味着 3306 端口是开放的，但 MySQL 限制了对它的访问。

创建用户时，请避免在任何地方提供访问权限（%选项）。既要限制对 IP 范围或子域名的访问，也要限制用户只能访问所需的数据库。例如，employees 数据库的 read_only 用户应该无法访问其他数据库：

```
mysql> CREATE user 'employee_read_only'@'10.10.%.%' IDENTIFIED BY
'<Str0ng_P@$$word>';
Query OK, 0 rows affected (0.00 sec)

mysql> GRANT SELECT ON employee.* TO 'employee_read_only'@'10.10.%.%'; Query
OK, 0 rows affected (0.01 sec)
```

employee_read_only 用户将只能从 10.10.%.%子网访问，并且只能访问

employee 数据库。

14.4 使用 mysql_config_editor 进行无密码认证

每次使用命令行客户端输入密码时，你可能会看到以下警告：

```
shell> mysql -u dbadmin -p'$troNgP@$$w0rd'
mysql: [Warning] Using a password on the command line interface can be
insecure.
Welcome to the MySQL monitor.  Commands end with ; or \g.
Your MySQL connection id is 1345
Server version: 8.0.3-rc-log MySQL Community Server (GPL)
~
mysql>
```

如果你未在命令行中传递密码，而是在系统提示时输入密码，则不会收到该警告：

```
shell> mysql -u dbadmin -p
Enter password:
Welcome to the MySQL monitor.  Commands end with ; or \g.
Your MySQL connection id is 1334
Server version: 8.0.3-rc-log MySQL Community Server (GPL)
~
mysql>
```

但是，当你通过客户端实用程序开发某些脚本时，是很难使用密码提示的。避免这种情况的一种方法是将密码存储在主目录中的.my.cnf 文件中。默认情况下，mysql 命令行实用程序读取.my.cnf 文件，并且不要求输入密码：

```
shell> cat $HOME/.my.cnf
[client]
user=dbadmin
password=$troNgP@$$w0rd

shell> mysql
Welcome to the MySQL monitor.  Commands end with ; or \g.
Your MySQL connection id is 1396
```

```
Server version: 8.0.3-rc-log MySQL Community Server (GPL)
~
mysql>
```

请注意，你可以在不提供任何密码的情况下连接服务器，但这会导致安全问题。有密码的话也是明文形式。为了解决这个问题，MySQL 引入了 mysql_config_editor，它以加密格式存储密码。该文件可以由客户端程序解密（仅在内存中使用）从而连接到服务器。

14.4.1 如何操作

使用 mysql_config_editor 创建.mylogin.cnf 文件：

```
shell> mysql_config_editor set --login-path=dbadmin_local --host=localhost
--user=dbadmin --password
Enter password:
```

你可以通过更改登录路径来添加多个主机名和密码。如果密码已经更改，则可以再次运行 mysql_config_editor，它将更新文件中的密码：

```
shell> mysql_config_editor set --login-path=dbadmin_remote --
host=35.186.157.16 --user=dbadmin --password
Enter password:
```

如果你想以用户 dbadmin 的身份登录到 35.186.157.16，只需要执行：

```
mysql --login-path = dbadmin_remote:
shell> mysql --login-path=dbadmin_remote
Welcome to the MySQL monitor.  Commands end with ; or \g.
Your MySQL connection id is 215074
~
mysql> SELECT @@server_id;
+-------------+
| @@server_id |
+-------------+
|         200 |
+-------------+
1 row in set (0.00 sec)
```

要连接到 localhost，只需执行 mysql 或 mysql --login-path = dbadmin_local：

```
shell> mysql
Welcome to the MySQL monitor. Commands end with ; or \g.
Your MySQL connection id is 1523
~
mysql> SELECT @@server_id;
+-------------+
| @@server_id |
+-------------+
|           1 |□
+-------------+□
1 row in set (0.00 sec)

shell> mysql --login-path=dbadmin_local
Welcome to the MySQL monitor.  Commands end with ; or \g.
Your MySQL connection id is 1524
~
mysql> SELECT @@server_id;
+-------------+
| @@server_id |
+-------------+
|           1 |
+-------------+
1 row in set (0.00 sec)
```

如果 dbadmin 的密码在所有服务器上都是相同的，则可以通过指定主机名来连接到其中的任何一台服务器。你不需要指定密码：

```
shell> mysql -h 35.198.210.229
Welcome to the MySQL monitor.  Commands end with ; or \g.
~
mysql> SELECT @@server_id;
+-------------+
| @@server_id |
+-------------+
|         364 |
```

```
+-------------+
1 row in set (0.00 sec)
```

如果要打印所有登录路径，可以执行以下操作：

```
shell> mysql_config_editor print --all
[dbadmin_local]
user = dbadmin
password = *****
host = localhost
[dbadmin_remote]
user = dbadmin
password = *****
host = 35.186.157.16
```

你会注意到 mysql_config_editor 遮住了密码。如果你尝试读该文件，则只会看到乱码字符：

```
shell> cat .mylogin.cnf
 ?-z???|???-B????dU?bz4-?W???g?q?BmV?????κ?I??
h%?+b???_??@V???vli?J???X`?qP
```

mysql_config_editor 只会帮助你避免以明文形式存储密码并简化连接到 MySQL 的过程。有许多方法可以解密存储在.mylogin.cnf 文件中的密码。因此，就算你使用了 mysql_config_editor，也不要认为密码就是安全的。你还可以将此文件复制到其他服务器，而不是每次创建.mylogin.cnf 文件（只有在用户名和密码相同的情况下才有效）。

14.5 重置 root 密码

如果你忘记了 root 密码，可以通过两种方法重置它。

14.5.1 如何操作

让我们看一看细节。

使用 init-file

在类 UNIX 系统上，通过指定 init-file 来停止服务器的运行再启动它。你可以在该文件中保存 ALTER USER 'root'@'localhost' IDENTIFIED BY 'New$trongPass1'SQL 代码。MySQL 在启动时执行 init-file 文件的内容，更改 root 用户的密码。

1. 停止服务器的运行：

```
shell> sudo systemctl stop mysqld
shell> pgrep mysqld
```

2. 将 SQL 代码保存在/var/lib/mysql/mysql-init-password 中，使其仅对 MySQL 可读：

```
shell> vi /var/lib/mysql/mysql-init-password
ALTER USER 'root'@'localhost' IDENTIFIED BY 'New$trongPass1';

shell> sudo chmod 400 /var/lib/mysql/mysql-init-password

shell> sudo chown mysql:mysql /var/lib/mysql/mysql-init-password
```

3. 使用--init-file 和其他必要的选项来启动 MySQL 服务器：

```
shell> sudo -u mysql /usr/sbin/mysqld --daemonize --pid-
file=/var/run/mysqld/mysqld.pid --user=mysql --init-
file=/var/lib/mysql/mysql-init-password
mysqld will log errors to /var/log/mysqld.log
mysqld is running as pid 28244
```

4. 验证错误日志文件：

```
shell> sudo tail /var/log/mysqld.log
~
2017-11-27T07:32:25.219483Z 0 [Note] Execution of init_file
'/var/lib/mysql/mysql-init-password' started.
2017-11-27T07:32:25.219639Z 4 [Note] Event Scheduler: scheduler
thread started with id 4
2017-11-27T07:32:25.223528Z 0 [Note] Execution of init_file
'/var/lib/mysql/mysql-init-password' ended.
```

```
2017-11-27T07:32:25.223610Z 0 [Note] /usr/sbin/mysqld: ready for
connections. Version: '8.0.3-rc-log' socket:
'/var/lib/mysql/mysql.sock' port: 3306 MySQL Community Server (GPL)
```

5. 验证你是否能够使用新密码登录：

```
shell> mysql -u root -p'New$trongPass1'
mysql: [Warning] Using a password on the command line interface can
be insecure.
Welcome to the MySQL monitor.  Commands end with ; or \g.
Your MySQL connection id is 15
Server version: 8.0.3-rc-log MySQL Community Server (GPL)
~
mysql>
```

6. 这一步最重要！删除/var/lib/mysql/mysql-init-password 文件：

```
shell> sudo rm -rf /var/lib/mysql/mysql-init-password
```

7. 或者，你也可以停止服务器的运行，然后不带--init-file 选项正常启动。

使用--skip-grant-tables

在这种方法中，通过指定--skip-grant-tables 停止服务器的运行然后再启动，MySQL 不会加载授权表。你可以以 root 身份不输入密码而连接到服务器，然后设置密码。由于服务器未经授权运行，因此其他网络的用户可能会连接进来。因此从 MySQL 8.0.3 开始，--skip-grant-tables 会自动启用--skip-networking，不允许远程连接。

1. 停止服务器的运行：

```
shell> sudo systemctl stop mysqld
shell> ps aux | grep mysqld | grep -v grep
```

2. 使用--skip-grant-tables 选项启动服务器：

```
shell> sudo -u mysql /usr/sbin/mysqld --daemonize --pid-
file=/var/run/mysqld/mysqld.pid --user=mysql --skip-grant-tables
mysqld will log errors to /var/log/mysqld.log
mysqld is running as pid 28757
```

3. 不使用密码连接到 MySQL，执行 FLUSH PRIVILEGES 重新加载授权，并更改用户以修改密码：

```
shell> mysql -u root
Welcome to the MySQL monitor.  Commands end with ; or \g.
Your MySQL connection id is 6
Server version: 8.0.3-rc-log MySQL Community Server (GPL)
~
mysql> FLUSH PRIVILEGES;
Query OK, 0 rows affected (0.04 sec)

mysql> ALTER USER 'root'@'localhost' IDENTIFIED BY
'New$trongPass1';
Query OK, 0 rows affected (0.01 sec)
```

4. 用新密码测试与 MySQL 的连接：

```
shell> mysql -u root -p'New$trongPass1'
mysql: [Warning] Using a password on the command line interface can
be insecure.
Welcome to the MySQL monitor.  Commands end with ; or \g.
Your MySQL connection id is 7
~
mysql>
```

5. 重启 MySQL 服务器：

```
shell> ps aux | grep mysqld | grep -v grep
mysql    28757  0.0 13.3 1151796 231724 ?     Sl   08:16   0:00
/usr/sbin/mysqld --daemonize --pid-file=/var/run/mysqld/mysqld.pid
--user=mysql --skip-grant-tables
shell> sudo kill -9 28757
shell> ps aux | grep mysqld | grep -v grep
shell> sudo systemctl start mysqld
shell> ps aux | grep mysqld | grep -v grep
mysql    29033  5.3 16.8 1240224 292744 ?     Sl   08:27   0:00
/usr/sbin/mysqld --daemonize --pid-file=/var/run/mysqld/mysqld.pid
```

14.6 使用X509设置加密连接

如果客户端和MySQL服务器之间的连接没有加密，任何有权访问网络的人就都可以查看数据。如果客户端和服务器位于不同的数据中心，建议使用加密连接。默认情况下，MySQL 8使用加密连接，但如果加密连接失败，将返回到非加密连接。你可以通过检查Ssl_cipher变量的状态来进行测试。如果连接是通过localhost建立的，则不会加密：

```
mysql> SHOW STATUS LIKE 'Ssl_cipher';
+---------------+--------------------+
| Variable_name | Value              |
+---------------+--------------------+
| Ssl_cipher    | DHE-RSA-AES256-SHA |
+---------------+--------------------+
1 row in set (0.00 sec)
```

如果你未使用SSL，则Ssl_cipher将是空白的。

可以强制某些用户只能使用加密连接（通过REQUIRE SSL子句来指定），而对于其他用户，则可以自由选择使用加密或者非加密连接。

根据MySQL文档：

MySQL支持采用TLS（传输层安全性）协议的客户端和服务器之间的加密连接。TLS有时被称为SSL（安全套接字层），但MySQL实际上不使用SSL协议来加密连接，因为其加密功能较弱。TLS使用加密算法来确保通过公共网络接收的数据可以被信任。它具有检测数据的更改、丢失或重放的机制。TLS还包含了使用X509标准提供身份验证的算法。

在本节中，你将了解如何使用X509来设置SSL连接。

所有与SSL（X509）相关的文件（ca.pem、server-cert.pem、server-key.pem、client-cert.pem和client-key.pem）都由MySQL在安装过程中创建并保存在数据目录下。服务器需要ca.pem、server-cert.pem和server-key.pem文件，客户端使用client-cert.pem和client-key.pem文件连接到服务器。

14.6.1 如何操作

1. 验证数据目录中的文件，更新 `my.cnf`，重启服务器并检查与 SSL 有关的变量。在 MySQL 8 中，默认情况下，设置下列值：

```
shell> sudo ls -lhtr /var/lib/mysql | grep pem
-rw-------. 1 mysql mysql 1.7K Nov 19 13:53 ca-key.pem
-rw-r--r--. 1 mysql mysql 1.1K Nov 19 13:53 ca.pem
-rw-------. 1 mysql mysql 1.7K Nov 19 13:53 server-key.pem
-rw-r--r--. 1 mysql mysql 1.1K Nov 19 13:53 server-cert.pem
-rw-------. 1 mysql mysql 1.7K Nov 19 13:53 client-key.pem
-rw-r--r--. 1 mysql mysql 1.1K Nov 19 13:53 client-cert.pem
-rw-------. 1 mysql mysql 1.7K Nov 19 13:53 private_key.pem
-rw-r--r--. 1 mysql mysql 451 Nov 19 13:53 public_key.pem

shell> sudo vi /etc/my.cnf
[mysqld]
ssl-ca=/var/lib/mysql/ca.pem
ssl-cert=/var/lib/mysql/server-cert.pem
ssl-key=/var/lib/mysql/server-key.pem

shell> sudo systemctl restart mysqld

mysql> SHOW VARIABLES LIKE '%ssl%';
+---------------+--------------------------------+
| Variable_name | Value                          |
+---------------+--------------------------------+
| have_openssl  | YES                            |
| have_ssl      | YES                            |
| ssl_ca        | /var/lib/mysql/ca.pem          |
| ssl_capath    |                                |
| ssl_cert      | /var/lib/mysql/server-cert.pem |
| ssl_cipher    |                                |
| ssl_crl       |                                |
| ssl_crlpath   |                                |
| ssl_key       | /var/lib/mysql/server-cert.pem |
+---------------+--------------------------------+
```

```
9 rows in set (0.01 sec)
```

2. 将 client-cert.pem 和 client-key.pem 文件从服务器的数据目录复制到客户端的位置：

```
shell> sudo scp -i $HOME/.ssh/id_rsa /var/lib/mysql/client-key.pem
/var/lib/mysql/client-cert.pem <user>@<client_ip>:
# change the ssh private key path as needed.
```

3. 通过传递--ssl-cert 和--ssl-key 选项连接到服务器：

```
shell> mysql --ssl-cert=client-cert.pem --ssl-key=client-key.pem -h
35.186.158.188
Welcome to the MySQL monitor. Commands end with ; or \g.
Your MySQL connection id is 666
Server version: 8.0.3-rc-log MySQL Community Server (GPL)
~
mysql>
```

4. 强制用户只能通过 X509 连接：

```
mysql> ALTER USER `dbadmin`@`%` REQUIRE X509;
Query OK, 0 rows affected (0.08 sec)
```

5. 测试连接：

```
shell> mysql --login-path=dbadmin_remote -h 35.186.158.188 --ssl-
cert=client-cert.pem --ssl-key=client-key.pem
Welcome to the MySQL monitor. Commands end with ; or \g.
Your MySQL connection id is 795
Server version: 8.0.3-rc-log MySQL Community Server (GPL)
~
mysql> ^DBye
```

6. 如果未指定--ssl-cert 或--ssl-key，你将无法登录：

```
shell> mysql --login-path=dbadmin_remote -h 35.186.158.188
ERROR 1045 (28000): Access denied for user
'dbadmin'@'35.186.157.16' (using password: YES)
```

```
shell> mysql --login-path=dbadmin_remote -h 35.186.158.188 --ssl-
cert=client-cert.pem
mysql: [ERROR] SSL error: Unable to get private key from 'client-
cert.pem'
ERROR 2026 (HY000): SSL connection error: Unable to get private key

shell> mysql --login-path=dbadmin_remote -h 35.186.158.188 --ssl-
key=client-key.pem
mysql: [ERROR] SSL error: Unable to get certificate from 'client-
key.pem'
ERROR 2026 (HY000): SSL connection error: Unable to get certificate
```

在默认情况下，所有与 SSL 相关的文件都保存在数据目录中。如果你想将它们保存在其他地方，可以在 my.cnf 文件中设置 ssl_ca、ssl_cert 和 ssl_key，并重新启动服务器。你可以通过 MySQL 或 OpenSSL 生成一组新的 SSL 文件。要了解更详细的步骤，请参阅 https://dev.mysql.com/doc/refman/8.0/en/create-ssl-rsa-files.html。还有许多其他可用的认证插件。你可以参考 https://dev.mysql.com/doc/ refman/8.0/en/authentication-plugins.html 来了解更多细节。

14.7 设置 SSL 复制

如果启用 SSL 复制，主从设备之间的二进制日志传输将通过加密连接发送。这与上一节中介绍的服务器/客户端连接类似。

14.7.1 如何操作

1. 如上一节所述，在主库上，需要启用 SSL。
2. 在主库上，将 client* 证书复制到从库：

```
mysql> sudo scp -i $HOME/.ssh/id_rsa /var/lib/mysql/client-key.pem
/var/lib/mysql/client-cert.pem <user>@<client_ip>:
```

3. 在从库上，创建 mysql-ssl 目录以保存与 SSL 相关的文件，并正确地设置权限：

```
shell> sudo mkdir /etc/mysql-ssl
shell> sudo cp client-key.pem client-cert.pem /etc/mysql-ssl/
shell> sudo chown -R mysql:mysql /etc/mysql-ssl
shell> sudo chmod 600 /etc/mysql-ssl/client-key.pem
shell> sudo chmod 644 /etc/mysql-ssl/client-cert.pem
```

4. 在从库上，执行 CHANGE_MASTER 命令，并进行与 SSL 相关的更改：

```
mysql> STOP SLAVE;

mysql> CHANGE MASTER TO MASTER_SSL=1, MASTER_SSL_CERT='/etc/mysql-
ssl/client-cert.pem', MASTER_SSL_KEY='/etc/mysql-ssl/client-
key.pem';

mysql> START SLAVE;
```

5. 验证从库的状态：

```
mysql> SHOW SLAVE STATUS\G
    *************************** 1. row ***************************
               Slave_IO_State: Waiting for master to send event
                  Master_Host: 35.186.158.188
~
             Slave_IO_Running: Yes
            Slave_SQL_Running: Yes
~
                 Skip_Counter: 0
          Exec_Master_Log_Pos: 354
              Relay_Log_Space: 949
              Until_Condition: None
               Until_Log_File:
                Until_Log_Pos: 0
           Master_SSL_Allowed: Yes
           Master_SSL_CA_File: /etc/mysql-ssl/ca.pem
           Master_SSL_CA_Path:
              Master_SSL_Cert: /etc/mysql-ssl/client-cert.pem
            Master_SSL_Cipher:
               Master_SSL_Key: /etc/mysql-ssl/client-key.pem
```

```
         Seconds_Behind_Master: 0
 Master_SSL_Verify_Server_Cert: No
~
                   Master_UUID: fe17bb86-cd30-11e7-bc3b-42010a940003
              Master_Info_File: mysql.slave_master_info
                     SQL_Delay: 0
           SQL_Remaining_Delay: NULL
       Slave_SQL_Running_State: Slave has read all relay log;
waiting for more updates
~
1 row in set (0.00 sec)
```

6. 在所有从库上进行了与 SSL 相关的更改之后，在主库上强制复制用户（replication user）使用 X509：

```
mysql> ALTER USER `repl`@`%` REQUIRE X509;
Query OK, 0 rows affected (0.00 sec)
```

请注意，这可能会影响其他复制用户。还有一种方法，你可以创建一个使用 SSL 的复制用户和一个正常的复制用户。

7. 验证所有从库的从库状态。